PETROLEUM EVALUATIONS AND ECONOMIC DECISIONS

PETROLEUM EVALUATIONS AND ECONOMIC DECISIONS

Arthur W. McCray

School of Petroleum
and Geological Engineering
The University of Oklahoma

Prentice-Hall, Inc. *Englewood Cliffs, New Jersey*

Library of Congress Cataloging in Publication Data

McCray, Arthur White.
Petroleum evaluations and economic decisions.

Includes bibliographies.
1. Petroleum industry and trade—Finance. 2. Risk.
3. Oil fields—Valuation. I. Title.
HD9560.6.M26 338.2′3 74–13088
ISBN 0–13–662213–6

10 9 8 7 6 5 4

Printed in the United States of America

PRENTICE-HALL INTERNATIONAL, INC., *London*
PRENTICE-HALL OF AUSTRALIA, PTY. LTD., *Sydney*
PRENTICE-HALL OF CANADA, LTD., *Toronto*
PRENTICE-HALL OF INDIA PRIVATE LIMITED, *New Delhi*
PRENTICE-HALL OF JAPAN, INC., *Tokyo*

Dedicated to

THE PETROLEUM INDUSTRY

CONTENTS

3

PROPERTIES OF PROBABILITY DISTRIBUTIONS

4

APPLICATIONS OF PROBABILITY DISTRIBUTIONS

5

APPRAISAL OF UNCERTAIN VENTURES

6

DECISION TREES AND ECONOMIC MODELS

7

SIMULATION—THE MONTE CARLO METHOD

8

EVALUATION OF EXPECTED DISCOVERIES IN MATURE REGIONS

9

BAYES STRATEGIES AND ESTIMATES OF VALUE

10

CONTROL OF OPERATIONS: CPM AND PERT METHODS

11

EVALUATION OF FUTURE PRODUCTION BY PERFORMANCE TRENDS

APPENDIX TABLES

PREFACE

Within the past fifteen years, new ways have been recognized for handling uncertainty. Appraisal methods for new drilling or development ventures and other high-risk uncertain ventures are the subject of this book. Historically, appraisals have been limited to consideration of definite values. Even the exploratory well was given a definite success ratio and a precise reward if successful. However, in many cases it was not so much that variations between estimates and actual results were unexpected, rather there was no way of handling or combining estimates which included variance.

Ways of interpreting estimates that include uncertainty are developed and explained. The methods permit converting a probabilistic description into an appraisal value. Proper combining of all unresolved ventures yields the total speculative position. These methods are fairly simple to apply and they are firmly based on mathematical principles of probability and statistics. Several such methods are developed and many variations are illustrated by examples. Most of the illustrations apply directly to the petroleum-producing industry.

A fundamental part of the evaluation process is the probabilistic description. This permits recognizing the unknown state of nature rather than insisting upon the precision of estimator. If any factors are not known precisely, the limits within which they may vary can be included and the effects of such variations on values become part of the estimate. Probabilistic descriptions can be made in a number of ways. Different situations lend themselves more readily to some types of descriptions rather than others and a number of examples of different methods are provided.

This book is concerned with applications of economics, probability, and statistics in the petroleum-producing industry. It makes no attempt to cover any one of these areas. Some areas of probability and statistics which are most useful and fundamental to later developments have been described for introductory purposes. The reader is encouraged to pursue further the studies of these areas.

An introductory chapter provides background on methods of allowing for uncertainty and common petroleum-evaluation processes. Next is a review of several methods used for estimating financial return. Subsequent chapters develop the applications of probability and statistics in evaluations. Several techniques, including simulation and critical path methods, are presented. The projection of future performance through production-decline curves is included to provide guidance for using the present value tables, which are helpful in estimating the present value of future oil or gas production. Commonly used probability tables and present value discount factors comprise the Appendix Tables.

Tribute is paid to the many engineers and others who have contributed so much to evaluation and decision-making theory and practice. Tribute is due also to the eminent schools and societies that sponsor programs and publications in these fields. Special recognition is given to the Society of Petroleum Engineers of AIME whose publications include many primary references in these as well as many other areas.

ARTHUR W. MCCRAY

PETROLEUM EVALUATIONS AND ECONOMIC DECISIONS

1

INTRODUCTION

1.1 INCLUDING UNCERTAINTY IN EVALUATIONS

1.1-1 Degrees of Uncertainty

People generally are more comfortable when they are dealing with certainty. The world seems more secure, and we can plan ahead with confidence. Why then, should we bring up uncertainty? Simply because there is a lot of it around. When the value per item is relatively small, when no great inconvenience results from a failure, when a slightly defective item serves almost as well, or when the probability of a defect or a failure is extremely small, then commonly the uncertainty is simply ignored. Many of the things that we do fall into such routine categories. On the other hand, the situation is greatly different when we try to develop new processes, better methods, new discoveries, and new resources. The drilling of an exploratory oil well is only one example where both the cost and the uncertainty regarding success are so great that they cannot be ignored.

1.1-2 Informal Ways of Making Allowances

Where uncertainties are great, and where there is no recognized way of taking them into account, decision making becomes difficult. It is also difficult to explain. In some cases experience can be called upon. Hunches and rumors play their part, as well as the prevailing degree of personal optimism. Groups of people are at least as susceptible as individuals to emotional reactions. Judgments of some kind must be made, the desirability of various consequences must be visualized, the future predicted, and the whole thing mentally integrated. One is forced to make assumptions and predictions; otherwise there is no basis for a decision.

Rules of thumb are often adopted. A payout time of three years, or of two years, may be insisted upon and the number of times paid out may be specified as at least twice the investment. Presumably such criteria are judged in terms of the "expected" performance. The factors used may be adjusted upward or downward according to the feelings about particular ventures. Indeed, in this book we shall attempt to adjust factors in individual cases in a more formal and presentable manner.

1.2 DECISION METHODS AND YARDSTICKS

1.2-1 Financial Measures

The order in which various financial performance standards have been developed and applied is probably unknown. Commonly their application is based on the assumption of a stable currency. Payout time, the number of times paid out, present values, and various discounted-cash-flow methods are all discussed in Chapter 2. In general, no one of these criteria is by itself a sufficient basis for judgment, and usually two or three are used in combination. That is, a prospect must be acceptable by more than one standard.

These financial criteria are developed under the assumption that the return cash-flow stream from a prospective venture can be precisely described both in total amount and by time pattern. In those cases where the results may vary widely, this is impossible. This has led to the use of rules of thumb and of fictitiously adjusted financial criteria, as previously noted. It may, for example, be specified that development wells should give at least an expected 20% rate of return, but it is known that a number of these will actually turn out to be dry. Thus the fictitiously high rate of return figure is essentially an attempt to adjust to uncertainty. It would be better, of course, to judge the uncertainty within each prospective venture on its own merits rather than to use a blanket figure on all prospects.

1.2-2 Statistical Approaches

Probably the first statistical concept which is mastered is that of expected value, the expectancy. If the income from a successful Ox Bow Sandstone well is $600,000 and only one-third are successful, the expected income is $200,000. The common mistake is then to assume that every well will have the certain income of $200,000, which simply is not true. A person might easily go broke before averaging out. Too much reliance is placed on the expectancy.

Statistics has very little effect on the outcome of an individual venture. It has much more to do with the outcomes of groups of ventures. Studies can reveal such things as 90% of some groups being above a particular value. If we keep our investment below that point we will be 90% confident of a positive outcome. The total number of trials in which we participate enters strongly into the picture. The number of trials is limited by the investment per venture and the total investment capital available. So eventually the financial strength of the investor enters into the calculations, as evidently it must.

Application of statistics and probability in evaluation follows rules that govern groups of individual trials. The outcomes of the groups are much more predictable than are the outcomes of the individual trials. These must be combined with the appropriate financial measures. The desired answers are then in the form of maximum investment that will still permit satisfying the necessary financial return. This must be expressed in terms of 95% confidence, for example. These methods are pursued in subsequent chapters.

1.2-3 Utility Measures

Utility is an old and respected concept in the science of economics. It recognizes that the second increment of a substance may have generally less value (*utility*) to a person than the first increment. Different people will of course have different regard or desire for the same substance. Thus each person would have his individual utility curve, or function. The increase in utility per each additional small increment of the substance (slope of the curve) is called *marginal utility*. This generally becomes smaller with each additional increment, and it may become negative, since too much of some substances becomes an annoyance or inconvenience.

More recently the term "utility" has been consigned to a quite different purpose. Attempts have been made to incorporate the informal, the intangible, and the personal factors that enter an evaluation into a *utility function*. This has been done by describing the information and possible returns from a prospective venture and then having an operator or manager state how much he would invest in such a venture. These results are plotted with negative dollars invested and positive dollars expected return both on the horizontal axis. The vertical scale, called *utiles*, is arbitrary. However, the height of the curve, at a point above the positive expected return, is equal to the distance below the axis, at the stated investment in such a venture. The resulting smoothed curve is generally logarithmic in shape. Presumably it goes to negative infinity somewhere near the total financial resources. Again the total financial strength of the investor is a factor. Thus different individuals would have different utility functions.

In this book we shall not attempt to develop or to use utility measures to guide investments. Where one controlling personal decision must be made, the intangible emotional and utility concepts are sometimes relied upon. However, it appears that formal ways for combining individual utilities of a number of different ventures are either lacking or suspect because of nonlinear functions. On the other hand, the combining of the variances of individual ventures to obtain the characteristics of the whole is the very essence of statistics. Thus the thrust of this book is away from the use of utility and toward the application of the mathematical principles of probability and statistics.

1.3 PETROLEUM EVALUATIONS

1.3-1 Evaluation Methods

An overview of common types of evaluations will permit us to place in perspective the methods discussed in this book. In different situations, entirely different methods are sometimes used for reasons of convenience, because of personal preference, or because the information available is more suited to one method than another. The method used must suit the purpose of the evaluation and the parties for whom the evaluation is made. Needless to say, it is often presumed that a precise evaluation can be made, whether it is possible or not. Following is a list of methods:

1. *Actual cost, less depreciation.* This concept is applied to tangible equipment for taxation purposes. It would seldom be applied to ordinary oil and gas producing property.

2. *Estimated replacement cost, less depreciation.* This concept might be used by an integrated oil company when they consider their crude oil resources. It would seldom, if ever, apply to an individual oil or gas producing property.

3. *Actual sales of similar properties.* Records of the same or similar properties have been used by the courts to establish values for legal purposes. Bona fide offers to buy or sell, as well as the opinions of qualified operators, have been accepted for establishing values. The opinions and offers must be those which prevailed on the basic date for which the evaluation was made, and not influenced by later information which might sway the price up or down. Another consideration is the *fair market value*, defined as acceptable to both a willing buyer and a willing seller, neither of whom are forced to make the transaction. Davis and Stephenson note that graduated income tax at high rates has distorted market forces and relationships which formerly were considered normal.[8] They define the fair market value of a property as the price at which a willing and informed owner would sell to a willing and informed buyer, with full consideration given to their relative tax positions.

4. *Capitalization of income derivable from the property.* Chapter 2 is devoted to the methods for such evaluations. Later chapters incorporate statistical and probabilistic estimates in evaluations. Income taxes and other taxes must be deducted from income, similar to any expense, in arriving at a true value.

1.3-2 Purposes of Evaluations

Evaluations are interpreted and modified according to their purpose. Particularly when used as security for loans, properties are counted at less than full value. Evaluations made for tax purposes must be in accordance with regulations that cover the particular cases. Changes in tax laws can, of course, result in considerable change in property values, particularly in the attractiveness of speculative ventures. Petroleum properties are commonly evaluated for one or more of the following purposes.

1. *Security for loans.*[6] When oil or gas properties are used as security, the amount of the loan is always less than the appraised value. Usually not more than 80% of the purchase price will be loaned. Practice at one time was to make no allowance for shallow formations which had not been perforated or otherwise brought into production, and no value was allowed for possible deeper production. Generally the size of the loan is limited to one-half of one of the following: (a) future gross revenue, (b) future net revenue before capital investments, (c) present worth of future net revenue before capital investments, (d) present worth of future pledged revenue. Bank loans usually run for periods from 5 to 10 years, and insurance companies and pension and investment funds make loans that run from 10 to 15 years. The lenders are concerned primarily with the security of their loans and the schedules at which they will be repaid. As a precaution against their being forced into the oil producing business, they want the operator to have a sizable interest in the property over and above the amount of the loan. Lenders prefer that their risk be spread over more than one property, one field, one horizon. The character of the operator is of course most important. Loans are made to small as well as to major oil companies. Banks that make oil and gas loans have on their staffs engineers and geologists to make the necessary evaluations.

2. *Transactions with other companies.*[7] Reasons for buying all or part of an oil or gas property include (a) financial profit; (b) acquiring petroleum reserves for refining and sales; (c) reducing incremental overhead and operating costs; (d) diversification, spreading risk; (e) acquiring management or technical personnel; and (f) adjusting tax position, investing earnings, acquiring a tax loss. Reasons for selling include (a) overzealous buyer offering high price; (b) retirement of principal owner; (c) estate problems; (d) disputes

among co-owners; (e) excessive overhead or operating costs; (f) marginal properties requiring excessive attention; (g) isolated properties; (h) lack of development funds; (i) debt liquidation; (j) loss of growth potential for reinvestment of earnings; (k) adjusting tax position for capital-gains treatment; and (l) expiring tax loss, suggesting merger with a company that can utilize this tax loss to its benefit.

3. *Management decisions.* Engineering studies are made of proved properties in order that they may be produced most efficiently. Workover and recompletion studies are performed routinely, and they must be economically justifiable. The remaining reserves are always a matter of importance. Studies are made on the feasibility of installing plants for treating, extracting, or liquidifying produced hydrocarbons and on the feasibility of secondary recovery operations. Engineering and geological studies of undeveloped properties are necessary for defining the probabilities of various sizes of deposits that may be discovered. In competitive bidding situations it is necessary to study both the possible resources that may be found and the behavior of competitors.

4. *Guiding investors.* Speculative capital that would otherwise be paid as taxes is often attracted into the petroleum industry. Where an investor is in the 50% tax bracket, money lost in an unfavorable venture would result in a net position change of only half of the apparent loss. The other part is composed of the "tax dollars." Such speculation has been very instrumental in maintaining reserves of petroleum. In addition to evaluations directed to specific individuals, there are others directed to the general investing public. The companies who discovered vast reserves of petroleum on the North Slope of Alaska published an evaluation of the discovery. The evaluation was made by an independent and highly reputable consulting firm. Such an evaluation could be presumed to attract investors.

5. *Government regulations and taxes.* These include unitization of oil fields, combining properties into drilling units to comply with well spacing rules, proration, and other rules. Income tax laws particularly specify how depreciation and other matters must be handled. Evaluations are also required for settling estate, inheritance, and gift taxes.

1.3-3 Forms of Oil Property in the United States

Before proceeding to a more detailed description of financial and probabilistic methods and standards, we shall consider briefly various forms of oil property in the United States. Following are brief descriptions of terms used in oil field development.

Ownership in fee means that the oil or gas producer, the operator, owns both the surface and the mineral rights of a property.

Mineral rights. Most landowners are not oil producers. They prefer to lease the mineral rights to oil companies. For this they receive a payment, called a *bonus*, and retain a royalty interest in the produced minerals.

Surface rights. This is the right to use the surface for normal purposes, such as agriculture. Payments to the holders of surface rights are customary to compensate for inconveniences caused by drilling operations and the laying of roads and pipelines.

Lease. This is the instrument that conveys the exclusive right to drill for and produce oil and/or gas. Standard forms, whose clauses have been interpreted by the courts, are generally used. The lessor, the landowner, customarily receives an initial bonus, rental, and a royalty. The most common amount for the landowner's royalty is one-eighth, although higher amounts have been paid. An example is found in the school lands of West Texas, where the royalty is one-sixth. Rental is in lieu of drilling. Usually it must be paid in advance on a yearly basis for the entire acreage leased, and amounts of $1 per acre are common for unproved lands. In Louisiana the mineral rights cannot remain separated indefinitely from the surface rights unless oil or gas is being produced.

Bonus is a cash payment made as an inducement to grant a lease. The amount depends upon the apparent value of production to be found, and has varied from $1 to well over $1,000 per acre.

Royalty generally refers to the landowner's royalty, which very commonly is one-eighth of the value of the oil and gas produced. It is calculated on the gross value and bears no part of the development, operating, or treating expenses. However, it is subject to taxation by both state and federal governments.

Overriding royalty is a royalty interest in addition to the landowner's royalty. These economic interests, which are often about one-sixteenth of the total, are retained by a lease owner when he assigns the remainder to an operator who is interested in drilling and producing the property. Such a transaction is called a *farm-out*.

Participating royalty is based on participating in development costs, such as furnishing casing or other equipment needed for drilling.

Term royalty expires, as upon the condition that oil production has not been obtained by a specified date.

Sliding royalty contains provisions for adjusting the fractional amount of the royalty upward if high production rates are attained, downward for lower production rates.

Off-set royalty is not royalty, but an agreed-upon monthly payment made in lieu of drilling an offset well on the property under lease. The offset well would prevent drainage of oil or gas across property lines by a well brought into production on adjacent property.

Oil production payment, *Oil payment*, or *Production payment* refer to payments that terminate upon payment of a total sum of money. Barrels of oil rather than dollars are sometimes specified. A fractional part of the production, such as 60%, may be assigned to a bank until such time as the principal and interest on a loan have been repaid. Payment of the total amount extinguishes the economic interest.

The *working interest* bears the cost of all drilling, development, and operating expenses. The fractional amount of the gross income that is assigned to the working interest is a factor in the economic limit of production. *Participating interest* and *fractional interest* generally refer to ownership of a fraction of the working interest.

A *carried interest* usually does not share in any revenue from a property or a well until a specified amount has been repaid to the carrying party. Often the carried party does not participate in the management of the property, but retained this type of interest when the lease changed hands or when a drilling deal was being promoted.

Dry hole money is a way of sharing exploration costs. It is a payment toward the cost of drilling a test well near acreage held by an operator. Usually geological and test information is made available to the contributor. Payment is made only if the well is drilled to a certain depth and is dry.

Bottom hole money, similarly, is a way of sharing exploration costs. This money is paid if the well is drilled to a certain depth, whether it is dry or not.

REFERENCES

1. Bross, I. D. F., *Design for Decision*. New York: Macmillan Publishing Co., Inc., 1953.
2. Von Neuman, John, and Oskar Morgenstern, *Theory of Games and Economic Behavior*. Princeton, N.J.: Princeton University Press, 1944.
3. Swalm, R. O., "Utility Theory—Insights into Risk Taking." *Harvard Business Review*, Nov.–Dec. 1966.
4. Grayson, C. J., Jr., *Decisions Under Uncertainty*. Boston: Division of Research, Harvard University, 1960.
5. Newendorp, P. D., "Applications of Utility Theory to Drilling Investment Decisions." D. Eng. Dissertation, University of Oklahoma, 1967.
6. Sherrod, G. E., "Financing Through Production Payments." Presented at Oklahoma City Section of SPE of AIME, Oct. 17, 1967.

7. JACKSON, E. G., "Cost of Capital and Financing of Undeveloped Secondary Recovery Acquisitions." Presented at Oklahoma City Section of SPE of AIME, Oct. 24, 1967.

8. DAVIS, R. E., and E. A. STEPHENSON, "Valuation of Natural Gas Property." *Journal of Petroleum Technology*, July 1953.

9. PAINE, PAUL, *Oil Property Valuation.* New York: John Wiley & Sons, Inc., 1942.

10. HUGHES, R. V., *Oil Property Valuation.* New York: John Wiley & Sons, Inc., 1967.

11. CAMPBELL, J. M., *Oil Property Evaluation.* Englewood Cliffs, N.J.: Prentice-Hall, Inc., 1959.

12. MEGILL, R. E., *Exploration Economics.* Tulsa, Okla.: Petroleum Publishing Company, 1971.

2

RETURN ON INVESTMENT

2.1 INTRODUCTION

Several methods used for estimating the worth of proposed economic ventures are described in this chapter. The viewpoint taken corresponds to ideal situations. That is, no probability of error is admitted to be present in any of the estimates. Accordingly, these would be called *deterministic methods*, or *deterministic models*. Sometimes the condition of little or no error is essentially true in real life, as when money is borrowed and repaid to a bank or mortgage loan company. In other cases this is not true, and a deterministic model should not be assumed. However, for these other cases, the methods presented in this chapter constitute a basis for the later development of other methods, where recognition is made of the probabilities of failure and the degrees of success.

The purpose of most evaluations is to make possible comparison of alternative choices of investments. Present values, profit/investment ratios, payout time, number of times paid out, and rates of return are indices that have been used for this purpose. They are discussed in the following sections. In general, any specific venture must be considered as one among several concurrent ventures, which interact with each other and with the general economic climate. No venture can be judged in a vacuum. Whether or not the returns are satisfactory depends largely upon what has been happening to like sums of money invested elsewhere in the same and in other industries and in other geographic areas; it depends upon competition and the changes in relative status among competing economic units; and it depends upon inflation, the rate of devaluation of money. The methods discussed in this chapter are intended to aid the judgment required in the economic environment.

2.2 INTEREST AND INFLATION

2.2-1 Interest

Interest is defined as a rate per unit time charged for the use of capital; it may also refer to the money so paid. For example, if the annual rate of interest were 8% ($i = 0.08$), a charge of \$8 interest would be made for using the amount of \$100 for 1 year. In compound interest, the interest is added periodically to the capital. The following relations apply where a sum A_0 accumulates to A_n at an annual rate of interest i over a period of n years.

$$A_n = A_0(1 + i)^n, \tag{2-1}$$

$$I = A_n - A_0. \tag{2-2}$$

The quantity I is the portion of A_n derived from interest. In general it is common to consider interest in the sense of an a priori quantity, as a predetermined amount or rate, much as occurs in the case of a bank loan.

2.2-2 Inflation

Inflation has to do with the decreased purchasing power of money. It is indicated by progressive increases in the amounts that must be paid for those things that are required to maintain households and to operate businesses. It is reflected in such things as commodity price indices and interest rates on municipal, state, and government bonds. It is perhaps least worrisome to those who use the savings of others for their operating capital. To many others it is a direct attrition against any values described in terms of currency. Although inflation would seem to result from a continuing series of individual actions, like interest it is generally expressed as a rate. It was generally conceded to be of the order of 3 or 4% annually in the 1940s and 1950s and had reached 6 or 7% annually by 1970. Table 2.1 contains figures published by government agencies.

Figure 2.1 illustrates the effect of inflation, expressed as an average rate over a period of time. Let A_1 be the number of dollars required after n years for purchasing the same amount of goods that could be purchased by amount A_0 at the reference time zero. Where the increase is expressed as an average rate f, one may write

$$A_1 = A_0(1 + f)^n.$$

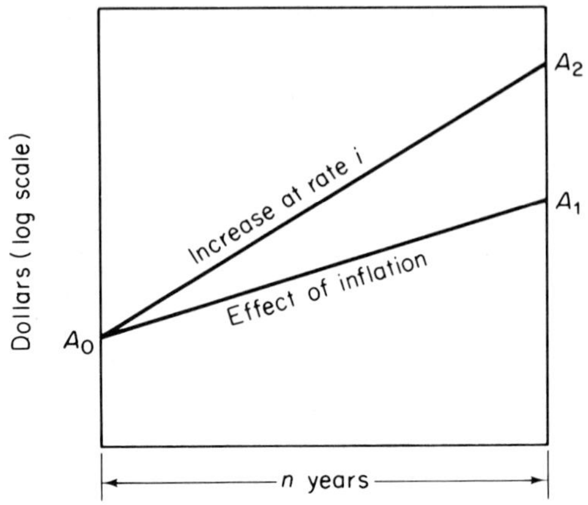

Figure 2.1. Effect of inflation.

Where the increase from A_0 to A_2 is expressed as an average interest rate over n years, one writes

$$A_2 = A_0(1 + i)^n.$$

Table 2.1
Purchasing Power of the Dollar

Year	*Purchasing Power*						
1900	1.00						
1910	0.83	1.00					
1920	0.40	0.47	1.00				
1930	0.47	0.57	1.20	1.00			
1940	0.57	0.68	1.43	1.19	1.00		
1950	0.33	0.39	0.83	0.69	0.58	1.00	
1960	0.27	0.32	0.68	0.56	0.47	0.81	1.00
1970	0.20	0.24	0.52	0.43	0.36	0.62	0.76

Source: Data from Bureau of Labor Statistics and Federal Reserve Bank of New York.

The real gain is only from A_1 to A_2, and if this occurs at a constant rate over n years, one has

$$A_2 = A_1(1 + s)^n.$$

By eliminating A_1 and A_2 from this equation, we obtain

$$1 + s = \frac{1 + i}{1 + f}, \tag{2-3}$$

$$s = \frac{i - f}{1 + f}. \tag{2-4}$$

By way of example, where one obtains a 10% interest rate and inflation proceeds at the rate of 6%, the actual gain is

$$s = \frac{0.10 - 0.06}{1.06} = 0.038, \quad \text{or} \quad 3.8\%.$$

Additional insight can be gained by observing Table 2.2. Here is indicated a series of yearly incomes e_n, received at the ends of the several years, whose total is E. But the dollars received in the future will not buy as much as they would today. Therefore, their value in today's dollars can be reckoned only by discounting according to the expected inflationary rate. If there is to be a true gain at rate s, the inflation-compensated dollars must be further discounted at rate s. The conventional manner of discounting at interest rate i is illustrated in the last column.

Table 2.2
Effect of Inflation on Reducing the Present Value of Future Income

Year	*Income (future dollars)*	*Present Value (today's dollars)*		
		Inflation Effect	*At True Rate*	*Apparent Rate*
1	e_1	$\frac{1}{1+f}e_1$	$\frac{1}{1+s}\frac{1}{1+f}e_1$	$\frac{1}{1+i}e_1$
2	e_2	$\left(\frac{1}{1+f}\right)^2 e_2$	$\left(\frac{1}{1+s}\right)^2\left(\frac{1}{1+f}\right)^2 e_2$	$\left(\frac{1}{1+i}\right)^2 e_2$
...	...	...	...	...
n	e_n	$\left(\frac{1}{1+f}\right)^n e_n$	$\left(\frac{1}{1+s}\right)^n\left(\frac{1}{1+f}\right)^n e_n$	$\left(\frac{1}{1+i}\right)^n e_n$
	E	$D_f E$	$D_s D_f E$	$D_i E$

The sums of the discounted columns indicate an additional relationship. The several D terms are simply composite present-value factors for reducing the totals of the dollars as received to their discounted values. Where both the inflationary rate and the true gain rate are recognized in the calculation, the following applies:

$$D_s D_f E = D_i E,$$

$$D_s D_f = D_i. \tag{2-5}$$

It is noted that both D_f and D_i are calculated to correspond to the series $e_1, e_2, \ldots, e_n$ of future dollar incomes. Thus their manner of calculation is rather conventional. Calculation of D_s, representing the reduction in present value due to true gain, would then be most easily accomplished through the use of Eq. (2-5).

Table 2.3 is an illustration of these relations using monetary values. The inflation rate used in the calculations is 6% and the apparent interest rate is 10%. The problem faced in this example is to make an estimate of how much could be paid, at the present time, for this series of payments, which will be received in the future.

Table 2.3
Monetary Example Accounting for the Effect of Inflation

Year	*Income at End of Year,* e_i (\$)	*Inflation Effect*		*Apparent Reductions*	
		Reduction Factor, 6%	*Value* (\$)	*Reduction Factor,* 10%	*Value* (\$)
1	1,000	0.9434	943.40	0.9091	909.10
2	1,000	0.8900	890.00	0.8264	826.40
3	1,000	0.8396	839.60	0.7513	751.30
4	1,000	0.7921	792.10	0.6831	683.10
5	1,000	0.7473	747.30	0.6207	620.70
	$E =$ 5,000		4,212.40		3,790.60

The following values can be extracted from Table 2.3:

$$D_f = \frac{4{,}212}{5{,}000} = 0.843, \qquad D_i = \frac{3{,}790}{5{,}000} = 0.758, \qquad D_s = \frac{D_i}{D_f} = \frac{0.758}{0.843} = 0.900.$$

If the series of payments of \$1,000 each were first reduced for effects of inflation at 6%, their total present value would be \$4,212.40. If it were desired to make an actual gain of 3.8% annually, then the most that could be paid now for this series of future payments can be calculated as follows:

$$\begin{aligned} \text{present value} &= (D_s)(\$4{,}212.40) \\ &= (0.8999)(4{,}212.40) \\ &= \$3{,}790.60. \end{aligned}$$

In practice, of course, this amount is found by using the 10% factors directly on each individual payment. The relations among the several rates are calculated by Eq. (2-3).

2.2-3 Effective, Nominal, and Continuous Interest Rates

Amount A_0 can be indicated as increasing to amount A_n by n yearly steps at yearly rate i:

$$A_n = A_0(1 + i)^n. \tag{2-1}$$

This same increase can be indicated by semiannual, quarterly, or monthly steps by the relation

$$A_n = A_0(1 + i_p)^{np}, \tag{2-6}$$

where i_p is the interest per period and p is 2, 4, 12, etc., periods per year. A comparison if Eq. (2-1) with (2-6) indicates that

$$1 + i = (1 + i_p)^p \tag{2-7}$$

or

$$1 + i_p = (1 + i)^{1/p}. \tag{2-7a}$$

For example, for $i = 0.10$, or 10%, yearly interest, the corresponding semiannual rate is found by

$$1 + i_{6\text{mo}} = (1.10)^{1/2} = 1.0488.$$

Thus one may say that interest rates of 10% annually or 4.88% semiannually are equivalent. They provide for the same accumulation from one amount to another over the same period of time.

Development of the continuous interest concept is conveniently approached by changing the frame of reference.[1] A start can be made by taking a nominal rate j. For example, let the nominal rate be 8% per year. A corresponding nominal rate would be 4% semiannually; another would be 2% quarterly. These are different rates, but the limiting value is of interest. The relation that describes the various rates is

$$A_n = A_0\left(1 + \frac{j}{p}\right)^{pn}. \tag{2-8}$$

If interest is compounded continously, p equals infinity. From series expansion, the following identity is obtained:

$$\lim_{p \to \infty}\left(1 + \frac{j}{p}\right)^{pn} = e^{jn}. \tag{2-9}$$

By substitution into Eqs. (2-1) and (2-8), the following results are obtained:

$$A_n = A_0 e^{nj}, \tag{2-10}$$

$$1 + i = e^j, \tag{2-11}$$

$$j = \ln(1 + i), \tag{2-12}$$

where j is the continuous interest rate. Table 2.4 shows some equivalent interest rates.

Table 2.4
Equivalent Interest Rates

i, Yearly	*i, Semiannual*	*j, Continuous*
0.05	0.0247	0.0489
0.10	0.0488	0.0955
0.15	0.0724	0.1400
0.20	0.0955	0.1826

Equations given in this section are relations between interest rates i, for various different time units, and the continuous rate j. Where only two amounts, A_0 and A_n, are concerned, it makes no difference which type of calculation is used. Cases where there is some basis for choice between different methods are discussed in the next section.

2.3 DISCOUNTED CASH FLOW

2.3-1 Present-Value Concepts

The concepts of *present value* are closely related to interest charges for the use of money. Some commonly used factors are as follows.

Where the payment for the nth year is received at the end of that year, Eq. (2-1) can be modified to give

$$\text{P.V.} = \left(\frac{1}{1+i}\right)^n \quad \text{(amount of the payment).} \tag{2-13}$$

The corresponding continuous-interest method is

$$\text{P.V.} = e^{-nj} \quad \text{(amount of the payment).} \tag{2-14}$$

Where the cash flow during the nth year is more-or-less uniformly spread over that year, a method commonly used in the petroleum industry is to assume that the entire amount is received at the midpoint of the year:

$$\text{P.V.} = \left(\frac{1}{1+i}\right)^{n-0.5} \quad \text{(cash flow in } n\text{th year).} \tag{2-15}$$

The corresponding continuous interest calculation is given by

$$\text{P.V.} = e^{-(n-1)j}\,\frac{1-e^{-j}}{j} \quad \text{(cash flow in } n\text{th year).} \tag{2-16}$$

In these relations the value of n is the difference, usually in years, between the time when an event takes place and the basic date of evaluation. Occasionally, these are the calendar dates when the evaluation is made. Frequently, however, the basic date of evaluation may be something like "date of well completion" or an equivalent event, which becomes a more or less accurately known calendar date only after the decision is made to proceed with a given project. Usually, many projects are evaluated but only the most promising selected. Those evaluations not selected are generally retained, however, and the date when the evaluation was made should be part of the record. These relations give the present values, which, if invested as indicated, would eventually yield the specified payments in the nth year. Thus in one sense they are appraisal methods, since the calculated present values are the most that could be paid now for the future payment, under the condition that the money must accumulate over a period of n years at the specified rate.

Table 2.5 shows a calculation of present value of a series of anticipated future payments. The payments are received at the ends of various years and are discounted at an annual rate of 8%.

Table 2.5
Calculation of Present Value

Year	*Sum Received at Year End* (\$)	*Present-Value Factor,* $(1.08)^{-n}$	*Present Value* (\$)
1	4,000	0.92593	3,704
2	7,000	0.85734	6,001
3	3,000	0.79383	2,381
4	5,000	0.73503	3,675
5	2,000	0.68058	1,361
	21,000		17,122

In general, an overall present-value factor, or *deferment factor*, is defined as follows:

$$D = \frac{\text{total present value}}{\text{sum of the undiscounted values}}. \tag{2-17}$$

For the example of Table 2.5,

$$D = \frac{\$17,122}{\$21,000}$$

$$= 0.81533.$$

In all cases the value of D, the discount factor, depends upon the rate of interest and the time pattern of the payments. For many cases where the amounts follow a regular pattern, values of D can be calculated by using an appropriate formula. A familiar example is the case of a series of uniform payments. Tables of various deferment factors are given in the appendixes at the end of Chapter 11.

2.3-2 Present Values and Present-Value/Cost Ratios

Interest rates are not ordinarily set at the pleasure of the individual bank or borrower. Rather, individuals are greatly influenced by the alternative opportunities, both their own and those of their counterparts. For this reason a lending institution can be assured that its invested funds will accumulate at the going rate of interest. Likewise, the borrower can accept the fact that he must pay back an amount greater than that which he originally received.

Appraisal of an individual investment venture can be made according to the same principles. Present-value concepts are widely used. Two present-value models, which aid in making selections between alternative investment opportunities, will be discussed. In both of these it is assumed that the interest rate, or discount rate, is predetermined. The first is based upon using an interest rate that corresponds to the cost of capital, which will actually be repaid to a lender as the venture progresses. The second model is based upon using an interest rate that corresponds to the average investment opportunity of the firm, and the flow of capital returned from the venture is presumed to be reinvested.

The first model has long been used, and has been referred to variously as the *present-value method*, the *engineer's method*, and the *accountant's method.* Essentially, the method sets the maximum appraisal value as the present value of the net operating income. This can be expressed as

$$C \leq DE, \tag{2-18}$$

where

C = appraisal value or cost,
D = present-value factor for the earnings,
E = total of the net operating income.

The time pattern of the net earnings E is known or estimated. The interest rate used in computing the present-value factor D corresponds to the rate at which capital can be borrowed. The question of whether to use annual, semiannual, or continuous interest methods can be determined by the manner in which the loan will be repaid.

An additional look at the first model reveals its significance. The present value is calculated, and the present value of the profits/investment ratio is readily available also. If the total net operating earnings E are divided into amortization C, interest I, and profit P, one has

$$E = C + I + P. \tag{2-19}$$

The present value of amortization plus interest is by definition the initial investment. Hence I can be eliminated by using

$$C = D(C + I). \tag{2-20}$$

After eliminating I, the equation can be arranged as

$$DP = DE - C. \tag{2-21}$$

That is, the present value of the net earnings minus the initial investment gives the present value of the anticipated profits. This might be accepted, without the above formulation, as an obvious conclusion, since the difference between two present values must itself be a present value. There is, however, the tacit assumption that the same present-value factor applies to both the profit fraction and the entire total stream of net earnings, which generally is a very good assumption. In selecting investments, maximizing the quantity $(DE - C)/C$, or the quantity DE/C, amounts to maximizing the present value of the profit per dollar invested.

An illustration can be taken from the data shown in Table 2.3. Let 8% be the cost of capital, and on that basis the present value of the future income is \$16,989. If this venture can be purchased for \$12,000, then

$$\frac{DE}{C} = \frac{\$16{,}989}{\$12{,}000} = 1.42.$$

This indicates that, after taking into account the cost of using capital, the \$12,000 invested in the venture has increased in present worth by 42%. The present-value profit/investment ratio is 0.42.

Example: A well that is estimated to cost \$140,000 is to be drilled on 40-acre spacing. The productive formation averages 30 ft in thickness. The time pattern of the production will give a present-value factor of 0.680 based on 10% cost of capital. Operating profit will average \$1.80 per barrel. If one could be absolutely certain of a recovery of 150 barrels per acre-foot, what would be the breakeven price of the bonus that could be paid to the landowner for the lease?

Solution:

$$\begin{aligned} \text{recovery} &= (40 \text{ acres})(30 \text{ ft})(150 \text{ bbl/acre-ft}) \\ &= 180{,}000 \text{ bbl}, \\ \text{P.V. of income} &= (0.680)(180{,}000 \text{ bbl})(\$1.80 \text{ per bbl}) \\ &= \$220{,}000, \\ \text{breakeven price} &= \$220{,}000 - \$140{,}000 = \$80{,}000. \end{aligned}$$

In this example it should be recognized that no provision is made for uncertainties and contingencies. Thus normally it would not be possible to pay the full indicated amount and continue to survive through a series of ventures, some of which might be failures. However, if the lease could be purchased for \$10,000, the following analysis results:

$$\begin{aligned} C &= \text{cost of well} + \text{cost of lease} \\ &= \$140{,}000 + \$10{,}000 \\ &= \$150{,}000, \end{aligned}$$

$$\begin{aligned} (\text{P.V. profit/investment ratio}) &= \frac{DE - C}{C} \\ &= \frac{220{,}000 - 150{,}000}{150{,}000} \qquad (2\text{-}22) \\ &= 0.47. \end{aligned}$$

This result indicates that the present value of the profits, as discounted at 10% per year, amounts to 47% of the investment. This criterion, by itself, appears favorable in this case. However, one would also like to know how long it will be before the initial investment is returned (payout time) and the rate of return that would reduce the profits to zero (discounted-cash-flow rate of return).

2.3-3 Projected Reinvestment of Earnings

Particularly in larger companies, the monies that are returned from one venture, in the form of operating profits, are regularly reinvested in other ventures. This gives rise to the concept that any one venture is best judged by its influence on the entire flow of capital through all ventures owned by the company.[2] One approach to the judging of prospective new ventures is to use the average rate of return achieved by the company.

The second model, which results from the above considerations, can be used to rank ventures according to their beneficial effects within the total investment stream of a company. It uses the present-value method as a convenient way to make comparisons among ventures. However, the basis of interpretation derives from projecting values into the future. An example is shown in Fig. 2.2, where a venture will return cash flows at the ends of years 1 through 4 and expire at the end of the fourth year. The present value of these cash flows is certainly not something that one has at hand, even though other monetary funds might be available. If an investment is made to purchase the venture, it will create a void in the total investment fund, which hopefully will be adequately shored up by the returning cash flow.

The second model utilizes the actual paths followed by the individual

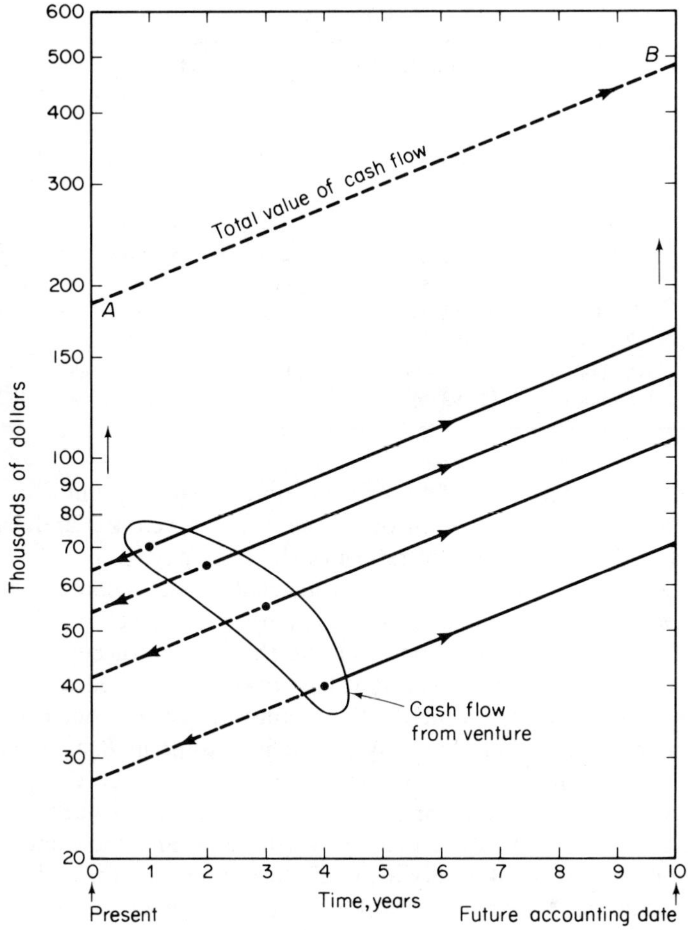

Figure 2.2. Accrued value of cash flow at company's average rate of earnings.

return cash-flow amounts as shown in Fig. 2.2 by solid lines. Corresponding data are given in Table 2.6. The figure is based upon the concept of reinvestment opportunity. A total future value will be reached by this method, and this offers a criterion for evaluation. As each yearly net earning comes in, it is reinvested at an earning rate of 10%, which is here assumed to be the average reinvestment rate available to the company. The parallel solid lines projecting to the right show how these amounts increase in value as time passes. A future accounting date was arbitrarily selected at the end of the tenth year. The individual values as of that time were added together, and their sum is shown at point *B*. This point represents a part of the total ulti-

Table 2.6
Calculations of Present Value and Value at end of 10 Years of a Cash Flow, Based on Average Rate of 10%

Year	*Present Value* ($)	$\left(\frac{1}{1+i}\right)^n$	*Net Cash Flow* ($)	$(1+i)^{10-n}$	*Value at End of Tenth Year* ($)
0				2.59374	
1	63,640	0.90909	70,000	2.35795	165,060
2	53,720	0.82645	65,000	2.14359	139,330
3	41,320	0.75131	55,000	1.94872	107,180
4	27,320	0.68301	40,000	1.77156	70,860
	186,000		230,000		482,430
	(186,000)	(2.59374) =			482,430

mate financial objective as of that future time. That is, point *B* would be attained by funds that flow through the present venture and then are reinvested at the average rate of return of the company.

There is also shown in Fig. 2.2 an alternative mathematical route for reaching point *B*. This is the present-value path. The dashed parallel lines to the left of each yearly net earning extend to the present, or beginning of the venture, and the individual present values have been added together. The sum of the present values, which is the quantity ordinarily determined in a present-value analysis, can obviously be extended to point *B*, as indicated by the upper dashed line. If two or more ventures are to be compared, the comparisons can be made as present values or as future values at an arbitrarily selected date, and their relative worths will rank the same. Clearly, this type of analysis is valid only if the discount rate used throughout corresponds to the rate of return that will prevail in the future. It depends upon the quality of investments that become available for the return cash flow from the present investment. An approximation would be to use the average rate of earnings of the company. Again, in the second model, as in the first, the quantity to be maximized is DE/C, the chief difference between the two models residing in the discount rate used to calculate the present value.

Another question that is answered in this analysis concerns whether the discount methods should correspond to an annual, semiannual, or continuous interest method. From Fig. 2.2, it can be seen that the answer to this question lies in the manner in which the return cash flow is reinvested. In the figure the return cash flow is depicted as being reinvested at the ends of the yearly periods. Thus end-of-the-year discount methods should be used for that case. If the monies would normally be reinvested semiannually, that type of calculation should be used. If the return cash flow from the present venture is comingled with cash flows from several similar ventures and reinvestments are made on a monthly, weekly, or essentially continuous basis, the use of continuous-interest-type calculations offers a good approximation.

One of the principles involved in causing different investors to prefer different investments is illustrated in Fig. 2.3 and Table 2.7. Investment *A* will yield \$90,000 at the end of 5 years. Investment *B* will yield \$160,000 at the end of 9 years. Company *X* enjoys an average rate of return of 10%. After discounting the two cash returns, company *X* prefers investment *B*. Company *Y* has been more fortunate and has achieved an average rate of return of

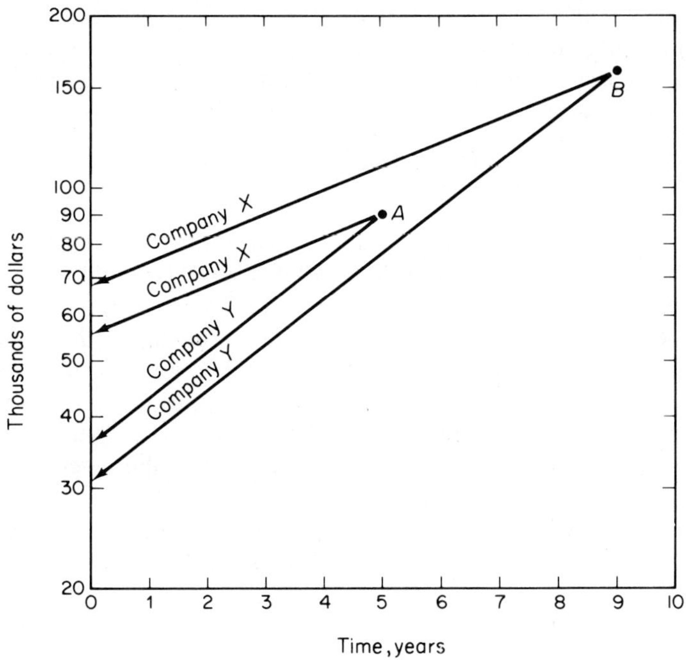

Figure 2.3. A case where different discount rates change investment preference.

Table 2.7
Reduction of Two Future Amounts to Present Values at Two Discount Rates

Company	*Discount Rate (%)*	*Venture*	*Cash Flow (\$)*	*Year*	$\left(\frac{1}{1+i}\right)^n$	*Present Value (\$)*	*Choice*
X	10	*A*	90,000	5	0.6207	55,860	
		B	160,000	9	0.4241	67,860	×
Y	20	*A*	90,000	5	0.4019	36,170	×
		B	160,000	9	0.1938	31,010	

20%. After discounting the two returns, company *Y* prefers investment *A*. Or, more explicitly, company *Y* can pay about \$5,000 more for investment *A* than it can pay for investment *B*, and still make 20%. But, of course, company *Y* must pay less than company *X*, in any case, if it is to realize a greater rate of return on the investment.

An alternative interpretation, and perhaps a more realistic one, can be placed upon this illustration. If may be that the average earnings of both companies are about the same. However, the managements are both attuned and influenced by probable future trends. It may well be that the management of company *Y* anticipates the arrival of investment opportunities that will yield about 20%. It, therefore, wishes to rank present investment opportunities on that basis. The shorter time until investment *A* expires tends to make it more attractive.

2.3-4 Present-Value Profile

Any proposed venture can be discounted at several interest rates and corresponding different present values will be found. Graphical plots of such data were illustrated by Woody and Capshaw.[2] These authors recommended this method for presenting condensed information to management that would aid in selecting between possible alternative investments. The technique eliminates the necessity of immediately establishing a discount rate for making an evaluation. Use of the average rate of return of all company investments was recommended as the proper discount rate for ranking ventures for the reinvestment of company-generated funds.

Figure 2.4 is a plot of the present-value profiles of two prospective oil wells. Both have a present value of \$7,000 at a discount rate of 16%. By this

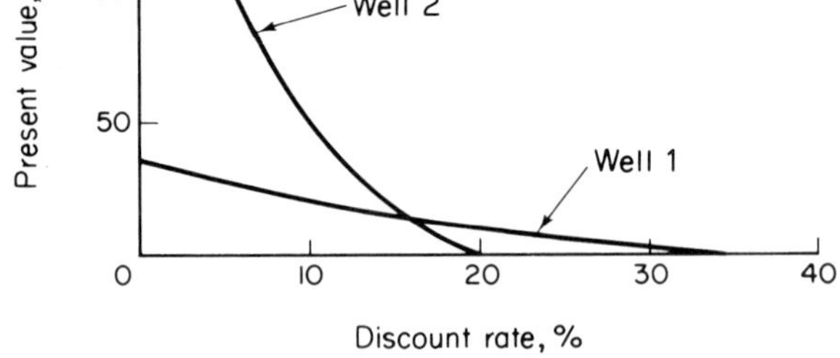

Figure 2.4. Present-value profiles of two drilling prospects.

method of ranking ventures, a company averaging less than 16% on its investments would prefer to drill well 2. A company averaging greater than 16% on its investments would prefer to drill well 1. Whether or not there is any appreciable advantage or significance one way or the other must be read along the vertical axis, the left margin. To make a comparison, one uses the present values of the two prospects, as discounted at the reinvestment rate available to the company.

The concealed reliance of this method on adequate and favorable reinvestment opportunities is shown in Table 2.8. Quite obviously, venture

Table 2.8
Comparison of Two Projects with Same Rate of Return

	Venture X	*Venture Y*
Investment ($)	100	100
Income first year ($)	150	0
Income second year ($)	0	225
Rate of return (%)	50	50

Source: After Woody and Capshaw.[2]

X is not as good as venture *Y*, unless by some rare chance the $150 received at the end of the first year could be reinvested at 50%. If it were reinvested at 10% then at the end of 2 years one would be comparing $165 from choosing venture *X* to $225 from choosing venture *Y*.

2.3-5 Payout Time and Number of Times Paid Out

One of the earlier criteria used to judge the desirability of an oil-well-drilling venture was payout time. For example, if it were estimated that a well would pay out in not more than 3 to 5 years, the venture often was judged to be acceptable. This was a reflection of the limited capital resources of the operators and of the probability that the flush production usually would be over by then. Thus it would be time to make a new move and the capital would be needed.

Another early criterion used to judge drilling ventures was the number of times that a venture would pay out. Where the operating profits would amount to twice the initial investment, one would ostensibly double his money. Although early usage of these two criteria generally neglected any reduction to present values, they can be used together with discounted-cash-flow concepts. Indeed, the discount rate, which reduces the future operating income so that the venture pays out exactly one time, has been called the *internal rate of return* or perhaps more properly the *discounted-cash-flow (DCF) rate of return.*

Advantages associated with a short payout period include the early return of the capital for reinvestment and minimizing the effect of uncertainties that increase with time. Among the latter are the possibilities of increased inflation, a decrease in crude oil price, increased operating costs, possible equipment failures, and increased pressure from competitors. Advantages associated with the larger number of times paid out include a hedge against uncertainties in geologic information and reservoir recovery factors. Also, the larger number of times paid out would generally be associated with larger hydrocarbon reserves, and often this is an important factor.

Two examples are shown in Fig. 2.5. The number of times paid out, calculated as the present value of the operating profits divided by the initial

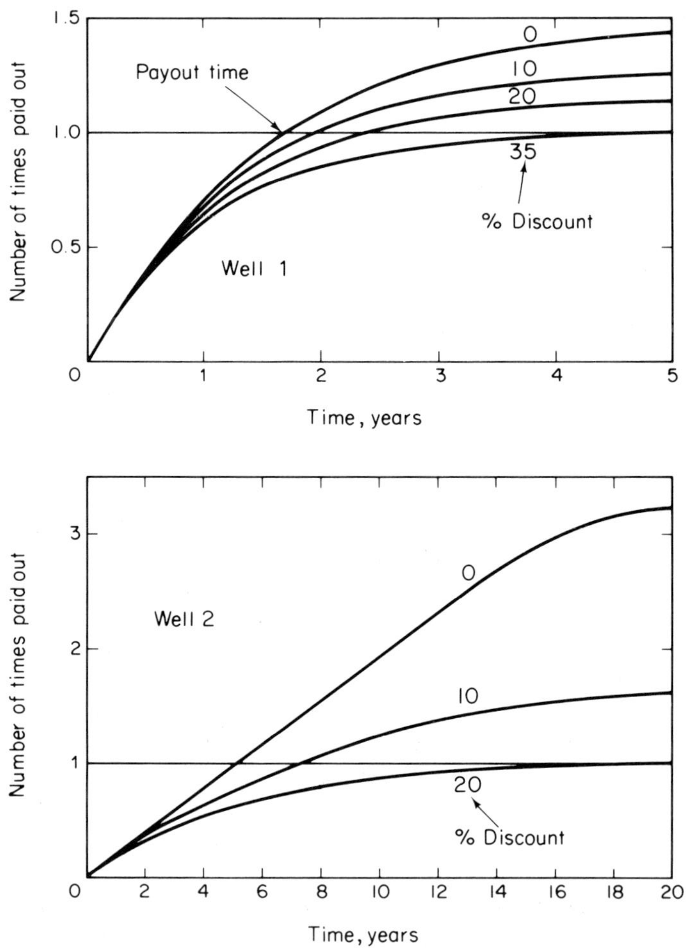

Figure 2.5. Number of times paid out versus time.

investment, is plotted versus time. The predicted costs and productions are as follows:

Well 1. Initial investment, $C = \$80{,}000$. Total reserves of 57,700 bbl, produced over 5 years on exponential decline, $q_0 = 96$ bbl/day, final rate of 6 bbl/day. Operating profit taken as $2 per bbl.

Well 2. Initial investment, $C = \$83{,}600$. Total reserves of 135,600 bbl, produced at rate of 20 bbl/day for 14 years, followed by exponential decline for 6 years to a final rate of 4 bbl/day. Operating profit taken as $2 per bbl.

Table 2.9 shows the manner in which the data were prepared for plotting in Fig. 2.5. Details of the manner in which each year's production can be calculated are explained in Chapter 11. The present value of the operating profit for each year was calculated using the midyear convention. That is, the income for each year was assumed to be received at the midpoint of the

Table 2.9
Payout Time and Number of Times Paid Out,
Example Well 1

		Operating Profits ($)					
		Undeferred			*Present Value at* 10%		
Year	*Production (bbl)*	*Yearly*	*Cumulative*	*Cumulative/C*	*Yearly*	*Cumulative*	*Cumulative/C*
1	28,800	57,600	57,600	0.72	54,920	54,920	0.69
2	15,000	30,000	87,600	1.10	26,000	80,920	1.01
3	7,800	15,600	103,200	1.29	12,290	93,210	1.17
4	4,000	8,000	111,200	1.39	5,730	98,940	1.24
5	2,100	4,200	115,400	1.44	2,740	101,680	1.27

		Operating Profits ($)					
		Present Value at 20%			*Present Value at* 35%		
Year	*Production (bbl)*	*Yearly*	*Cumulative*	*Cumulative/C*	*Yearly*	*Cumulative*	*Cumulative/C*
1	28,800	52,580	52,580	0.66	49,580	49,580	0.62
2	15,000	22,820	75,400	0.94	19,140	68,720	0.86
3	7,800	9,890	85,290	1.07	7,380	76,100	0.95
4	4,000	4,230	89,520	1.12	2,810	78,910	0.99
5	2,100	1,850	91,370	1.14	1,090	80,000	1.00

year, thereby approximating continuous receipts and disbursements of monies. Incidently, these are the same two wells whose present-value profiles are plotted in Fig. 2.4.

Payout time can be read from Fig. 2.5 by noting where the cumulative present-value lines cross the paid-out one-time line. Thus a range of values of payout time begins where the cumulative undeferred value line, indicated as discounted at zero interest rate, crosses the one-time paid-out line and extends to the end of the venture. In most instances the payout time that corresponds to the cost of capital, such as 8 or 10%, would be of more significance than the 0% interest line. On the right-hand edge of the diagrams it can be seen that the 35% line in the case of well 1, and the 20% line in the case of well 2, intersect the paid-out one-time line at the ends of the ventures. These are the DCF rates of return for the respective ventures.

2.3-6 Discounted-Cash-Flow Rates of Return

Impartial discussion of the DCF rate of return method requires the presentation of both the best and the worst features of the concept. The best features appear in those cases where there is one investment that is made at the beginning, and there is only one apparent solution, one rate of return. The worst features appear in cases where there are several alternate periods of investments and profits, and there appear to be several solutions, several rates of return. Another name for this same method is the *internal rate of return.*

The discounted-cash-flow rate of return may be defined as the rate of interest that will reduce future net incomes so that the sum of their present values equals the initial investment. This may be written as

$$0 = -C + \frac{S_1}{(1+i)^{0.5}} + \frac{S_2}{(1+i)^{1.5}} + \cdots + \frac{S_n}{(1+i)^{n-0.5}}, \qquad (2\text{-}23)$$

where

C = initial investment, dollars,

S_1 = net cash flow, considered to be spread throughout the first year,

i = interest rate, per year.

The defining equation is sometimes written as

$$0 = \sum_{k=0}^{n} \frac{S_k}{(1+i)^k}. \qquad (2\text{-}24)$$

In this form all transfers can be considered to occur at the midpoint of each year. Multiplying all terms by $(1 + i)^{0.5}$ would, of course, be redundant and would produce no new or different answers. Quite obviously, all transfers can as well be considered to occur at the end of each year. Equation (2-23) indicates how to provide for cases where lump-sum investments are interspersed with more-or-less continuous disbursements and receipts. Nor does this exhaust the possibilities, since continuous-interest-type calculations are available. One needs, however, to examine the character of the results.

For there to be any solutions to Eq. (2-24) there must be some positive and some negative terms on the right-hand side. Where there is only one investment, the S_0 term is negative and the remaining terms positive. This involves one change of sign, and there is only one root, one value of i, which satisfies the equation. Suppose that the terms were $S_0(-)$, $S_1(+)$, $S_2(-)$, $S_3(+)$, and $S_4(+)$. Here there are three changes of sign, and there may be as many as three values of i that satisfy the equation. The possible number of solutions is governed according to *Descartes Rule*.

An example of a venture that requires only a single investment at the beginning is shown in Tables 2.10 and 2.11, and its present-value profile is given in Fig. 2.6. Table 2.10 shows the calculations for three points on the present-value profile. The solution of Eq. (2-24) is indicated to be $i = 0.148$, or 14.8%, for the DCF rate of return. Table 2.11 shows the amount of the remaining (unamortized) balances as of the beginning of each year. A profit amounting to 14.8% of the current investment is taken out of the total net cash flow at the end of each year, and the remainder of the cash flow is used

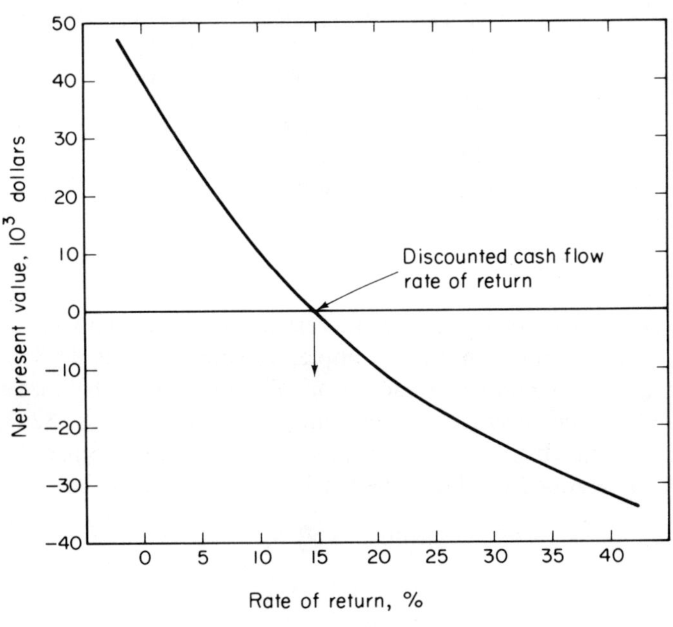

Figure 2.6. Present-value profile of venture shown in Tables 2.10 and 2.11.

Table 2.10
Present Values of Return Cash Flow, End-of-Year Method

		$i = 0.05$		$i = 0.30$		$i = 0.148$	
Year	*Net Cash Flow* ($)	$\left(\frac{1}{1+i}\right)^n$	*Present Value* ($)	$\left(\frac{1}{1+i}\right)^n$	*Present Value* ($)	$\left(\frac{1}{1+i}\right)^n$	*Present Value* ($)
0	−80,000	1.0000	−80,000	1.0000	−80,000	1.0000	−80,000
1	20,000	0.9524	19,050	0.7692	15,380	0.8711	17,420
2	30,000	0.9070	27,210	0.5917	17,750	0.7588	22,750
3	25,000	0.8638	21,600	0.4552	11,380	0.6610	16,510
4	20,000	0.8227	16,450	0.3501	7,000	0.5757	11,480
5	15,000	0.7835	11,750	0.2693	4,040	0.5015	7,500
6	10,000	0.7462	7,460	0.2072	2,070	0.4369	4,350
	40,000		23,520		−22,380		0

Table 2.11
Yearly Profits and Amortization Payments ("Proof of Return on Investment"), End-of-Year Method

Year	*Investment Balance, First of Year* ($)	*Net Cash Flow, End of Year* ($)	*Profit at* $i = 0.148$ ($)	*Amortization, End of Year* ($)	*Remaining Investment, End of Year* ($)
1	80,000	20,000	11,840	8,160	71,840
2	71,840	30,000	10,630	19,370	52,470
3	52,470	25,000	7,940	17,060	35,410
4	35,410	20,000	5,240	14,760	20,650
5	20,650	15,000	3,060	11,940	8,710
6	8,710	10,000	1,290	8,710	0
	269,080	120,000	40,000	80,000	

to reduce the investment. In this manner the investment is extinguished by the end of the venture.

In the DCF method profit and interest are considered to be the same thing. The yearly profit, in this example, is calculated as 14.8% of the investment at the beginning of each year. The investment balance is commonly called the *unamortized balance*, and the sum of these, $269,080 in this example, is called the *sum of the unamortized balances*. Since the yearly equations can be added to obtain the total profit, one can write

$$\text{profit} = i \sum B, \tag{2-25}$$

where

$\sum B$ = sum of the unamortized balances,

i = rate of return on investment.

In this example (0.148)($269,080) = $40,000. Any degree of numerical accuracy required can be achieved by increasing the number of significant figures.

In considering this example it should be noted that the $80,000 does not remain invested in this venture for the 6-year life of the venture. If it did, it would accumulate to a total of $183,000, giving a profit of $103,000, as compared to the $40,000 actually realized. Therefore, once again attention is inevitably drawn to the necessity of considering the reinvestment opportunities. Other investments, which yield approximately the same rate of return, are required to fill out the total investment pattern. Provided that equal reinvestment opportunities will be available, as the venture returns cash flow, the DCF rate of return provides a method for comparison of investment opportunities.

Another example, one that brings out the less satisfactory aspects of the DCF method, is shown in Tables 2.12 and 2.13. The present-value profile is shown in Fig. 2.7. For this example there are three solutions of Eq. (2-24); roots are found at values of i of 8, 19, and 41% annual interest rates. Not more than one of these could be used to characterize the venture, and concern about this prompts an inquiry into the meaning of these solutions.

Table 2.13 attempts to answer some of the questions regarding the meaning of solutions of Eq. (2-24). It uses two investment funds for this purpose. Both of these funds accumulate at a specified rate of interest, and for this example they are shown in successive illustrations as accumulating at 0, 8, 19, and 41% annual interest rates. Fund A serves as the source and sink for the venture; it supplies funds when needed and it receives cash flows shed by the venture. All balances in fund A accumulate at the specified rate

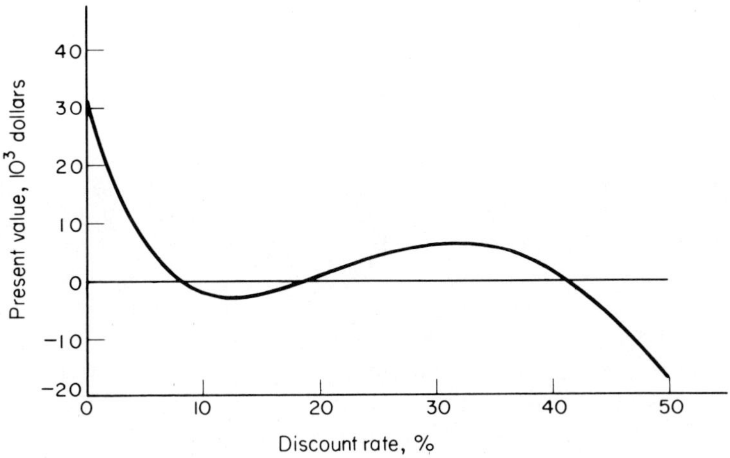

Figure 2.7. Present-value profile showing three rates of return. Intercepts are at 8, 19, and 41%.

Table 2.12
Cash-Flow Pattern Giving Three Discounted-Cash-Flow Rates of Return

Year	*Cash Flow, End of Year* ($)
0	—5,000
1	18,409
2	—22,452
3	9,075

Table 2.13
Investment-Fund Method of Analyzing Discounted-Cash-Flow Rates of Return

i	*Year*	*Fund A, Partially Invested in a Venture* ($)						*Fund B, Without Venture* ($)
		Balance, First of Year	*Transfer to Venture, First of Year*	*Balance, During Year*	*Interest on Balance*	*Transfer from Venture, End of Year*	*Balance, End of Year*	*Balance, End of Year*
0	1	10,000	5,000	5,000	0	18,409	23,409	10,000
	2	23,409	0	23,409	0	0	23,409	10,000
	3	23,409	22,452	957	0	9,075	10,032	10,000
0.08	1	10,000	5,000	5,000	400	18,409	23,809	10,800
	2	23,809	0	23,809	1,904	0	25,713	11,664
	3	25,713	22,452	3,261	261	9,075	12,597	12,597
0.19	1	10,000	5,000	5,000	950	18,409	24,359	11,900
	2	24,359	0	24,359	4,628	0	28,987	14,161
	3	28,987	22,452	6,535	1,242	9,075	16,852	16,852
0.41	1	10,000	5,000	5,000	2,050	18,409	25,459	14,100
	2	25,459	0	25,459	10,438	0	35,897	19,881
	3	35,897	22,452	13,445	5,512	9,075	28,032	28,032

of interest. Fund B simply accumulates at the specified rate of interest and is otherwise undisturbed. It can be observed that when the interest rates correspond to the roots of the equation, in this case 8, 19, and 41% the two funds accumulate to exactly the same amounts by the end of the venture. This result is foretold by the fact that the present value of the venture is zero when it is discounted at these particular interest rates.

Solutions of Eq. (2-23) and (2-24) and their like indicate interest rates that will discount cash flows from a venture to zero present value. By extension, the solutions indicate characterizing rates of interest-accumulation fields in which the cash flows to and from the venture will have an ultimate zero effect, where they will not influence the final amounts. These mathematical solutions do not guarantee that the firm has access to any such interest-accumulation fields or opportunities in the real world. Such solutions are intercepts on the horizontal axis of present-value profile plots. They serve as criteria for judging the worth of ventures when other characteristics are understood as being satisfactory also.

Appraisals by the DCF method were discussed previously. Equation (2-22) or (2-23) is used for the purpose. It might be well to mention that many times a fictitiously high rate of return, 20%, for example, is used to calculate present worth. This figure has been used to evaluate the development drilling that is necessary when extending the limits of an oil field. On the average, one of every four or five of these wells is not productive. Thus the fictitiously high rate provides sufficient profit on the good wells to pay also for the bad wells. It is not necessarily recommended as the best procedure; rather, it is noted as a common practice.

2.3-7 Acceleration Projects

In many operations, the opportunity often arises where a supplementary investment will augment the total amount of the return cash flow or decrease the required time. These situations are known as *acceleration projects*. The features are

1. An original cash-flow stream is assured by presently existing investments.
2. An accelerated cash-flow stream can be obtained by making an additional investment.

Analysis of the attractiveness of the original investment would have been made at the outset of the venture. Analysis of the attractiveness of the acceleration investment is made at a later time, when the second investment is contemplated. Two basic evaluation dates are evident. Total knowledge available would be greater at the later date, and uncertainties of the outcome would generally be greatly reduced. Evaluation of an acceleration project reduces to an evaluation of the difference between the accelerated and the original cash-flow streams.[2,8]

A case where an additional investment increased the undiscounted profit is shown in Table 2.14. The accelerating investment of $10,000 was made in the fourth year. When the values are undiscounted, the differences between the sums of the original and accelerated streams directly check the sum of the differences, \$32,000 − \$21,000 = \$11,000. When they are discounted, the differences between the sums check after correcting to the same basic date. The difference column is discounted to the beginning of the fourth year (the

Table 2.14
Undiscounted and Present Values of Net Cash-Flow Streams of an Acceleration Project

Year	*Undiscounted Values (\$)*			*Present Values at 10% (\$)*		
	Original	*Accelerated*	*Difference*	*Original*	*Accelerated*	*Difference*
0	−20,000	−20,000		−20,000	−20,000	
1	10,000	10,000		9,090	9,090	
2	7,000	7,000		5,780	5,780	
3	5,000	5,000		3,760	3,760	
4	4,000	−6,000	−10,000	2,730	−4,100	−9,090
5	4,000	16,000	12,000	2,480	9,930	9,120
6	3,000	12,000	9,000	1,690	6,770	9,760
7	3,000	8,000	5,000	1,540	4,110	3,410
8	2,000		−2,000	930		−1,240
9	2,000		−2,000	850		−1,130
10	1,000		−1,000	390		−510
	21,000	32,000	11,000	9,240	15,340	8,120
Evaluated as of beginning of:				Year 1	Year 1	Year 4

end of the third year). A check is obtained as follows: $(1/1.10)^3$(\$8,120) = \$6,100; \$15,340 − \$9,240 = \$6,100. Thus the equivalence between evaluations made at different basic dates is readily demonstrated.

Analysis of this project is shown in Fig. 2.8. The curves for the complete project, both the original and the accelerated versions, are evaluated as of the beginning of the original investment. The curve for the difference is evaluated as of the beginning of the fourth year. Considering the values as reckoned on the horizontal axis, the curve for the accelerated venture lies above the original in the discount-rate range from −38 to 80%. Whether or not there is any appreciable significance attached to this must be determined from examining the images projected on the vertical axis, from considering the actual differences in the present worths. At 0% discount, the net worth of the acceleration investment is \$11,000. At 10% discount, the net present value is \$8,120, so the acceleration investment is to be preferred where the yields of other available investments is in the 0, 10, or 20% range. At higher discount rates, the advantage of the acceleration investment decreases. If other currently available investments yield anything over 80%, they would be preferred.

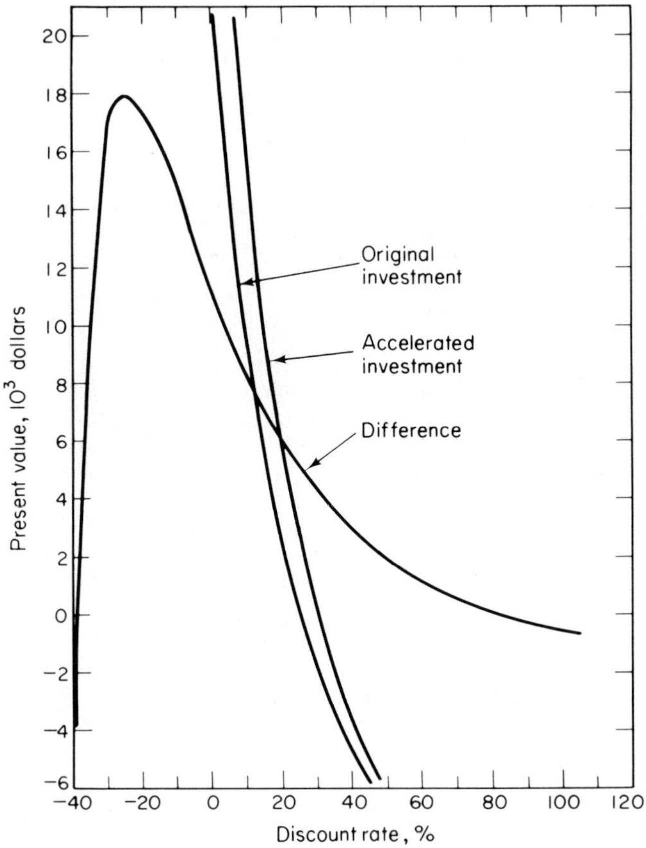

Figure 2.8. Present-value versus discount rate for an acceleration project. Data from Table 2.14.

Table 2.15 and Fig. 2.9 illustrate an acceleration project whose undiscounted net value is negative but which becomes a positive advantage at higher discount rates. This feature of showing positive present values in a limited discount range is the characterizing feature of most acceleration projects. In this illustration the present values are positive between 8 and 38 %. Interpretation may be made as follows. Unless the income from the project can be reinvested at rates greater than 8 %, the loss in undiscounted income will not be compensated by the faster rate of return cash flow. On the high side of the range, the reasoning is that if other investments are available that will yield greater than 38 %, they should be preferred to the acceleration investment. However, by noting that the present value at 20 % discount rate is \$1,260 from an acceleration investment of \$15,000, the relative gain in present-value position would be \$1,260/\$15,000 or about 8 %. If it is assumed

Table 2.15
Analysis of Acceleration Project by Discounting Differences Between the Original and the Accelerated Versions

Year	*Undiscounted Values* ($)			*Present Values** ($)	
	Original	*Accelerated*	*Differences*	$i = 20\%$	$i = 50\%$
1	15,000	15,000			
2	14,000	14,000			
3	12,000	12,000			
4	10,000	10,000			
5	10,000	10,000			
6	8,000	−7,000	−15,000	−12,500	−10,000
7	8,000	20,000	12,000	8,330	5,330
8	7,000	17,000	10,000	5,790	2,960
9	7,000	15,000	8,000	3,860	1,580
10	6,000	10,000	4,000	1,610	530
11	6,000		−6,000	−2,010	−530
12	5,000		−5,000	−1,140	−290
13	5,000		−5,000	−1,160	−200
14	4,000		−4,000	−780	−100
15	3,000		−3,000	−480	−50
	120,000	116,000	−4,000	1,260	−770

*Present values are as of the beginning of the acceleration phase.

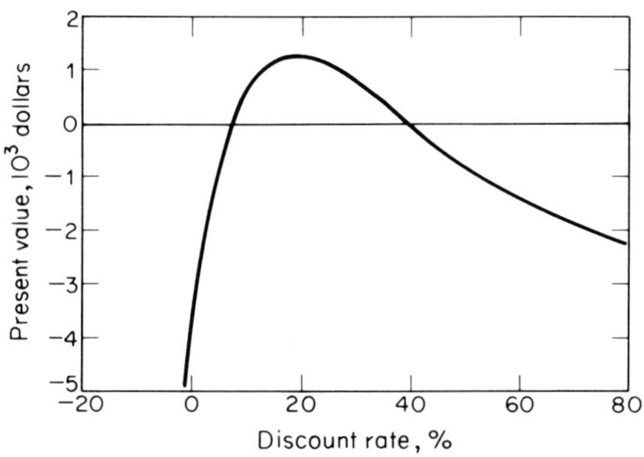

Figure 2.9. Present-value profile for an acceleration project. Data from Table 2.15.

that some risk and uncertainty is involved, the possible gain is small, and the project ordinarily might not be recommended in any case.

2.3-8 Proposed Modifications of the Discounted-Cash-Flow Method

The desire to rank and characterize the worth of investment projects with a single index, and the shortcomings of the DCF method, have spawned various attempts at modifications of the basic method. Two of these are discussed here. The first is a successful modification; the second is often difficult to apply and to interpret adequately.

Several evaluation equations have been formulated in terms of one initial investment. It is usually suggested that if any additional investments are required, they should be reduced to the time of the major initial investment at an interest rate of 6 or 8%, which would reflect the cost of borrowed capital. This approach can be used with Eq. (2-24). To illustrate, in the example shown in Table 2.12, the negative flow at the end of the second year can be combined with the initial investment. If it is reduced according to 8% (one of the original roots), then there remains only the one root at 8%. If that same term had been reduced according to 6%, then there would be only one root at approximately 7%. This procedure leaves only one change of sign, only one root. Thus it eliminates the problem of multiple roots, multiple rates of return. In general, it is to be recommended.

A second proposed modification turns out to be unsuited for application to a typical oil producing venture where the bulk of the cash flow occurs early, followed by a gradual decline period. The modification would be more appropriate where cash flows are uniform and end abruptly. A premise of the method is that return cash flow from a project is generally reinvested at the average rate of earnings of the firm. The values of the net earnings per time period can be projected according to the average rate and summed up as of the end of the venture. In a similar manner, all investment amounts can be discounted at the same average rate to the beginning of the venture. It is then possible to make a direct solution for the interest rate which would transform the initial amount into the final amount over the life span of the venture.[3,4] We shall next analyze the results so obtained.

In the notation of this chapter, the sum attained at the end of the venture can be expressed as its present value, as of the beginning of the venture, multiplied by the proper single-amount interest factor, or $DE(1 + i)^n$. If this same amount is to be attained by the present value of the investments, which accumulates at interest rate r, the amount can be expressed as $C(1 + r)^n$. Equating these future-value expressions gives

$$C(1 + r)^n = DE(1 + i)^n,$$

where D is the present-value factor for the future net earnings E. D is based upon i, the average earnings rate of the company, and r is a modified rate of return for the combination of the particular project and the average projects of the company. The equation can be arranged as

$$1 + r = \left(\frac{DE}{C}\right)^{1/n}(1 + i),$$

and in this form the value obtained for r can be seen to be very sensitive to the value of n, the lifetime assigned to the project. As n increases, the value of r is mathematically forced closer to the value of i. This characteristic is unsuited for judging ventures, such as many in the petroleum industry, in which the bulk of the cash flows occur early but may be followed by an extended life of very satisfactory, although declining, productivity.

2.4 AVERAGE-ANNUAL-RATE-OF-RETURN METHOD

The name "average annual rate of return" was given to a method proposed by J. J. Arps.[5] Rates of return calculated by the *average annual method* are usually very close to those obtained by the DCF method. Both methods use the total of the investment balances over the life of the venture as a basis for calculating rate of return. Trial-by-error calculations, required for the DCF method, are eliminated in the average annual method. An assigned interest rate equal to the cost of using capital permits determining the basis. Most, if not all, methods do find average rates that can be used throughout the life of a venture.

In the average annual rate of return method, the total undeferred cash flow is divided as follows:

$$E = C + I + P, \tag{2-26}$$

where

E = total undiscounted net cash flow as received over the life of the venture,

C = initial investment,

I = interest paid at rate i on the invested capital,

P = profit,

i = interest rate charged for the use of capital.

It was observed in discussing Table 2.11 that if the sum of the unamortized balance can be established, the rate of return can be calculated very easily. Where the rate of return reflects both interest on capital and profit, Eq. (2-25) can be rewritten as

$$r = \frac{I + P}{\sum B}, \tag{2-27}$$

where

r = average annual rate of return,

$\sum B$ = sum of the unamortized balances.

Using Eq. (2-26) to eliminate I and P permits writing

$$r = \frac{E - C}{\sum B}. \tag{2-28}$$

Attention is next focused on finding a replacement for $\sum B$ in terms of quantities that are more readily available. The first step is given by the following relation, which can be observed to be true in Table 2.11:

$$I = i \sum B. \tag{2-29}$$

Next, I can be eliminated. By definition, the overall deferment factor must reduce the future repayments of capital plus interest to a present value equal to the investment. Thus

$$C = D(C + I) \tag{2-30}$$

or

$$I = C\frac{1 - D}{D}, \tag{2-31}$$

where

D = deferment or present-value factor based on interest rate i and usually based on the time pattern of the total earnings E.

Using the foregoing relations to eliminate $\sum B$ from Eq. (2-28) produces the following:

$$r = i\frac{D}{1 - D}\left(\frac{E}{C} - 1\right). \tag{2-32}$$

In the form of an appraisal equation, this becomes

$$C \leq \frac{DE}{\frac{r}{i} - D\left(\frac{r}{i} - 1\right)}. \tag{2-33}$$

Example: A well is to be drilled on 40-acre spacing. Net pay thickness is 20 ft, giving a total of 800 acre-ft. A recovery of 150 bbl/acre-ft is anticipated, a total of 120,000 bbl. Initial rate will be about 24 bbl/day, and the economic limit is 3 bbl/day. The deferment factor, based on 8% cost of capital, is

0.51392. If the net operating profit averages \$2 per barrel, what is the value of the future production, if a 20% profit is required as a cushion against unfavorable outcomes?

Solution:

$$C \leq \frac{DE}{\frac{r}{i} - D\left(\frac{r}{i} - 1\right)}$$

$$\leq \frac{(0.51392)(120{,}000 \text{ bbl})(\$2 \text{ per bbl})}{\frac{0.20}{0.08} - 0.51392\left(\frac{0.20}{0.08} - 1\right)}$$

$$\leq \$71{,}300.$$

This is the maximum amount that could be paid for buying the lease and drilling and completing the well.

2.5 AVERAGE-BOOK-RATE-OF-RETURN METHOD

Another method that bases rate of return on the total of the investment balances is the *average book method.* This method gives answers that are usually very close to the DCF rate of return. The sum of the investment balances is calculated directly. An amortization procedure must be assumed for this purpose. In many cases the value of an asset is depreciated in proportion to the units it has produced, taken as a fraction of the estimated total production. Rephrasing, one would say that amortization of the investment was made in proportion to production. The concept is straightforward and rather easily explained. In most cases it gives a very good answer.

Establishing a rate of return relationship for this method requires only the estimating of the sum of the unamortized balances. This quantity was observed in the DCF and average annual rate of return methods. Two developments of this method have been published.[6,7] Both can be modified to suit the cases where money transfers are made at year end or those where average conditions can be assumed in order to better approximate continuous flow of funds. The corresponding amortization patterns are illustrated graphically in Fig. 2.10. First, the year-end conventions will be considered.

There are several ways of arriving at an expression for the sum of the unamortized balances. The easiest way is to recognize that the amount amortized at the end of the first year was invested for 1 year, the amount amortized at the end of the second year was invested for 2 years, and so on to the end of the venture. One can then directly write

$$\sum B = \sum_{k=1}^{n} k a_k, \tag{2-34}$$

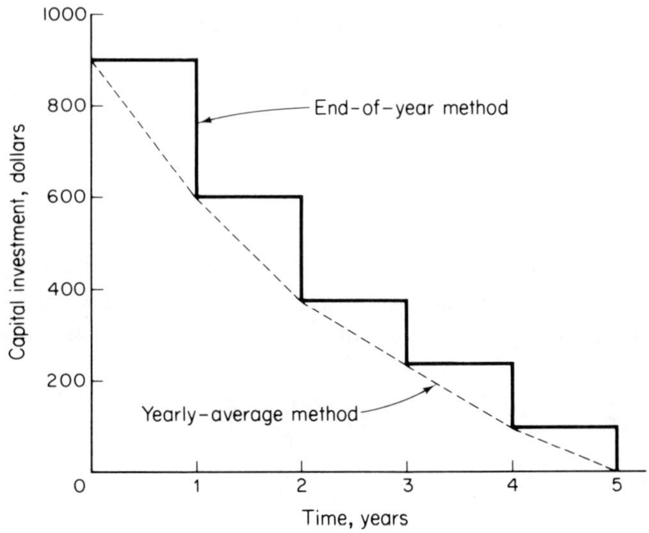

Figure 2.10. Amortization patterns for the average book method.

where

$\sum B$ = sum of the unamortized balances, dollars,

n = life of the venture, years,

C = initial investment, dollars,

a_k = amortization in kth year, dollars.

The following relation indicates that the total amortization will equal the initial investment:

$$\sum_{k=1}^{n} a_k = C. \tag{2-35}$$

Where the amortization is set proportional to the yearly productions, one can write

$$\sum B = \frac{C}{N} \sum_{k=1}^{n} kQ_k, \tag{2-36}$$

where

Q_k = oil produced during kth year, barrels,

N = total recoverable oil, barrels.

The relation can as easily be expressed in terms of net cash flow per year. To accomplish this, one would replace barrels of oil with net-cash-flow dollars.

The average balance in any year is the balance at the first of the year less half of the amortization effected during the year. It follows that the sum of the average balances must be half of the initial investment smaller than the sum of the first-of-the-year balances. For the midyear convention, one has

$$\sum B = \left(\frac{C}{N}\sum_{k=1}^{n} kQ_k\right) - \frac{C}{2}. \tag{2-37}$$

The other development is in terms of total oil produced from the beginning of the venture. In this notation the total amortization is proportional to the cumulative production. The sum of the unamortized balances can be written as

$$\begin{aligned}\sum B &= (C) + \left(C - \frac{N_1}{N_p}C\right) + \left(C - \frac{N_2}{N_p}C\right) + \cdots + \left(C - \frac{N_{n-1}}{N_p}C\right) \\ &= nC - \frac{C}{N_p}[N_1 + N_2 + \cdots + N_{n-1} + (N_n - N_n)] \\ &= C(n+1) - \frac{C}{N_p}\sum_{k=1}^{n} N_k\end{aligned} \tag{2-38}$$

where

N_k = cumulative oil produced by end of kth year, barrels,

N_p = cumulative oil produced by end of nth year, barrels.

Subtracting half of the initial investment to obtain the midyear convention relationship gives

$$\sum B = C\left(n + \frac{1}{2}\right) - \frac{C}{N_p}\sum_{k=1}^{n} N_k. \tag{2-39}$$

Example: *Calculating Rate of Return by the Average Book Method*
The estimated production of a well is shown by year in the second column of Table 2.16. The operating profit will average \$2 per barrel. Cost of drilling and completing the well will be \$144,000. Estimate the rate of return, assuming that amortization of the initial investment is proportional to the relative amount of oil produced.
Solution: The sum of the unamortized balances is determined in two different ways in Table 2.16. The total amount to be amortized is \$144,000. The amount amortized during the first year is (40,000 bbl/120,000 bbl) × (\$144,000) = \$48,000. In similar manner, the other amortizations were calculated. The sum of the unamortized balances is calculated explicitly as \$348,000. In the preceding column it can be seen that the sum of the ka_k products is also \$348,000.

Table 2.16
Calculation of Sum of the Unamortized Balances, Average Book Method

Year	*Production,* Q (10^3 *bbl*)	*Cumulative Production,* N_k (10^3 *bbl*)	*Amortization,* a_k (10^3 *dollars*)	ka_k (\$)	*Balance, First of Year* (10^3 *dollars*)	*Average Balance During Year* (10^3 *dollars*)
1	40	40	48	48	144	120
2	30	70	36	72	96	78
3	20	90	24	72	60	48
4	20	110	24	96	36	24
5	10	120	12	60	12	6
	120	430	144	348	348	276

The same value of the sum of the unamortized balances can be calculated by Eq. (2-38) as follows:

$$\sum B = C\left[(n + 1) - \frac{1}{N}\sum N_k\right]$$

$$= 144{,}000\left[(5 + 1) - \frac{1}{120}(430)\right]$$

$$= \$348{,}000.$$

For the midyear convention calculation, the sum would be

$$\sum B = \$348{,}000 - \frac{144{,}000}{2}$$

$$= \$276{,}000.$$

The rate of return is calculated as follows:

$$r = \frac{E - C}{\sum B}$$

$$= \frac{(120{,}000 \text{ bbl})(\$2 \text{ per bbl}) - \$144{,}000}{\$348{,}000}$$

$$= 0.276, \text{ or } 27.6\%.$$

Using the midyear convention, the rate calculates as follows:

$$r = \frac{(120{,}000 \text{ bbl})(\$2 \text{ per bbl}) - \$144{,}000}{\$276{,}000}$$

$$= 0.348, \text{ or } 34.8\%.$$

For this example the DCF method gives a rate of return of 25.8% for the end-of-year convention and a rate of 35.8% by the midyear convention. Since the future incomes and costs are estimates, these slight differences between the DCF and the average book methods would usually be considered insignificant.

2.6 HOSKOLD'S METHOD

Hoskold's method has been popular in mine evaluations.[5] It is suited to ventures that have a limited profitable life and where the original capital must be returned at the end of the venture. Thus it is a sinking-fund method wherein the money accumulated in the sinking fund will be available for reinvestment when the present venture has expired. The sinking fund is accumulated at a "safe rate of interest" in bank deposits or negotiable securities. A uniform series of profits is postulated, and these profits are expressed as a fraction of the initial investment.

In arriving at the method, the total net earnings over the life of the venture are divided into two parts, one of which will go to pay profits and the other into the sinking fund:

$$E = E_p + E_{sf}, \tag{2-40}$$

where

E = total undeferred net earnings,

E_p = portion that will satisfy the profit requirements,

E_{sf} = portion that will satisfy the sinking-fund requirements.

The present values of both sides of Eq. (2-40) will also be equal, provided that discounting is done at the same rate throughout. Here both sides are discounted at the safe rate of interest, i. The following relation holds true for the entire amount, by definition of D, the present-value factor:

$$\text{P.V.}(E) = DE. \tag{2-41}$$

The present value of the portion of total earnings which are disbursed as profits is the present value of a series of uniform payments. The amount of each payment is $r_H C$. The present value of the series is given by the following familiar relation:

$$\text{P.V.}(E_p) = r_H C \frac{1 - \left(\frac{1}{1+i}\right)^n}{i} \tag{2-42}$$

The sinking fund must accumulate by the end of the venture to a value equal to C, the initial investment. Its value at that point in time is thus known. This value can be reduced to the present value by the following fundamental relation:

$$\text{P.V.}(E_{sf}) = \left(\frac{1}{1+i}\right)^n C. \tag{2-43}$$

Equating the present values gives

$$DE = r_H C \frac{1 - \left(\frac{1}{1+i}\right)^n}{i} + \left(\frac{1}{1+i}\right)^n C, \tag{2-44}$$

where

C = cost of the venture, the initial investment,

r_H = fraction of the cost that is taken for yearly profits, commonly referred to as the *speculative rate*,

i = interest rate at which money accumulates in the sinking fund, commonly referred to as the *safe rate*,

n = life span of the venture, years,

D = present-value factor that reduces the net earnings E to a present value at interest rate i; D depends upon both the value of i and the time pattern of the earnings E.

The following solution may be obtained:

$$r_H = i \frac{\frac{DE}{C} - (1+i)^{-n}}{1 - (1+i)^{-n}}. \tag{2-45}$$

In the form of an appraisal equation, one obtains the following:

$$C \leq \frac{DE}{\frac{r_H}{i} - \left(\frac{r_H}{i} - 1\right)(1+i)^{-n}}. \tag{2-46}$$

Hoskold's equation is widely known in the following form. If the total earnings are in the form of a series of uniform payments, each of which has a value of S, then the present value of the total earnings is given by

$$\text{P.V.}(E) = S \frac{1 - (1+i)^{-n}}{i}. \tag{2-47}$$

Using the above with Eqs. (2-42) and (2-43) permits obtaining the following relations:

$$r_H = \frac{S}{C} - \frac{i}{(1+i)^n - 1}, \tag{2-48}$$

$$C \leq \frac{S}{r_H + \dfrac{i}{(1+i)^n - 1}}, \tag{2-49}$$

where

S = total net earnings, received at the end of each year.

Example: A company attempts to pay 10% on its capital funds. When capital is not used within a venture, it is accumulated at 5% in a bank deposit. A well will produce 50,000 bbl over the next 10 years. The present-value factor, based on 5% interest, is 0.80. If the operating profit is \$1.80 per bbl, how much can be paid for the well?

Solution:

$$C \leq \frac{DE}{\dfrac{r_H}{i} - \left(\dfrac{1}{1+i}\right)^n \left(\dfrac{r_H}{i} - 1\right)}$$

$$\leq \frac{(0.80)(50{,}000\text{ bbl})(\$1.80)}{\dfrac{0.10}{0.05} - \left(\dfrac{1}{1.05}\right)^{10}\left(\dfrac{0.10}{0.05} - 1\right)}$$

$$\leq \$51{,}940.$$

Example: An addition to a school will cost \$500,000. Bonds will be sold at 7%, interest payable annually, to be retired after 10 years. A sinking fund will be created from annual taxes, which will accumulate in a bank at 4%. How much annual taxes must be collected?

Solution: Equation (2-47) is rearranged to the following form:

$$S = C\left[r_H + \frac{i}{(1+i)^n - 1}\right]$$

$$= \$500{,}000\left[0.07 + \frac{0.04}{(1.04)^{10} - 1}\right]$$

$= \$76,670$ annual taxes required to pay interest and retire the bonds.

NOMENCLATURE

A = amount of money,

a = amortization during a particular year,

$\sum B$ = sum of the unamortized balances, over the life of a venture,

C = initial investment, cost of a venture,

D = composite present-value factor,

DCF = discounted cash flow,

E = total undiscounted net operating profit or earnings received over the life of a venture,

$e_1, e_2, \ldots$ = undiscounted earnings in years 1, 2, . . . ,

e = 2.718, base of natural logarithms,

f = rate of inflation per time period,

I = the total amount of interest,

i = interest rate per time period,

j = continuous interest rate,

k = index designating a particular year,

N_p = total recoverable oil,

N_k = total recovery through the kth year,

n = number of time periods, years, separating two events,

P = total undiscounted profit over the life of a venture,

P.V. = present value,

p = number of time periods per year,

Q_k = oil produced during the kth year,

r = rate of return, similar to an interest rate,

S = net operating profit (or loss) during a time period (usually 1 year),

s = rate of gain or interest after compensating for inflation.

APPENDIX

2.A-1 Comparison of Various Rate-of-Return Methods

The following examples use the same initial investment, C = \$60,000, and the same stream of future net-cash-flow amounts. Different methods require different amortization patterns.

1. *Present value and present-value profit/investment ratio:*

Table 2.17
Present Value at 8%

Year	*Net Operating Profit* ($)	$(1.08)^{-n}$	*Present Value* ($)
1	20,000	0.926	18,520
2	25,000	0.857	21,430
3	20,000	0.794	15,880
4	15,000	0.735	11,030
5	10,000	0.681	6,800
	E = 90,000		73,660

$$D = \frac{73{,}660}{90{,}000} = 0.818,$$

$$\text{Present value/investment ratio} = \frac{DE}{C} = \frac{73{,}660}{60{,}000} = 1.23,$$

$$\begin{pmatrix}\text{present value of}\\ \text{profits/investment ratio}\end{pmatrix} = \frac{DE}{C} - 1 = 0.23.$$

2. *Discounted-cash-flow rate of return:*

Table 2.18
Discounted-Cash-Flow Rate* of Return

Year	*Net Operating Income* ($)	$(1.173)^{-n}$	*Present Value* ($)	*Investment, First of Year* ($)	*Interest at 17.3%* ($)	*Amortization* ($)
1	20,000	0.852	17,040	60,000	10,380	9,620
2	25,000	0.726	18,160	50,380	8,740	16,260
3	20,000	0.619	12,380	34,120	5,920	14,080
4	15,000	0.528	7,920	20,040	3,480	11,520
5	10,000	0.450	4,500	8,520	1,480	8,520
	90,000		60,000	173,060	30,000	60,000

*Rate of 17.3% obtained by previous trial-by-error calculations.

$$\text{Check on rate of return, } \frac{E - C}{\sum B} = \frac{90{,}000 - 60{,}000}{173{,}060} = 0.173.$$

3. *Average book rate of return:*

$$\text{Amortization fraction of net income} = \frac{C}{E} = \frac{60{,}000}{90{,}000} = 0.667.$$

Table 2.19
Average Book Rate of Return

Year	*Net Operating Income* ($)	*Outstanding Balance* ($)	*Amortization* ($)	*Interest at* 8% ($)	*Profit* ($)
1	20,000	60,000	13,330	4,800	1,870
2	25,000	46,670	16,670	3,730	4,600
3	20,000	30,000	13,330	2,400	4,270
4	15,000	16,670	10,000	1,330	3,670
5	10,000	6,670	6,670	530	2,800
	90,000	160,010	60,000	12,790	17,210

$$\text{Rate of return} = \frac{E - C}{\sum B} = \frac{90{,}000 - 60{,}000}{160{,}010} = 0.188.$$

4. *Average annual rate of return:*
(This example is based on 8% cost of capital.)

$$r = i\frac{D}{1 - D}\left(\frac{E}{C} - 1\right) = (0.08)\frac{0.8184}{0.1816}\left(\frac{90{,}000}{60{,}000} - 1\right) = 0.180,$$

$$\text{Profit fraction of net income} = \frac{P}{E} = 1 - \frac{C}{DE} = 1 - \frac{60{,}000}{73{,}660} = 0.185.$$

Table 2.20
Average Annual Rate of Return

Year	*Net Operating Income* ($)	*Outstanding Balance* ($)	*Interest at* 8% ($)	*Profit* (0.185) ($)	*Amortization* ($)
1	20,000	60,000	4,800	3,710	11,490
2	25,000	48,510	3,880	4,640	16,480
3	20,000	32,030	2,560	3,710	13,730
4	15,000	18,300	1,460	2,780	10,760
5	10,000	7,540	600	1,860	7,540
	90,000	166,380	13,300	16,700	60,000

$$\text{Rate of return} = \frac{E - C}{\sum B} = \frac{90{,}000 - 60{,}000}{166{,}380} = 0.180.$$

5. *Hoskold's Method—a sinking-fund method:*
(This example is based on 4% interest on the sinking fund.)

Table 2.21
Determining Present-Value Factor for the Cash Flow

Year	*Net Operating Income* ($)	$(1.04)^{-n}$	*Present Value* ($)
1	20,000	0.962	19,230
2	25,000	0.925	23,110
3	20,000	0.889	17,780
4	15,000	0.855	12,830
5	10,000	0.822	8,220
	90,000		81,170

$$D = \frac{81,170}{90,000} = 0.901,$$

$$r_H = i\frac{\frac{DE}{C} - (1+i)^{-n}}{1-(1+i)^{-n}} = 0.04\frac{\frac{81,170}{60,000} - 0.822}{1-0.822} = 0.119,$$

$$\text{Yearly profit} = r_H C = (0.119)(60,000) = 7150.$$

Table 2.22
Hoskold's Method

Year	*Net Operating Income* ($)	*Profit* ($)	*Deposit in Sinking Fund* ($)	*Year's Interest on Sinking Fund* ($)	*Value of Sinking Fund End of Year* ($)
1	20,000	7,150	12,850	0	12,850
2	25,000	7,150	17,850	510	31,210
3	20,000	7,150	12,850	1,240	45,300
4	15,000	7,150	7,850	1,810	54,960
5	10,000	7,150	2,850	2,190	60,000
	90,000	35,750	54,250	5,750	

REFERENCES

1. ESSLEY, P. L., Jr. "The Difference Between Nominal and Effective Interest Tables and Nominal and Effective Rates of Return." *Journal of Petroleum Technology*, Vol. 17, No. 8 (Aug. 1965), pp. 911–918.

2. WOODY, L. D., Jr., and T. J. CAPSHAW, "Investment Evaluation by Present-Value Profile." *Journal of Petroleum Technology*, Vol. 12, No. 6 (June 1960), p. 15.

3. BALDWIN, R. H., "How To Assess Investment Proposals." *Harvard Business Review*, May–June 1959, p. 98.

4. CAMPBELL, J. M., "Optimization of Capital Expendures in Petroleum Investments." *Journal of Petroleum Technology*, July 1962, p. 708.

5. ARPS, J. J., "Profitability of Capital Expendures for Development Drilling and Producing Property Appraisal." *Journal of Petroleum Technology*, July 1958, p. 13.

6. WADE, F. R., and D. E. EVERT, "Simplified Methods for Investment Analysis in the Petroleum Industry." *Society of Petroleum Engineers Paper 1527-G*, 1960.

7. BRONS, FOLKERT, and M. SILBERG, "The Relation of Earning Power to Other Profitability Criteria." *Journal of Petroleum Technology*, Mar. 1964, p. 269.

8. HENRY, A. J., "Appraisal of Income Acceleration Projects." *Journal of Petroleum Technology*, Apr. 1972, p. 393.

PROBLEMS

2-1. A certain country used "purple coupons" for money. In 1970 it required 140 coupons to purchase the same composite amount of goods which could be purchased for 100 coupons in 1960. What was the average rate of inflation per year during that period?

2-2. One hundred shares of common stock were purchased at $80 per share. The stock pays an annual dividend of $2 per share, paid quarterly. If the annual rate of inflation is taken as 5% and if the investor wishes to realize a true gain of 4% per year, what would be the necessary market value of the stock at the end of a 10-year period?

2-3. Calculate the value, as of Jan. 1, 1982, of a series of six payments of $1,000 each to be received on Jan. 1, 1987, through Jan. 1, 1992, inclusive. Use an interest rate of 6% per year. Provide the following two solutions. The required factors can be found in interest tables.

(a) Calculate the value of individual payments by using $(1 + i)^{-n}$ factors and add the resulting values.

(b) Check (exactly) the solution in part (a) by using a formula and factor, developed for a series of uniform payments, such as $P.V. = A(1 - (1 + i)^{-n})/i$, where A is the amount of each of the n payments, which are received at the end of each interest period. After the series factor reduces the series to a single-amount value as of the beginning (or end) of the series, the single-amount value can be transferred to some other time by a suitable $(1 + i)^n$, or reciprocal, factor.

2-4. Calculate the value, as of Jan. 1, 1980, of 50 payments of $1,000 each year, beginning Jan. 1, 1990. Use 8% annual interest.

2-5. A loan company in a neighboring state recently advertised the following loan schedule:

Cash You Get	*24 Payments*
$100	$ 5.93
200	11.80
300	17.49
400	22.75
500	27.69
600	32.49

For a \$500 loan repaid in 24 equal payments, calculate the discount factor D and the corresponding yearly interest rate. *Note:* D = (present value)/(total amount of the undiscounted payments). For a series of n uniform payments made at the end of each interest period,

$$D = \frac{1 - (1 + i)^{-n}}{ni}.$$

Also, use the following relation between yearly and monthly interest rates:

$$1 + i_y = (1 + i_m)^{12}.$$

2-6. Assume that the average loan in Problem 2-5 is \$500 and that this company borrows 30% of its total operating capital at 6% annual interest and then is able to employ 90% of its total capital, on the average. What annual interest rate does this company realize as a gross profit on its own money?

2-7. Calculate the effective annual interest rate i which corresponds to a continuous interest rate of $j = 8\%$ per year. The relation $i = e^j - 1$ applies.

2-8. A continuous stream of income is received at the rate of \$1,000 per month for a period of 5 years. Calculate the present value, as of the beginning of the 5-year period, at a continuous interest rate of $j = 0.09$ per year (9% per year). *Note:* The present-value factor for the entire undiscounted amount is

$$D = \frac{1 - e^{-nj}}{nj}.$$

2-9. (a) A continuous stream of income is received at the rate of \$12,000 per year for a period of 5 years. Calculate its present value, as of the beginning of the 5-year period, at the continuous rate of $\frac{3}{4}\%$ per month.

(b) If the income is received as five yearly payments at the ends of the years, calculate its present value at a yearly rate equivalent to the continuous rate.

2-10. (a) A project requires an investment of \$50,000. The return cash flow will occur in equal amounts at the ends of the next 10 years. If an apparent return of 20% is required, what must be the amount of each payment?

(b) What is the corresponding present value of the profit/investment ratio at a 0% interest rate? At a 10% discount rate?

2-11. (a) A project when completed will yield return payments of \$20,000 each at the ends of the next 10 years. If an apparent return of 25% is required, what is the maximum amount that can be invested in the completed project?

(b) What is the corresponding present value of profit to investment ratio at a 0% interest rate? At a 10% discount rate?

2-12. A certain project requires an investment of \$100,000. It will return the following cash flow:

Year	*Net Operating Profit* ($)
1	$50,000
2	40,000
3	30,000
4	30,000
5	10,000

Assume that the cash flows are received at the ends of each yearly period. Calculate the rate of return by the following methods:

(a) Discounted cash flow.
(b) Average annual, based on 8% cost of capital.
(c) Average book rate of return.
Assume that the cash flows are received at the midpoints of each yearly period. Recalculate the rate of return by the following methods:
(d) Discounted cash flow.
(e) Average annual, based on 8% cost of capital.
(f) Average book rate of return.

2-13. (a) A total of $750 interest was paid on a loan over a 5-year period. The yearly interest rate was 5%. Calculate the sum of the unamortized balances.
(b) If the loan was repaid so that the principal was reduced by equal amounts at the ends of each of the 5 years, what was the amount of the initial loan?

2-14. The table shows a contemplated acceleration project. The cash flows can be considered as occurring at the ends of the years. The net original investment, as indicated by the table, amounts to $100,000. Plot the present value of the differences between the accelerated and the original project at interest discount rates between 0 and 80%. What is the range of average return on investment which might be enjoyed by a company so that the particular company would be interested in considering this added investment?

Year	*Estimated Original Project* ($)	*Net Incomes Accelerated Project* ($)
0	180,000	80,000
1	160,000	250,000
2	140,000	260,000
3	120,000	200,000
4	90,000	—
5	80,000	—
6	50,000	—

3

PROPERTIES OF PROBABILITY DISTRIBUTIONS

3.1 INTRODUCTION

3.1-1 The Role of Probability

The efforts and works of man encompass fields where there are varying chances of success. At one end of the spectrum are those endeavors whose satisfactory completion is characterized by nearly absolute certainty. These can be analyzed by the use of deterministic models, wherein the result of an action is presumed to be known precisely. The construction of a building serves as an example, for almost all buildings can be completed essentially as planned, although minor variations commonly occur in materials and the performance of craftsmen. Any small risks that are inherent in such activities are accepted as part of ordinary existence and quite generally ignored. The attractiveness of economic participation in many such projects can be described adequately in terms of an expected return, and any variations in this quantity will be small.

A continuum of increasing relative uncertainty extends to those ventures where the probability of success of the individual trial is low. Among those endeavors characterized by high uncertainty, the probability of failure of an individual trial often exceeds the probability of its success. Producing a highly successful theatrical play would serve as an example of this group, as would the exploratory well drilled for discovery of petroleum. As a general class, such fields of endeavor are characterized by high profits in a relatively few fortunate cases, which serve to attract new entrants into the field. They are also characterized by losses of fortunes in many cases where individuals are eliminated from the field. Even though based upon accurate information, expected rates of return serve as an insufficient guide for participation in such areas. Decisions should allow for the variations in outcome. The study of probability, besides being fascinating in its own right, leads toward the making of better decisions.

Mathematical models quite often take the form of equations that express relationships between quantities. The friction horsepower required to run a particular air compressor might be expressed satisfactorily as a function of speed and temperature. Color of the machine and effects of small vibrations would not appear in the relationship, for they are not important in the operating range. Thus models do not completely represent the real-world situation, only those parts of it which are considered important. Consequently, results obtained from models generally need to be monitored and interpreted. However, by changing the value of one or more quantities in the model, the expected real-world result can be predicted rapidly and economically from the model. It is said that one manipulates the model to gain insight. In probabilistic models, provision must be made for different outcomes, whether or not a particular trial will result in failure or success, and the degree of success. Both the long-term averages and the extent of short-term fluctuations need to be studied.

3.1-2 Kinds of Probabilities

Relative frequency is one of the concepts used for explaining probability. It was advanced about 1920 by R. von Mises, an Austrian mathematician. Let there be two or more possible outcomes, call them A, B, C, . . ., of which only one can occur in each trial. Repeated trials indicate a long-run ratio, which has been expressed as

$$p(A) = \lim_{n \to \infty} \frac{n_A}{n}, \tag{3-1}$$

where n_A is the number of times event A occurred in a total of n trials, or experiments. A familiar example is the tossing of a fair coin, where the probability of obtaining a head, p(H), approaches 0.5, as measured by a large number of trials. Attempts to define probability by means of Eq. (3-1) encounter theoretical difficulties, since in the real world an experiment cannot be performed an infinite number of times.[1] It is interesting to note in coin-tossing experiments that, although the ratio of heads may approach closer to 0.5, as the number of trials increases the difference between the total number of heads and the total number of tails invariably increases.

While the notion of relative frequency remains a concept, it is nevertheless a very useful one and is commonly used whenever data are available. Gambling wheels and other devices exhibit long-run frequencies that permit the gambling houses to set their payoff schedules. Insurance companies establish rates based upon long-run frequencies. Machines that produce such things as bolts or electronic parts generally produce a small number of defectives, possibly due to combinations of undetectable causes, yet the relative frequencies of the defective items are predictable.

A group, or a set, of equally likely outcomes, only one of which shall occur, is the concept that served as a basis for the early theoretical developments of probability. Early interest in probability stemmed from an interest in games of chance, such as the draw of a card or the throw of a die. The French mathematicians Fermat and Pascal in the seventeenth century and Laplace in the eighteenth, developed much of the theory. In the nineteenth century Whitworth based an appraisal equation on the number of tickets sold in a lottery.[2] Where there is a group of n equally likely outcomes, if a total of n_A of these is favorable to event A, then it is taken that

$$p(A) = \frac{n_A}{n}. \tag{3-2}$$

For example, there are four aces in a deck of 52 playing cards. If the deck is well shuffled, we say that the probability of drawing an ace is

$$p\,(\text{ace}) = \tfrac{4}{52}.$$

Similarly, for a fair die whose six sides are numbered one to six, the probability of any one of these numbers, such as five, coming up, is

$$p(5) = \tfrac{1}{6}.$$

Unfortunately, this concept does not always apply. Suppose that a die is loaded or weighted so that some of the individual sides are less likely to come up than others. Then there is unfortunately no way of selecting a group or set of equally likely events.

Although not always applicable, the concept of a set of equally likely events is a powerful tool for developing insight. Suppose that in a particular area the probability of drilling a dry well is 0.5, and the probability of a producer is then also 0.5. If a total of three wells were to be drilled, what are the expected outcomes? There are a total of eight equally likely outcome events, which can be displayed as shown in Table 3.1. The results are summarized in Table 3.2. When all possible outcomes have been listed for the group or set or system of interest, their individual probabilities must sum to 1, or unity.

Personal, or subjective, probabilities are required to interpret many

Table 3.1
Ways for Outcomes of Drilling Three Wells

Well A	*Well B*	*Well C*
D	D	D
D	D	P
D	P	D
D	P	P
P	D	D
P	D	P
P	P	D
P	P	P

Table 3.2
Probabilities of Combinations of Producers and Dry Wells

p(3 dry)	0.125
p(2 dry, 1 producer)	0.375
p(1 dry, 2 producers)	0.375
p(3 producers)	0.125
	1.000

situations, particularly those which cannot be repeated. For example, a farm manager might feel that there is a 40% chance that a disastrous flood will occur in a certain valley during the coming year. A person familiar with the geology of a region might feel that there is a 30% chance that a certain geologic structure contains a rich oil pool. Such probabilities were used by Grayson to make drilling decisions.[3] Although such situations, or experiments, cannot be repeated, some basis must be adopted for making a decision that events will prove to be right or wrong. Hadley has explained such probabilities in terms of a *degree of rational belief.*[1] There is obviously a limit to how closely such probabilities can be stated. Their worth depends upon the experience and perception of the individual.

There is a fourth way of considering probabilities which is compatible with all the foregoing. It is the concept of *multiple similar experiments.* Situations are considered as though occurring in a number of different universes. In some of the universes the outcomes would be favorable, in others unfavorable. As far as relative frequency is concerned, it does not matter, for example, whether one tosses the same coin 100 times or whether under equivalent conditions he simultaneously tosses 100 different coins. The results will be equivalent. It rather obviously makes no difference whether one draws repeatedly from the same well-shuffled full deck or from different decks of playing cards. However, the idea of similar situations occurring in different universes, thereby permitting multiple outcomes, does give credibility for formulating subjective probabilities, necessary for many decisions, where clearly the identical case, or experiment, cannot be repeated. In a practical sense it may be noted that although no two geologic basins are identical, they are similar enough so that experience gained in one is valuable in others.

There is no universal agreement at present regarding exactly what probability is. Perhaps most generally it is considered as a fundamental concept that is aligned with many if not all natural phenomena and thus can be used to interpret them. The approach taken by Savage leads toward assuming that an individual behaves so as to satisfy certain axioms and therefore has a personal probability function.[4] The problem faced by decision makers in petroleum exploration and in industry generally is to obtain probabilities that can serve as guides both for survival and for making necessary progress in the real world. In the following paragraphs it is initially assumed that probabilities are known, at least to an accuracy sufficient for estimation and decision purposes. Ways of testing and improving probability estimates follow in later developments.

3.2 FREQUENCY DISTRIBUTIONS

3.2-1 Summarizing Data

Interpretation of even moderately large amounts of data requires that they be summarized. This is commonly accomplished by means of frequency distributions. Rather than attempt to interpret meanings concealed within a large number of individual values, the data are gathered into groups or classes. This facilitates interpretation as well as numerical computations. Measurements and estimates generally contain errors, so it would often be difficult to justify not using a frequency distribution. An example of raw data is given in Table 3.3.

Class boundaries, as given in Tables 3.4 and 3.5, coincide, so the upper boundary of one class is the same as the lower boundary of the next class. Common convention is to place members that fall on a boundary into the higher class. Labeling them nominal class boundaries implies that they may not be the true class boundaries. In general, the difference between nominal and true class boundaries is related to the precision of the measurements. In the present case one notes that the measurements are given to the second decimal. It can be presumed that a measurement of 0.115 would be rounded to 0.12, for example, whereas a measurement of 0.1149 would be rounded to 0.11. The true class boundaries, based upon this reasoning, are shown in

Table 3.3
Porosity Measurements (20 values)

0.12	0.14	0.15	0.18
0.19	0.16	0.17	0.15
0.16	0.14	0.15	0.13
0.14	0.15	0.13	0.13
0.14	0.12	0.10	0.11

Table 3.4
Forming a Frequency Distribution

Nominal Class Boundaries	*Members*	*Number of Members*
0.10–0.12	0.10, 0.11	2
0.12–0.14	0.12, 0.13, 0.13, 0.13, 0.12	5
0.14–0.16	0.14, 0.15, 0.15, 0.14, 0.15, 0.14, 0.15, 0.14	8
0.16–0.18	0.16, 0.17, 0.16	3
0.18–0.20	0.18, 0.19	2

Table 3.5
Frequency Distribution

Nominal Class Boundaries	*True Class Boundaries*	*Class Mark,* X_i	*Frequency,* f_i	f_iX_i
0.10–0.12	0.095–0.115	0.105	2	0.210
0.12–0.14	0.115–0.135	0.125	5	0.625
0.14–0.16	0.135–0.155	0.145	8	1.160
0.16–0.18	0.155–0.175	0.165	3	0.495
0.18–0.20	0.175–0.195	0.185	2	0.370
			20	2.860

Table 3.5. The difference between upper and lower boundaries is referred to as the *class interval.* In this example all the class intervals are equal. This practice is common, but it is not a necessary condition.

The average porosity can be found from the data in Table 3.3 as follows:

$$\text{mean} = \frac{\sum x}{n} = \frac{2.86}{20} = 0.143. \tag{3-3}$$

These same data are arranged in a frequency distribution in Table 3.5. When data are in this form, the average porosity can be found as follows:

$$\begin{aligned}\text{mean} &= \frac{\sum f_i X_i}{\sum f_i} \\ &= \frac{2.860}{20} = 0.143.\end{aligned} \tag{3-4}$$

The class mark is the value assigned, for convenience, to all members within a class. Ideally it should be the weighted average value of all members within each class. A common convention is to set a class mark at the arithmetic average of the upper and lower true class boundaries. This convention was followed in Table 3.5.

3.2-2 Graphical Presentations

Histograms and frequency polygons are used to show probability density. Figure 3.1 includes both methods of presentation for the porosity data contained in the foregoing tables. Both are based upon true class limits, which is more evident in the case of the histogram, where each horizontal segment extends between true class limits for each class. Segments of the frequency polygon are drawn between midpoints of true class boundaries. By convention, the first and last segments are each extended half of an interval beyond the upper and lower true class limits of the histogram.

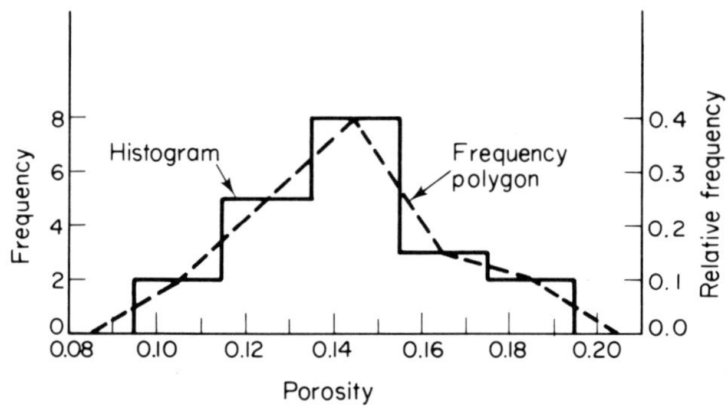

Figure 3.1. Histogram and frequency polygon.

Cumulative frequency polygons are shown in Fig. 3.2. The "less than" type can be drawn directly from the data given in Table 3.5. It is shown as a solid line, from which one can read directly that 20% (cumulative relative frequency = 0.2) of the cases have a porosity of less than about 12.4%. The "more than" type is obtained for each line by subtracting from 1.0 the

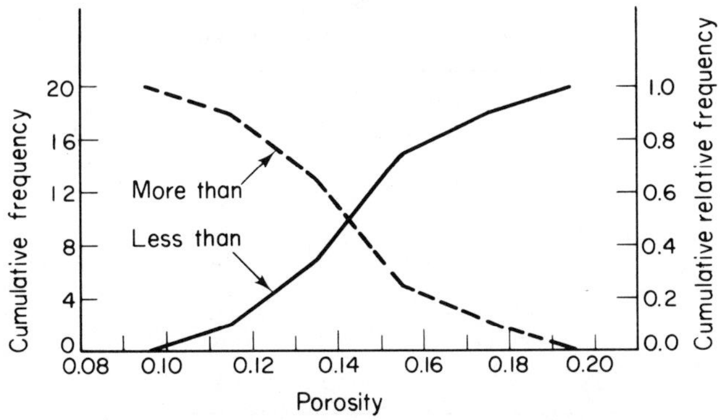

Figure 3.2. Cumulative frequency polygons.

cumulative relative frequency. Both cumulative values then correspond to the upper true class boundary given in each line in the table. Using the dashed line, one can read directly from the graph, for example, that 40% of the cases have a porosity greater than about 14.6%.

3.3 GROUPS OF DATA: STATISTICAL PROPERTIES

3.3-1 Statistical Measures

Data, particularly when there are many of them, are not handled as individual points or values, but rather their salient features are noted or determined for comparative purposes. These characterizing features present a summary that can be used for analysis, for control, or for evaluation. Many times it can be recognized that a series of individual measurements or other data are following the pattern of a familiar distribution. When this occurs, the interpretation of the data is greatly facilitated, for the important features of several very useful distributions have been established and demonstrated many times. Thus one becomes interested in distributions. First, however, several of the important characteristics of groups of data points will be noted.

The *range* is the difference between the highest and lowest values. In Table 3.6 the range of W is $14 - 3 = 11$; the range of X is $44 - 12 = 32$.

The *median* has as many values above it as below. The median of the W-values is 7; the median of the X-values is $(26 + 29)/2 = 27.5$. The idea of a median is extended to *quartiles*. For the X-values, the first quartile is $(15 + 19)/2 = 17$; the second quartile is the mean; the third quartile is $(31 + 38)/2 = 34.5$. The idea is extended to *deciles* and to *percentiles* in the same manner. When the data have been grouped into a frequency distribution, and a quartile occurs somewhere within the range covered by a class interval, it is customary to compute the quartile (X-value, for example) by using

Table 3.6
Two Groups of (unrelated) Data

Values of W	*Values of X*
3	12
4	15
6	19
7	26
9	29
11	31
14	38
	44

linear interpolation based on the ratio of the number of points needed to complete the quartile to the total number of points in the frequency class.[5]

A *mode* is a value that occurs with greater relative frequency. That is, measured values tend to cluster around a mode. Where a distribution is symmetric, the mean and the mode coincide. When a distribution is skewed, they do not coincide. When more than one mode appears, it is sometimes evident that more than one species is present within the group or population that is being measured.

The *mean* is the arithmetic average.* The symbols $\bar{X}$ and m are both commonly used for the mean.

$$\bar{X} = \frac{1}{N} \sum_{i=1}^{N} X_i. \tag{3-5}$$

For convenience, Eq. (3-5) will usually be written more simply as

$$\bar{X} = \frac{1}{N} \sum X, \tag{3-5a}$$

where it is understood that the summation is made over all the X-values. For the values listed in Table 3-6, the means are

$$\begin{aligned} \bar{W} &= \frac{3 + 4 + 6 + 7 + 9 + 11 + 14}{7} \\ &= 7.71, \\ \bar{X} &= \frac{12 + 15 + 19 + 26 + 29 + 31 + 38 + 44}{8} \\ &= 26.75. \end{aligned}$$

*Other means used for some purposes are the geometric mean G and the harmonic mean H:

$$G = (X_1 X_2 X_3 \ldots X_N)^{1/N},$$

$$\frac{1}{H} = \frac{1}{N} \sum \frac{1}{X}.$$

Two important properties are the *variance* and the *standard deviation.* These are measures of the tendency for the data to be spread out about the mean. Other measures of this tendency can be devised, but these two prove to be the most useful. Variance is the additive property, and after variances are added together, the square root is taken to give the standard deviation.

Two cases must be considered when calculating the variance and standard deviation of a group of data points. When the group contains all points or members of interest, so that the entire population of interest is used in the calculation, the following apply:

$$\sigma^2 = \frac{\sum (X_i - \bar{X})^2}{N}, \tag{3-6}$$

$$\sigma = \sqrt{\frac{\sum (X_i - \bar{X})^2}{N}}. \tag{3-7}$$

When the group does not contain all the members of thc population, but is indeed a sample used to estimate the properties of the entire population, the variance should be multiplied by Bessel's correction of $N/(N - 1)$ before taking the square root, so that the following obtain[6]:

$$s^2 = \frac{\sum (X_i - \bar{X})^2}{N - 1}, \tag{3-8}$$

$$s = \sqrt{\frac{\sum (X_i - \bar{X})^2}{N - 1}}, \tag{3-9}$$

where

σ^2 = variance of the population,

σ = standard deviation of the population,

s^2 = estimated variance of the population,

s = estimated standard deviation of the population,

N = sample size.

Equations (3-8) and (3-9) are generally used when dealing with a series of measurements, because generally one works with a sample rather than with all possible cases. As the sample size becomes larger, it matters little whether N or $N - 1$ is used in the calculation.

The following calculations use the data in Table 3.7:

$$\bar{X} = \frac{\sum X}{N} = \frac{40.0}{5} = 8.0.$$

Table 3.7
Calculation of Variance

X_i	$x = X_i - \bar{X}$	$x^2 = (X_i - \bar{X})^2$
2.5	−5.5	30.25
5.5	−2.5	6.25
9.5	1.5	2.25
10.0	2.0	4.00
12.5	4.5	20.25
40.0	0.0	63.00

For the case where the table includes all members of the population,

$$\sigma^2 = \frac{\sum (X_i - \bar{X})^2}{N} = \frac{63.00}{5} = 12.60,$$

$$\sigma = \sqrt{\sigma^2} = \sqrt{12.60} = 3.55.$$

For the case where the data in the table represent a sample from a (larger) population,

$$s^2 = \frac{\sum (X_i - \bar{X})^2}{N - 1} = \frac{63.00}{5 - 1} = 15.75,$$

$$s = \sqrt{s^2} = \sqrt{15.75} = 3.97.$$

It is very convenient to use the following notation:

$$x_i = X_i - \bar{X}. \tag{3-10}$$

From the definition of $\bar{X}$ it follows that

$$\sum x = 0. \tag{3-11}$$

The expression $\sum (X_i - \bar{X})^2$ appears in several of the equations. Any time it is more convenient to do so, the following equivalent method may be used in the calculation of this quantity:

$$\sum x^2 = \sum (X_i - \bar{X})^2 = \sum X^2 - \frac{(\sum X)^2}{N}. \tag{3-12}$$

This latter method is generally more convenient on a desk calculator and where it is convenient to proceed with calculations before the mean is deter-

Table 3.8
Alternative Method of Calculation of Sum of Squares About the Mean

X_i	X_i^2
2.5	6.25
5.5	30.25
9.5	90.25
10.0	100.00
12.5	156.25
40.0	383.00

mined. The data in Table 3.8 are used for the following solution:

$$\sum x^2 = \sum X^2 - \frac{(\sum X)^2}{N}$$

$$= 383.00 - \frac{(40)^2}{5}$$

$$= 63.00.$$

3.3-2 Transferring and Combining Variances

Many times it is convenient to calculate the variance of a portion of the data about its own mean. This can occur when part of the data follow a familiar distribution. Remaining is the problem of how to combine the several parts to obtain the variance of the whole distribution, which must be calculated about the mean of the entire distribution. Figure 3.3 illustrates the problem schematically. In this case the data are divided into two parts.

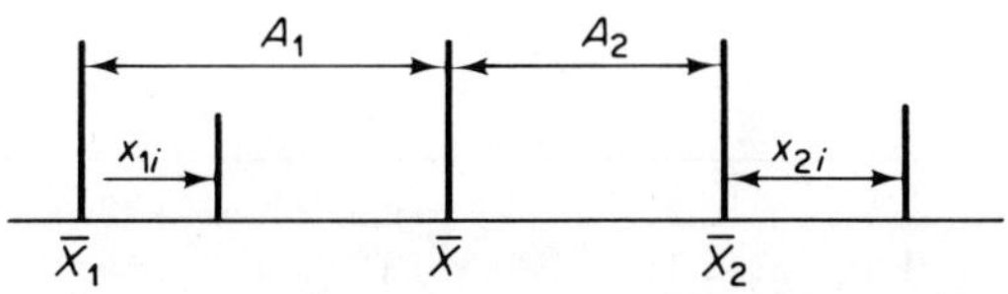

Figure 3.3. Transferring variance of individual parts to the mean of the entire distribution.

Part I has a mean value of $\bar{X}_1$, and part II has a mean of $\bar{X}_2$. Only one individual data point is illustrated for each part. The following relations hold. For part II,

$$X_i = \bar{X} + A_2 + x_{2i},$$

and a similar relation holds for part I. The following development makes use of the fact that $\sum x = 0$.

$$\sigma^2 = \frac{\sum (X_i - \bar{X})^2}{N}$$

$$= \frac{\sum (x_1 + A_1)^2 + \sum (x_2 + A_2)^2}{N}$$

$$= \frac{\sum x_1^2 + 2A \sum x_1 + N_1 A_1^2 + \sum x_2^2 + 2A_2 \sum x_2 + N_2 A_2^2}{N}$$

$$= \frac{N_1}{N}\left(\frac{\sum x_1^2}{N_1} + A_1^2\right) + \frac{N_2}{N}\left(\frac{\sum x_2^2}{N_2} + A_2^2\right)$$

$$= \frac{N_1}{N}(\sigma_1^2 + A_1^2) + \frac{N_2}{N}(\sigma_2^2 + A_2^2), \qquad (3\text{-}13)$$

where

N_1 = number of data points in part I,

N_2 = number of data points in part II,

$N = N_1 + N_2$ = total number of data points,

σ_1^2 = variance of part I about its own mean,

σ_2^2 = variance of part II about its own mean,

A_1 = difference between mean of part I and mean of the total,

A_2 = difference between mean of part II and mean of the total.

Table 3.9 illustrates the use of Eq. (3-13), where the same data used pre-

Table 3.9
Combining Variances of Individual Parts of a Distribution

Part I			*Part II*		
X_i	x_{1i}	x_{1i}^2	X_i	x_{2i}	x_{2i}^2
2.5	−1.5	2.25	9.5	−1.17	1.36
5.5	1.5	2.25	10.0	−0.66	0.45
8.0	0.0	4.50	12.5	1.83	3.36
			32.0	0.00	5.17

$\bar{X}_1 = \frac{8.0}{2} = 4.0$

$\bar{X} = \frac{8.0 + 32.0}{2 + 3} = 8.0$ $\qquad$ $\bar{X}_2 = \frac{32.0}{3} = 10.67$

$\sigma_1^2 = \frac{4.50}{2} = 2.25$ $\qquad$ $\sigma_2^2 = \frac{5.17}{3} = 1.72$

$A_1^2 = (4.0 - 8.0)^2 = 16.00$ $\qquad$ $A_2^2 = (10.67 - 8.0)^2 = 7.11$

viously in Table 3.8 are divided into two parts and subsequently recombined. Equation (3-13) is used to calculate the variance of the total as follows:

$$\begin{aligned}\sigma^2 &= \frac{N_1}{N}(\sigma_1^2 + A_1^2) + \frac{N_2}{N}(\sigma_2^2 + A_2^2) \\ &= \tfrac{2}{5}(2.25 + 16.00) + \tfrac{3}{5}(1.72 + 7.11) \\ &= 12.60.\end{aligned}$$

This result is of course the same as was obtained from Table 3.8, where all the data were used in a single calculation. It can be noted that Bessel's correction factor, $N/(N - 1)$, should be used only after combining the variances of the several parts of the distribution.

3.4 DISCRETE PROBABILITY DISTRIBUTIONS

3.4-1 Fundamental Characteristics

A common die is a cube whose six faces are numbered one to six. When a die is thrown, it will come to rest with one of these faces up. Other possibilities are ruled out—the result cannot be something like 2.4 or 5.1. Only the following are possible: $X = 1, 2, 3, 4, 5$, or 6. This is a discrete distribution, and if the die is fair, the probability associated with each number is $\frac{1}{6}$. The probability of the distribution sums to 1 unity, since one of the several possible events must occur.

The prospects for discovery of petroleum in an exploratory well might be described as shown in Table 3.10. This indicates a 60% chance of a dry well, a 10% probability that the producible reserves will amount to 20,000 bbl of oil, and so on. The several probabilities sum to 1.0 (see also Table 3.11). This same distribution is pictured in Fig. 3.4, where the lengths of the vertical lines are proportional to the probabilities. Interpretation in this case is somewhat different from that where a die is thrown, for the background understanding is, of course, that the outcome could occur at points other

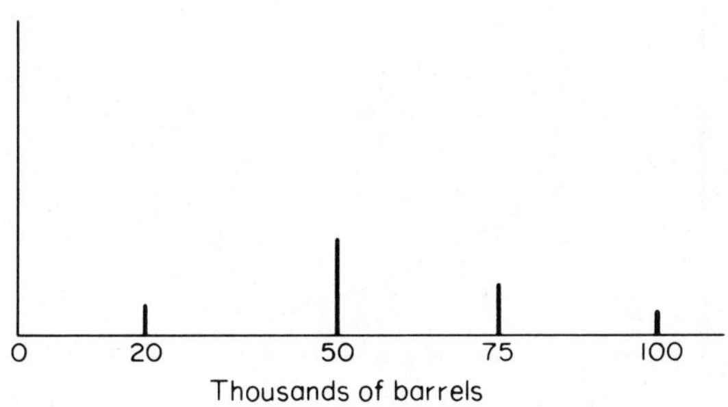

Figure 3.4. A discrete distribution summarizing possible outcomes from drilling an exploratory well. A probability density function plot.

Table 3.10
Discovery Prospects of an Exploratory Well

Probability, p	*Reserves (M bbl)*
0.60	0
0.10	20
0.15	50
0.10	75
0.05	100

Table 3.11
Cumulative Probabilities of Discovery Prospects
Data of Table 3.10

Reserves (M bbl)	*Probability, p*	*Cumulative Probability*	*Probability of a Greater Value*
0	0.60	0.60	0.40
20	0.10	0.70	0.30
50	0.15	0.85	0.15
75	0.10	0.95	0.05
100	0.05	1.00	0.00

than those where the vertical lines are drawn. That is, each vertical line is, in effect, serving as a classmark for the adjacent possibilities. Information in this form serves very well for the determination of the mean and variance of the distribution, which are the properties of greatest interest. Table 3.11 and Fig. 3.5 illustrate the cumulative probability of this same distribution.

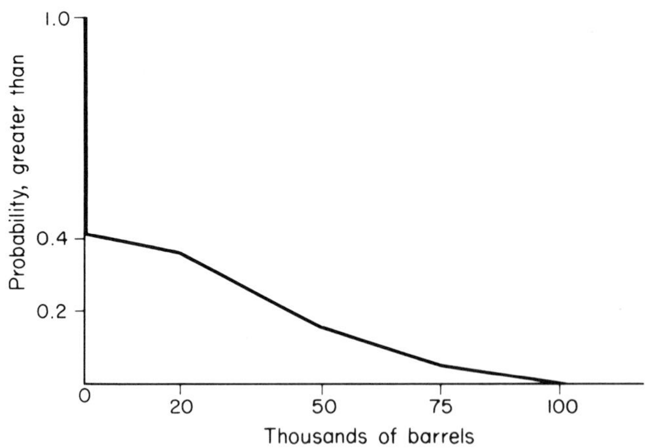

Figure 3.5. Cumulative probabilities arranged as "probability of greater than."

3.4-2 Statistical Properties

Finding the mean and variance of a discrete distribution is analogous to methods used for groups of data. However, the probabilities assigned to the several parts can be used and the calculations simplified somewhat. The mean of the whole is given by

$$m = \bar{X} = \sum p_i X_i. \tag{3-14}$$

Variance of a discrete distribution is given by

$$\sigma^2 = \sum p_i (X_i - \bar{X})^2. \tag{3-15}$$

In Table 3.12 these equations are used to calculate the mean, variance,

Table 3.12
Calculation of Mean, Variance, and Standard Deviation of a Discrete Distribution

Probability, p_i	*Reserves,* X_i *(M bbl)*	$p_i X_i$	$X_i - \bar{X}$	$p_i(X_i - \bar{X})^2$
0.60	0	0.0	−22	290.4
0.10	20	2.0	−2	0.4
0.15	50	7.5	28	117.6
0.10	75	7.5	53	280.9
0.05	100	5.0	78	304.2
1.00		22.0		993.5

and standard deviation of the data contained in Table 3.10:

$$\text{mean} = \sum p_i X_i = 22.0 \text{ M bbl},$$

$$\text{variance} = \sigma^2 = \sum p_i (X_i - \bar{X})^2 = 993.5,$$

$$\text{standard deviation} = \sigma = \sqrt{\sigma^2} = \sqrt{993.5} = 30.3 \text{ M bbl}.$$

If again $x_i = X_i - \bar{X}$, then the following holds:

$$\sum p_i x_i = 0. \tag{3-16}$$

3.4-3 Combining Means and Variances

The mean of a total distribution can be found by combining means of its parts, weighted according to their respective probabilities. Table 3.13

Table 3.13
Combining Discrete Distributions

Probability, p_i	*Reserves,* X_i *(MM bbl)*	p_iX_i	$X_i - \bar{X}$	$p(X_i - \bar{X})^2$
		First Part		
0.857	0	0.00	−2.82	6.82
0.143	20	2.86	17.18	42.18
1.000		2.86		49.00
		Second Part		
0.500	50	25.00	−16.67	138.94
0.33	75	25.00	8.33	23.13
0.167	100	16.67	33.33	185.15
1.000		66.67		347.22

shows the distribution of Table 3.12 broken into two parts. Let the first part have a weight of 0.7, and the second part a weight of 0.3, as they did originally. Let X_i represent the mean of an individual part. The mean of the total is then found by proceeding as follows:

$$\begin{aligned} m &= \sum p_i X_i \\ &= (0.7)(2.82) + (0.3)(66.67) \\ &= 22.00, \text{ as before.} \end{aligned}$$

Variances of the several parts can be combined, but they first must be transferred to the mean of the whole distribution. Variance of any unit is least about its own mean, and generally these individual variances must be increased by the square of the displacement between the mean of the individual part of the mean of the whole. Let A_i represent this displacement. For the two parts shown in Table 3.13, these displacements are

$$\begin{aligned} A_1 &= 22.00 - 2.86 = 19.18, \\ A_2 &= 22.00 - 66.67 = -44.67. \end{aligned}$$

The variance about the mean of the total is then calculated as follows:

$$\begin{aligned} \sigma^2 &= \sum p_i(\sigma_i^2 + A_i^2) \\ &= (0.7)(49.00 + (19.18)^2) + (0.3)(347.22 + (-44.67)^2) \qquad (3\text{-}17) \\ &= 993.5, \qquad \text{as before.} \end{aligned}$$

3.5 CONTINUOUS PROBABILITY DISTRIBUTIONS

3.5-1 Fundamental Properties

One concept of a continuous distribution is that it is a continuous envelope conforming to a discrete distribution. Very often the mathematics associated with using continuous distributions are simpler and can serve as convenient methods for the determination of the properties of distributions which actually consist of discrete points. Figure 3.6 indicates this concept. The lengths of the individual vertical lines are proportional to the probabilities of occurrence of the corresponding X-values, and the sum of all such probabilities is defined as 1. Similarly, the total area under the continuous curve is defined as 1.

Figure 3.7 illustrates a triangular distribution. A mode occurs at $X = 6.5$, which indicates that more outcomes of trials will be near 6.5 than near $X = 4.0$, for example. Probabilities are proportional to areas. The probability is 0.23 that X will occur between $X = 3$ and $X = 5$. The probability is 0.28 that X will occur between $X = 5$ and $X = 6$. The cumulative probability

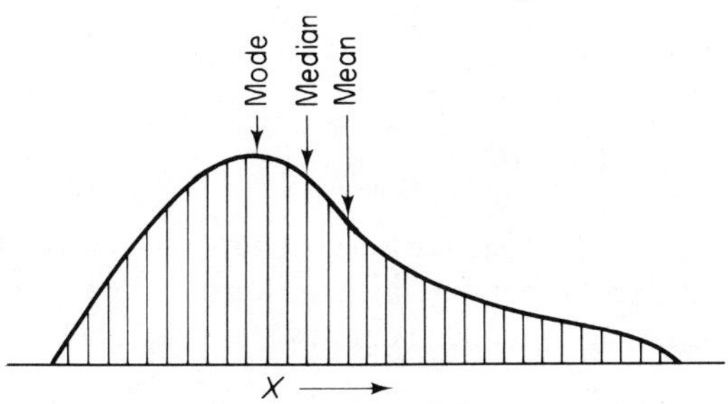

Figure 3.6. Continuous distribution curve. Vertical lines indicate a discrete distribution enclosed by the continuous curve. This distribution is skewed to the right.

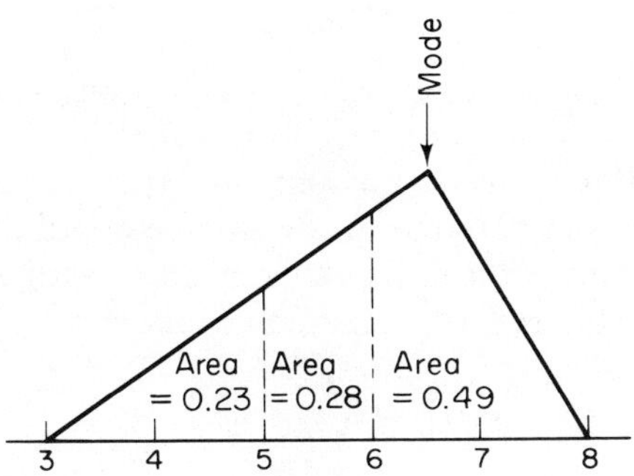

Figure 3.7. Probability and area relations.

to $X = 6$ is $0.23 + 0.28 = 0.51$. Accordingly, the probability is 0.51 that X will occur between $X = 3$ and $X = 6$. The remaining area, $1.00 - 0.51 = 0.49$, is the probability that X will occur at a value greater than $X = 6$, which in this case would be between $X = 6$ and $X = 8$. Since the total area under the curve is 1, the probabilities are zero of X occurring anywhere other than between 3 and 8.

Another name for the mean is the *expectancy*. Thus when X is the random variable, the mean equals the expectancy of X, written as

$$m = E[X]. \tag{3-18}$$

The variance can be obtained by the difference between the expectancy of X^2 and the mean squared, which general relation is written as

$$\sigma^2 = E[X^2] - m^2. \tag{3-19}$$

These expectancies are defined as follows:

$$E[X] = \int_{-\infty}^{\infty} f(X)X\,dX, \tag{3-20}$$

$$E[X^2] = \int_{-\infty}^{\infty} f(X)X^2\,dX \tag{3-21}$$

3.5-2 Uniform Distributions

A continuous uniform distribution is shown in Fig. 3.8. This type is sometimes referred to as a rectangular distribution. The randomly occurring X-values are evenly distributed throughout the range between a and b. Use

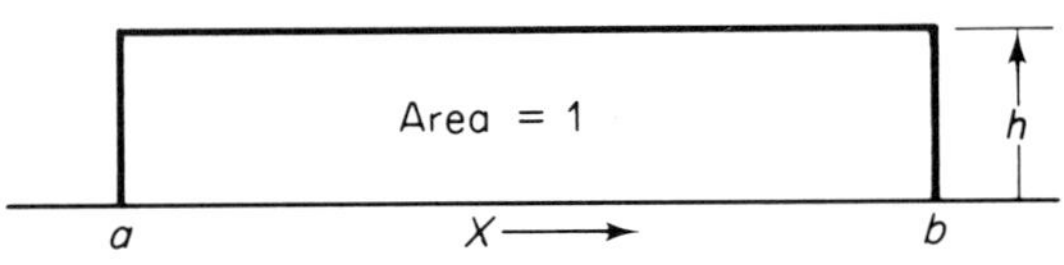

Figure 3.8. A continuous uniform distribution.

of the general equations given in the foregoing paragraphs will be illustrated by finding the mean and variance of this simple distribution. For this case, $f(X) = h =$ a constant in the range of interest and zero elsewhere. Since the total area is 1, the value of h is computed as follows:

$$1 = h(b - a),$$

$$h = \frac{1}{b - a}.$$

The mean is calculated by evaluating the integral between a and b, since $f(X)$, or h, is zero elsewhere.

$$E[X] = \int_a^b hX\,dX$$

$$= h\frac{b^2 - a^2}{2} = \frac{b^2 - a^2}{2(b - a)}$$

$$= \frac{b + a}{2}.$$

From the symmetry it is, of course, rather obvious that the mean should occur at the midpoint.

$$E[X^2] = \int_a^b hX^2\,dX$$

$$= h\frac{b^3 - a^3}{3}$$

$$= \frac{b^3 - a^3}{3(b - a)}.$$

The variance is then found by using Eq. (3-19).

$$\sigma^2 = E[X^2] - m^2$$

$$= \frac{b^3 - a^3}{3(b - a)} - \left(\frac{b + a}{2}\right)^2$$

$$= \frac{(b - a)^2}{12}, \qquad \text{for a uniform distribution.}$$

Final form of this expression for variance is of interest, for it repeats clearly that variance is related to the spread of the values and not to their magnitude.

Cumulative probability is defined by

$$F = \int_{-\infty}^{X} f(X)\,dX. \tag{3-22}$$

For the uniform distribution,

$$F = \int_a^X h\,dX$$

$$= \frac{x - a}{b - a}$$

for $a \leq X \leq b$. Thus $F = 0$ when $X = a$ and $F = 1.0$ when $X = b$. Figure 3.9 is a graph of cumulative probability for a uniform distribution.

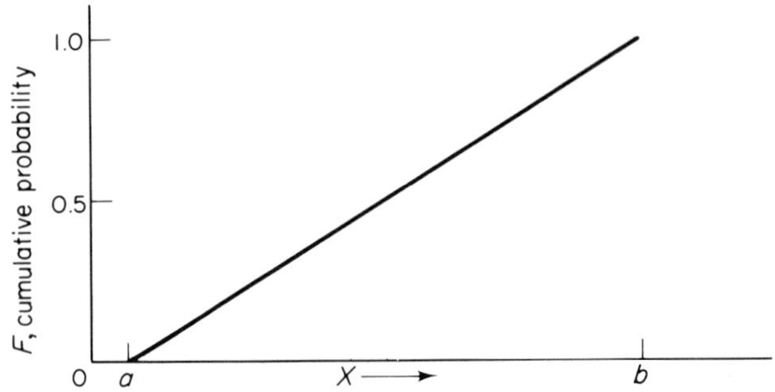

Figure 3.9. Cumulative probability of a continuous uniform distribution.

The methods used here to obtain the mean, variance, and cumulative probability can be used for triangular, parabolic, or other types of distributions. Where integration is not possible by analytic methods, resort may be had to numerical methods.

Combining the various portions of a distribution to find the mean and variance of the combined distribution constitutes a chief area of interest. The same methods used to combine discrete distributions are applicable. A first step is to determine the mean of each separate part. These are weighted according to their various probabilities and added to obtain the mean of the total distribution. Variances about these individual means, plus the transfer variances from these means to the mean of the total, are weighted according to their respective probabilities and added to obtain total variance.

Combining parts of continuous distributions will be illustrated here using the uniform distribution. The method is applicable to other distributions as well, the key being that one works from the individual means of the various parts. Figure 3.10 illustrates a uniform distribution that has been

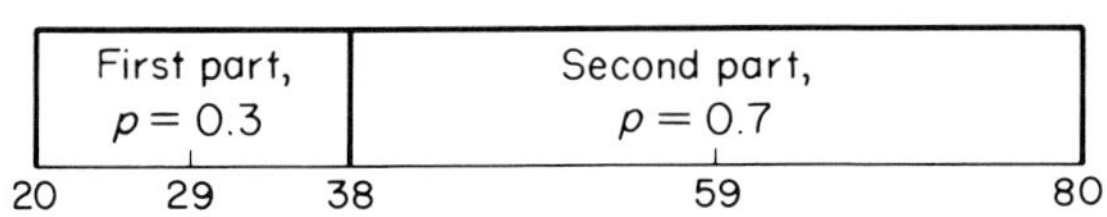

Figure 3.10. Uniform distribution divided into two parts.

broken into two separate parts. These are to be combined. Analysis is first made to determine the probabilities, means, and variances of the individual parts, and then they are combined.

For the *first part:*

$$m_1 = \frac{a_1 + b_1}{2} = \frac{20 + 38}{2} = 29.$$

Variance about the mean of the first part is

$$\sigma_1^2 = \frac{(b_1 - a_1)^2}{12} = \frac{(38 - 20)^2}{12} = 27.$$

Probability associated with the first part is

$$p_1 = \frac{38 - 20}{80 - 20} = 0.3.$$

For the *second part:*

$$m_2 = \frac{a_2 + b_2}{2} = \frac{38 + 80}{2} = 59.$$

Variance about the mean of the second part is

$$\sigma_2^2 = \frac{(b_2 - a_2)^2}{12} = \frac{(80 - 38)^2}{12} = 147.$$

Probability associated with the second part is

$$p_2 = \frac{80 - 38}{80 - 20} = 0.7.$$

Combining the first and second parts:

$$\begin{aligned} m &= \sum p_i \bar{X}_i = (0.3)(29) + (0.7)(59) = 50, \\ \sigma^2 &= \sum p_i[\sigma_i^2 + (m_i - m)^2] \\ &= 0.3[27 + (29 - 50)^2] + 0.7[147 + (59 - 50)^2] \qquad (3\text{-}23) \\ &= 0.3(468) + 0.7(228) \\ &= 300. \end{aligned}$$

Checking by calculating mean and variance of the whole distribution:

$$m = \frac{a + b}{2} = \frac{20 + 80}{2} = 50,$$

$$\sigma^2 = \frac{(b - a)^2}{12} = \frac{(80 - 20)^2}{12} = 300,$$

which is the variance about the mean of the whole distribution. The standard deviation is $\sqrt{300} = 17.3$.

A second example concerns the prospects of a proposed exploratory well. These have been described as shown in Fig. 3.11. The spike represents a 60% chance of drilling a dry well. It was estimated that there is a 30% chance of

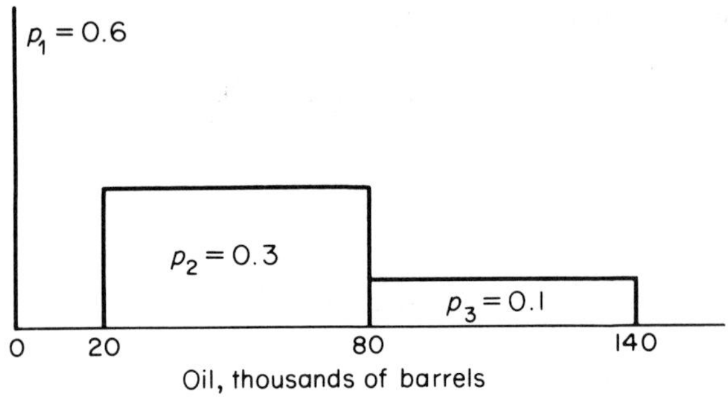

Figure 3.11. Discovery prospects of a proposed exploratory well.

obtaining between 20 and 80 M bbl and a 10% chance of obtaining between 80 and 140 M bbl.

The properties of the three parts are determined as follows:

First part (the spike):

$$m_1 = 0,$$
$$\sigma_1^2 = 0.$$

Second part:

$$m_2 = \frac{a_2 + b_2}{2} = \frac{20 + 80}{2} = 50,$$

$$\sigma_2^2 = \frac{(b_2 - a_2)^2}{12} = \frac{(80 - 20)^2}{12} = 300.$$

Third part:

$$m_3 = \frac{a_3 + b_3}{2} = \frac{80 + 140}{2} = 110,$$

$$\sigma_3^2 = \frac{(b_3 - a_3)^2}{12} = \frac{(140 - 80)^2}{12} = 300.$$

Combining the parts,

$$\begin{aligned} m &= \sum p_i m_i \\ &= (0.6)(0) + (0.3)(50) + (0.1)(110) \\ &= 26 \text{ M bbl}, \\ \sigma^2 &= \sum p_i[\sigma_i^2 + (\bar{X}_i - m)^2] \\ &= 0.6[0 + (0 - 26)^2] + 0.3[300 + (50 - 26)^2] + 0.1[300 + (110 - 26)^2] \\ &= 405.6 + 262.8 + 735.6 \\ &= 1{,}404, \\ \sigma &= \sqrt{1{,}404} = 37.47 \text{ M bbl}, \qquad \text{standard deviation.} \end{aligned}$$

3.5-3 Triangular Distributions

Properties of triangular distributions and of a triangle with a spike added are given in Table 3.14. Mean and variance of a right triangle are readily calculated by means of Eqs. (3-24) and (3-25 *a* or *b*), and properties of others are obtained by adding and transferring properties in the usual manner.

An example can be drawn from the probable average water saturation in a particular oil reservoir. It is believed that the lowest value would be 20%, the most probable 30%, and the highest value at 50% water saturation. This

Table 3.14
Properties of Triangular Distributions

				Eq. No.
	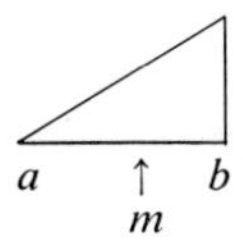	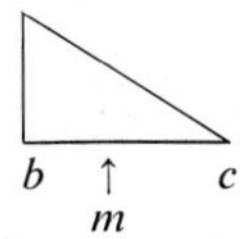	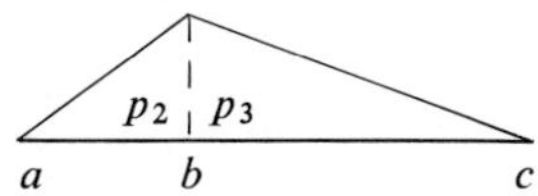	
Mean, m	$\frac{a+2b}{3}$	$\frac{2b+c}{3}$	$\frac{a+b+c}{3}$	(3-24)
Probabilities, p	1	1	$p_2 = \frac{b-a}{c-a}$ $p_3 = \frac{c-b}{c-a}$	
Variance, σ^2	$\frac{(b-a)^2}{18}$	$\frac{(c-b)^2}{18}$	$\frac{(c-a)^2}{18}(1-p_2p_3)$	(3-25)
Cumulative probability, F	$a \leq X \leq b$ $b \leq X \leq c$		$F = p_2\left(\frac{x-a}{b-a}\right)^2$ $F = 1 - p_3\left(\frac{c-X}{c-b}\right)^2$	(3-26) (3-27)
	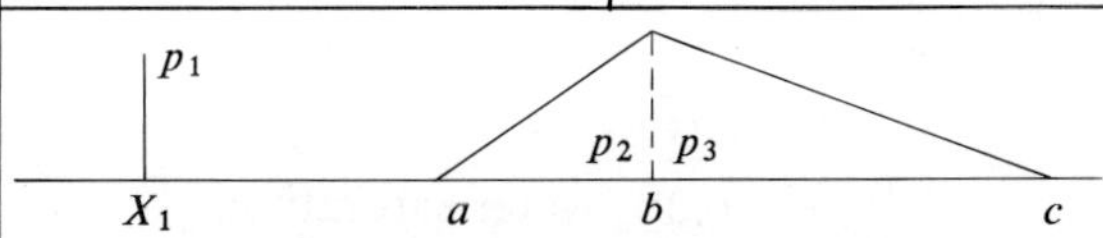			
Mean, m	$m = p_1X_1 + (1-p_1)\frac{a+b+c}{3}$			(3-28)
Probabilities, p	$p_2 = (1-p_1)\frac{b-a}{c-a}$ $p_3 = (1-p_1)\frac{c-b}{c-a}$			
Variance, σ^2	$\sigma^2 = \frac{p_1}{1-p_1}(m-X_1)^2 + \frac{(c-a)^2}{18}\left(1-p_1-\frac{p_2p_3}{1-p_1}\right)$			(3-29)

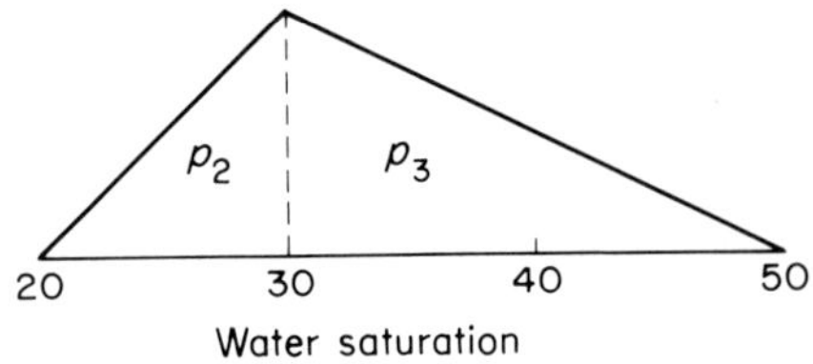

Figure 3.12. An estimate of the probable average water saturation in a particular oil reservoir.

estimate is illustrated in Fig. 3.12, and the following quantities are calculated according to a triangular distribution:

Mean:

$$m = \frac{a+b+c}{3} = \frac{20+30+50}{3}$$
$$= 33\% \text{ water saturation.}$$

Probabilities:

$$p_2 = \frac{b-a}{c-a} = \frac{30-20}{50-20} = 0.33,$$
$$p_3 = \frac{c-b}{c-a} = \frac{50-30}{50-20} = 0.67.$$

Variance:

$$\sigma^2 = \frac{(c-a)^2}{18}(1-p_2p_3)$$
$$= \frac{(50-20)^2}{18}[1-(0.33)(0.67)]$$
$$= 38.88.$$

Standard deviation:

$$\sigma = \sqrt{\sigma^2} = \sqrt{38.88}$$
$$= 6.2\% \text{ water saturation.}$$

The cumulative probability that the water saturation will be less than 40% is

$$F = 1 - p_3\left(\frac{c-X}{c-b}\right)^2 = 1 - 0.67\left(\frac{50-40}{50-30}\right)^2$$
$$= 0.83.$$

A second example is concerned with the reserves that may be discovered by a certain exploratory well. The chance of a dry well is estimated as 0.3.

There is little, if any, chance of the well being completed for any reserves less than 20,000 bbl. The most probable reserves discovered would be 80,000, and there is a decreasing probability of a discovery up to the limit of 200,000 bbl. This is indicated in Fig. 3.13. The following quantities are calculated for the triangular distribution with a spike.

Mean:

$$\begin{aligned} m &= p_1 X_1 + (1 - p_1)\frac{a + b + c}{3} \\ &= (0.3)(0) + (1 - 0.3)\frac{20 + 80 + 200}{3} \\ &= 70 \text{ M bbl oil.} \end{aligned}$$

Probabilities:

$$\begin{aligned} p_2 &= (1 - p_1)\frac{b - a}{c - a} = (1 - 0.3)\frac{80 - 20}{200 - 20} \\ &= 0.233, \\ p_3 &= (1 - p_1)\frac{c - b}{c - a} = (1 - 0.3)\frac{200 - 80}{200 - 20} \\ &= 0.467. \end{aligned}$$

Variance:

$$\begin{aligned} \sigma^2 &= \frac{p_1}{1 - p_1}(m - X_1)^2 + \frac{(c - a)^2}{18}\left(1 - p_1 - \frac{p_2 p_3}{1 - p_1}\right) \\ &= \frac{0.3}{0.7}(70 - 0)^2 + \frac{(200 - 20)^2}{18}\left[0.7 - \frac{(0.233)(0.467)}{0.7}\right] \\ &= 3{,}080. \end{aligned}$$

Standard deviation:

$$\begin{aligned} \sigma &= \sqrt{\sigma^2} = \sqrt{3{,}080} \\ &= 56 \text{ M bbl.} \end{aligned}$$

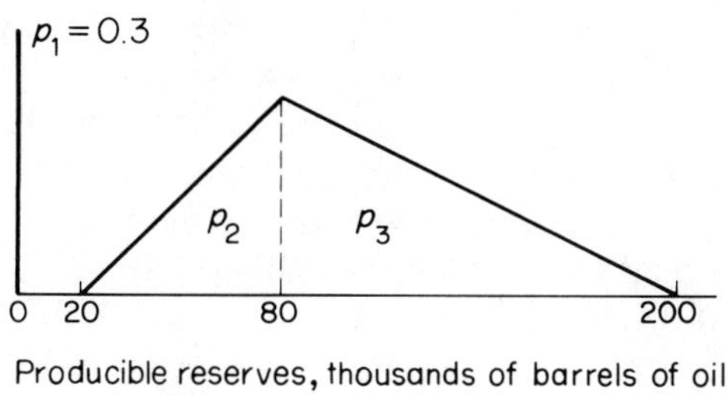

Figure 3.13. Estimated prospects of a certain exploratory well.

In summarizing the properties of this distribution, it should be noted that the range of possible outcomes is considered to be from 0 to 200 M bbl. The expectancy (the mean) is 70 M bbl, which must be interpreted as the average of many similar trials. Also, the variance and the standard deviation have but little meaning if only a single trial is to be made. However, these properties do become very useful when several prospects are considered together. Methods of evaluating the prospects of combinations of ventures will be described in subsequent chapters.

3.6 RELATIONS BETWEEN SAMPLING AND PARENT DISTRIBUTIONS

3.6-1 Characteristics of Sampling Distributions

Information about the quality of a shipment of manufactured articles can be obtained by selecting and testing several of the individual articles. Very rarely are they all tested, for usually the results obtained by sampling are sufficient. Tests made on the strengths of a few steel rods justify accepting a carload, and tests on a few light bulbs characterize a large quantity. For many reasons sampling is a necessary procedural method. It is therefore important to know the relations that exist between samples and the populations from which they are drawn.

Sampling, in effect, creates new distributions. More belief is placed in the average of several trials, and correctly so. Several samples can be taken, and their mean or average values will be found to differ from each other. The collection, or set, of such average values is known as a *sampling distribution.* Where a sample is comprised of several trials, the tendency is for high, low, and medium values to be mixed together. The greater the number of individual trials within a sample, the closer the sample mean will be to the mean of the parent distribution from which the sample was drawn. It is customary to state this as: The greater the sample size, the smaller will be the variance of the sampling distribution.

Sampling distributions tend to become normal distributions. This is in accordance with the *central limit theorem.* When the sample size is 30 or more, there is excellent agreement. Even when there are only 3 or 4 individual trials within a sample, there is often a close resemblance to the normal distribution. Thus whether samples are drawn from a triangular, or a normal, or a log-normal, or a rectangular distribution, the average or mean values of the samples will tend to form a normal distribution.

Important relations between parent and sampling distributions follow. For all cases, the mean of the sample is used to estimate the mean of the parent distribution:

$$m_{\bar{x}} = m. \tag{3-30}$$

Table 3.15
Parent Distribution: Values of X

(2)	(5)	(8)	(9)

Relations between parent and sampling distributions are illustrated by a simple example. The parent distribution given in Table 3.15 will be used. The mean and standard deviation of this parent distribution are as follows:

$$m = \frac{\sum X}{M} = \frac{2+5+8+9}{4} = \frac{24}{4} = 6,$$

$$\sigma^2 = \frac{\sum X^2 - [(\sum X)^2/M]}{M} = \frac{174 - [(24)^2/4]}{4} = \frac{30}{4},$$

$$\sigma = 2.74.$$

3.6-2 Sampling Without Replacement

For sampling without replacement, standard deviations are related by

$$\sigma_{\bar{X}} = \frac{\sigma}{\sqrt{N}}\sqrt{\frac{M-N}{M-1}}. \tag{3-31}$$

where

m = mean of the parent distribution,

$m_{\bar{X}}$ = mean of the sample distribution,

σ = standard deviation of the parent distribution,

$\sigma_{\bar{X}}$ = standard deviation of the sample distribution,

M = size of the parent population,

N = sample size, number within the sample.

[Where M, the size of the parent population, is large, $M - N \simeq M - 1$, and Eq. (3-31) reduces to Eq. (3-32).]

The case of sampling without replacement can be illustrated with samples of size $N = 2$. If one hypothesizes that a very large number of trials will be made, analysis can be made by setting up a table of equally likely samples. Table 3.16 illustrates this concept. This new distribution of sample means can be observed to be clustered more closely about the center. Its mean and standard deviation are as follows:

$$m = \frac{\sum \bar{X}}{n_{\bar{X}}} = \frac{3.5 + 5.0 + 5.5 + 6.5 + 7.0 + 8.5}{6} = \frac{36}{6} = 6,$$

$$\sigma_{\bar{X}}^2 = \frac{\sum \bar{X}^2 - [(\sum \bar{X})^2/n_{\bar{X}}]}{n_{\bar{X}}} = \frac{231 - [(36)^2/6]}{6} = \frac{15}{6} = \frac{5}{2},$$

$$\sigma_{\bar{X}} = 1.58.$$

Table 3.16
Sampling Distribution Without Replacement

Sample	*Sample Mean,* $\bar{X}$
(2, 5)	3.5
(2, 8)	5.0
(2, 9)	5.5
(5, 8)	6.5
(5, 9)	7.0
(8, 9)	8.5

From Eq. (3-31) the value of the standard deviation is

$$\sigma_{\bar{X}} = \frac{\sigma}{\sqrt{N}}\sqrt{\frac{M-N}{M-1}} = \frac{2.74}{\sqrt{2}}\sqrt{\frac{4-2}{4-1}}$$
$$= 1.58,$$

which indicates agreement between the methods.

3.6-3 Sampling With Replacement

For sampling with replacement, or from a large population with or without replacement, standard deviations are related by

$$\sigma_{\bar{X}} = \frac{\sigma}{\sqrt{N}}. \tag{3-32}$$

The case of sampling with replacement can be analyzed by again visualizing a very large number of trials. In these, each of the four values in Table 3.15 would be equally likely to be chosen first in the sample. Then the same value or any of the others would be equally likely to be chosen as the second value. This gives the group of samples listed in Table 3.17. The corresponding sample means, values of $\bar{X}$, which constitute a new distribution, are listed in Table 3.18. The mean and standard deviation of these values are as follows:

$$m = \frac{\sum \bar{X}}{n_{\bar{X}}} = \frac{2.0 + \cdots + 9.0}{16} = \frac{96}{16} = 6,$$

$$\sigma_{\bar{X}}^2 = \frac{\sum \bar{X}^2 - [(\sum X)^2/n_{\bar{X}}]}{n_{\bar{X}}} = \frac{636 - 576}{16} = \frac{15}{4},$$

$$\sigma_{\bar{X}} = 1.94.$$

A check for agreement with Eq. (3-32) is made as follows:

$$\sigma_{\bar{X}} = \frac{\sigma}{\sqrt{N}} = \frac{2.74}{\sqrt{2}} = 1.94,$$

which illustrates agreement between the two methods.

Table 3.17
Sampling Distribution with Replacement: Values of X

(2, 2)	(2, 5)	(2, 8)	(2, 9)
(5, 2)	(5, 5)	(5, 8)	(5, 9)
(8, 2)	(8, 5)	(8, 8)	(8, 9)
(9, 2)	(9, 5)	(9, 8)	(9, 9)

Table 3.18
Sampling Distribution with Replacement: Values of $\bar{X}$

2.0	3.5	5.0	5.5
3.5	5.0	6.5	7.0
5.0	6.5	8.0	8.5
5.5	7.0	8.5	9.0

These concepts can be summarized as follows. When samples, each consisting of two or more individual trials, are drawn from a parent distribution, the average values of such samples form a new distribution. The mean of the new sampling distribution tends to equal the mean of the original parent distribution. The standard deviation of the sampling distribution is smaller than the standard deviation of the parent distribution, and the two are related by Eq. (3-31) or (3-32). The form of the new sampling distribution is generally different from the original (unless the original was a normal distribution), and the sampling distribution tends to be similar to a normal distribution.

Quite evidently relations must and do exist between the characteristics of samples and the confidence that can be placed in properties estimated from such samples. The student's t distribution is widely used for this purpose when sampling is from a normal or near-normal distribution. The mean and variance of the sample are used to estimate the mean and variance of the parent population. In general, the mean of the parent population is estimated to be either above or below the mean of the sample, within certain confidence limits. Conversely, how much the estimate can be narrowed down depends upon the standard deviation of the sample, the size of the sample, and the degree of confidence required in the estimate. One branch of the science of statistics is entirely concerned with such estimates.

NOMENCLATURE

A = displacement between mean of a subgroup and mean of total population,
$= \bar{X}_i - \bar{X}$, for a discrete distribution,
$= m_i - m$, for continuous distributions,

$E[X]$ = expectancy of X,

$f(X)$ = function describing the value of the probability density in terms of X,

f_i = number of values, or outcomes, occurring in the ith class,

F = cumulative probability,

m = arithmetic mean or average value,

$m_{\bar{X}}$ = mean of a sample,

M = size of the parent population,

M = abbreviation for thousand (MM = 1 million),

n = number of trials,

N = sample size, number of trials,

$p(X_1)$ = probability that quantity X_1 will be chosen in a trial,

s = estimated standard deviation, based on a sample,

s^2 = estimated variance, based on a sample,

$x = X_i - \bar{X}$ = normalized X-values,

$\bar{X}$ = arithmetic average of X-values in the group,

$\sum X$ = sum of all X-values in the group or set,

σ = standard deviation, based on all members in a population,

σ^2 = variance, based on all members in a population,

$\sigma_{\bar{X}}$ = standard deviation of a sample.

REFERENCES

1. Hadley, G., *Introduction to Probability and Statistical Decision Theory.* San Francisco: Holden-Day, Inc., 1967, p. 10.

2. Whitworth, W. A., *Choice and Chance.* New York: Cambridge University Press, 1878.

3. Grayson, C. J., Jr., *Decisions Under Uncertainty.* Boston: Division of Research, Harvard University, 1960.

4. Savage, L. J., *The Foundations of Statistics.* New York: John Wiley & Sons, Inc., 1954.

5. Spegel, M. R., *Statistics* (Schaum Outline Series). New York: Schaum Publishing Co., 1961.

6. Neville, A. M., and J. B. Kennedy, *Basic Statistical Methods.* New York: Intext Publishers Group, 1964.

PROBLEMS

3-1. Four wells are to be drilled, *A*, *B*, *C*, and *D*. The probability of each being succesful is 0.5, independent of the outcomes of the other wells. Prepare a table that shows all possible results (permutations), such as *A*, dry; *B*, producer; *C*, dry; *D*, dry. Summarize the results in a table that shows the probabilities of the various combinations, such *p* (1 dry, 3 producers) = (?).

3-2. The following percentage porosity values were measured in a well; each has equal weight:

6	12	5	11
10	6	11	9
5	8	8	14
8	12	10	10
11	4	8	8
9	13	7	13
13	7	15	9

Calculate the average porosity. Make a frequency distribution, using nominal class boundaries such as 4–6, 6–8, etc. Show the number of cases in each class. Calculate the true class boundaries and class mark. Calculate the average porosity from the frequency and true-class-mark data.

3-3. Calculate the mean, variance, and standard deviation for the following distribution:

p:	0.4	0.3	0.2	0.1
X:	−50	40	60	80

3-4. Calculate the variance and standard deviation by first using an assumed mean of 100 for the following data, then transferring to the true mean. Check the result by conventional calculations.

X
102.4
103.7
99.2
104.1
102.6

3-5. The prospects of an exploratory well are summarized in the table. The possible outcomes are expressed as net present values. Calculate the mean, variance, and standard deviation.

Probability	*Outcome* ($)
0.5	−70,000
0.1	50,000
0.3	150,000
0.1	200,000

3-6. Referring to Problem 3-5, if the cost of the well were increased by \$10,000, such that all outcome values were reduced by \$10,000, what would the variance and standard deviation be?

3-7. A well treatment will cost \$50,000 whether or not it is successful. The following table gives a summary view of various possible increases in the present value of production which may be realized as a result of the proposed treatment. Calculate the net present value of each outcome listed in the table and the means and standard deviations of the possible net present-value outcomes.

Present Value of Increased Production (\$)	*Probability*
0	0.25
40,000	0.10
60,000	0.30
80,000	0.20
100,000	0.15

3-8. A financial venture was simulated on a computer. Individual outcomes in simulation depend upon random combinations of the several variables, controlled so as to simulate nature, the real world. The following table gives the number of trials that fell into each of several previously selected classes. In this case, since each result would be calculated to eight or more significant figures, the true class boundaries would be as given in the table. Calculate the mean and variance by two methods:

(a) Assume that all outcomes within each class fell at the class mark.

(b) Assume that the data points were dispersed so that they can be represented by a continuous uniform distribution within each class.

No. Trials in Each Class	*Class Limits* (*10^3 dollars*)
350	−20 to 0
200	0 to 20
150	20 to 40
100	40 to 80
100	80 to 100

3-9. The possible outcomes of drilling a particular exploratory well have been expressed in terms of two separate continuous distributions. The lower distribution has a probability of 0.6 and extends from −\$80,000 to −\$50,000. The favorable outcomes will lie between \$30,000 and \$250,000, with a probability of 0.4. The values are net present values in all cases. Calculate the mean, variance, and standard deviation of the combined distribution.

3-10. A well is given 0.4 probability of being dry. If successful, it may recover anywhere from 50,000 to 200,000 bbl of oil. Calculate the mean, variance, and standard deviation.

3-11. The ultimate recovery of a proposed development well is estimated at somewhere between 60,000 and 140,000 bbl. There is no reason to believe that the outcome is more likely to occur in one part of the range than another. Calculate the mean and standard deviation of this continuous uniform distribution.

3-12. Calculate the mean, variance, and standard deviation of a uniform distribution with a spike. The uniform distribution extends from 0 to 100,000 bbl of oil. The spike is located at 0 and has a probability of 0.4.

3-13. (a) Calculate the mean, variance, and standard deviation of the following uniform distribution with a spike. The uniform distribution extends from 20,000 to 60,000 bbl of oil. The spike is located at 0 and has a probability of 0.3.
(b) What is the probability that the results will be greater than 40,000 bbl?
(c) What is the probability that the results will be greater than 50,000 bbl?

3-14. The probability of a particular well being dry is 0.6. If successful, the lower limit of recovery is 75 bbl/acre-ft, the most likely recovery is 150 bbl/acre-ft, and the upper limit is 300 bbl/acre-ft. Calculate the mean, variance, and standard deviation of this triangular distribution.

3-15. The prospects of a certain exploratory well have been summed up as follows. Probability of a dry hole and attendant financial loss of \$80,000 is 0.6. If successful, the total present value, with initial investment deducted, could run somewhere between \$50,000 and \$350,000, with a most probable value of \$200,000. Calculate the mean, variance, and standard deviation of the possible outcomes.

3-16. The results that can be expected from a proposed water flood, in terms of additional oil recovered per acre-foot, are represented by a triangular distribution. The lower limit has been estimated to be 50, the upper limit 200, and the most likely result to be 100 bbl of oil per acre-foot. Calculate the mean and standard deviation of this distribution. What is the probability that the recovery will be less than 75 bbl/acre-ft? What is the probability that the recovery will be greater than 100 bbl/acre-ft? What is the probability that the recovery will be greater than 150 bbl/acre-ft?

3-17. One alternative in developing a certain water flood prospect would be to develop the entire field as rapidly as possible. This would require an investment of \$700,000. There is estimated to be a 25% chance of no response from the water flood. If the formation responds to water flooding, the present value of future operating profits is represented by a triangular distribution as follows: the lower limit is \$600,000, the upper limit is \$2,400,000, and the most likely outcome is \$1,500,000. Calculate the net present values of the spike and the three points that define the triangular distribution.
(a) Calculate the mean and standard deviation of the triangular distribution with a spike.
(b) What is the probability that the present value of the profit/investment ratio will be less than 50%?

3-18. In the west flank area of the Trimble Basin there are sufficient geologic features for the occurrence of five oil fields. From past experience in similar areas it is estimated that the field sizes will be about 2, 3, 4, 5, and 10 MM bbl. If the company discovers two of these fields, what would be the mean and standard deviation of the possible results? Note that this is a case of sampling without replacement.

3-19. For the prospective fields described in Problem 3-18, if operator *A* is able to discover three of the fields, what are the mean and standard deviation of the possible results?

3-20. In the central area of the basin, 20 fields have been discovered to date. In equal proportions these have sizes of 2, 3, 4, 5, and 10 MM bbl, so the distribution of field sizes is the same as in Problem 3-18. If the company discovers three fields in this area, what would be the mean and standard deviation of the possible outcomes? *Note:* There are many more fields to be discovered in this region, and the discovery of a few fields will not appreciably change the size distribution of the remaining fields.

4

APPLICATIONS OF PROBABILITY DISTRIBUTIONS

4.1 INTRODUCTION

4.1-1 The Character of Distributions

The notion of probability can be shown graphically by the ratio of areas. In the simplest such diagram, only the probability of occurrence of an event is illustrated. In Fig. 4.1 an area of unity is assigned to sample space S. If an event occurs in S, the probability of its being A equals the area enclosed by envelope A. Probability of failure, or not A, equals the area in S outside envelope A. This simplest of cases does not always provide an adequate answer. When an event occurs, its value may not always be the same. Rather, its value may vary enormously. It therefore becomes important to know

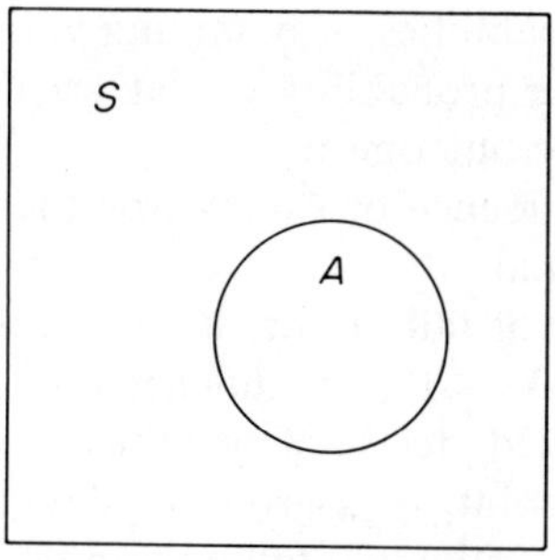

Figure 4.1. Simple Venn diagram.

which values are more likely to occur. This useful service is performed by distributions; they tell where events are more likely to occur and where they are very likely to occur.

A common form of distribution is sketched in Fig. 4.2. Along the horizontal base line are values of X. For example, X could be the number of

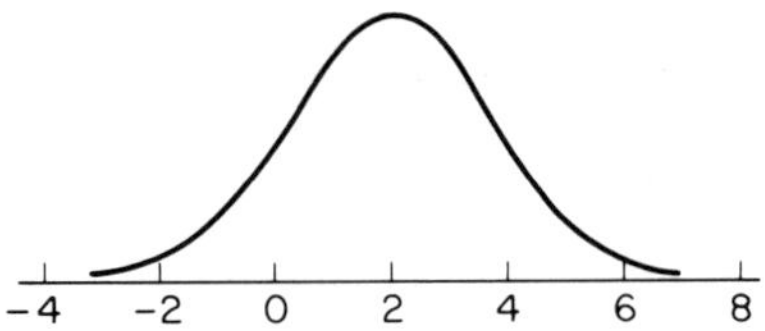

Figure 4.2. Common (normal) probability distribution.

barrels of oil discovered by an exploratory well, or X could be the number of heads obtained in, say, 10 tosses of a quarter, or X could be the measured diameter of a rod turned out by a production machine. Total area under the curved line is equal to unity, so the probability of the outcome occurring between two particular values is equal to the area under the curve between these two values.

4.1-2 Kinds of Applications

Applications of probability include formulating problems so that they can be solved by known mathematical methods. Formulation can sometimes be made so that the problem exactly fits a well known distribution whose solutions are available. Some of these solutions are developed in the following sections.

Types of applications of interest include the following:

1. Given an overall probability of success, determine the probabilities that various numbers of failures will occur before a successful outcome occurs.
2. Determine the probabilities of obtaining various economic outcomes, generally expressed as the probability of "at least" or "equal to or better than" the stated economic outcomes.
3. Determine the confidence of success, or the probability of achieving at least some minimum goal.
4. Determine the risk of failure, or the probability of failing to achieve some minimum goal. (Risk $= 1 -$ confidence.)
5. Formulate one-tailed tests of significance. Tests of significance concern whether a data point, or a group of data, do not belong in a particular distribution. If a point would fall only in the upper 5% or only in the lower 5% of the total probability space, the point is said to be significant at the 5% level. Similar considerations apply for a point to be significant at the

1% or other level of significance. Such points would be rejected at the stated level as not belonging to the distribution.

6. Formulate two-tailed tests of significance. If a value does not lie within the bounds corresponding to the middle 95% of the probability space of a particular distribution, the presumption (hypothesis) that it belongs to the distribution is rejected at the 5% level. Similar tests and rejections are made at 1% and other levels of significance.

Solutions of these general types are discussed in the following sections. They are developed for several probability distributions.

4.2 BINOMIAL AND NORMAL DISTRIBUTIONS

4.2-1 Binomial Distribution

Early exploratory efforts in a newly recognized geologic area provide an ideal application for the binomial probability distribution. Overall composition of the sediments—the proportions of shales, sandstones, and limestones present—could lead to an estimate, for example, that deposits of hydrocarbons will be scattered but will underlie 20% of the area. A total of 400 exploratory wells might be required to fully test the prospective area. For the first several years the probability of success for an exploratory well would be taken to be 0.2. In using the binomial, no consideration is given respecting how big or how good a discovery might prove to be. It is taken to be either good or bad, acceptable or unacceptable, dry or producer, with classification restricted to one or the other of two groups.

The binomial is a sampling distribution. In the above example, a group of 10 exploratory wells would represent a sample of size 10, and the sample would be drawn, or chosen, from the set consisting of all possible locations for exploratory wells within the confines of the geologic area. Eventually, as time would approach for the final stages of exploratory drilling in the area, the success ratio would be expected to shift considerably from its early value. Other methods can be employed at the later date, but initially the binomial serves to describe fluctuations in the number of productive versus dry wells. For the binomial to apply strictly, the probability of success must remain constant.

There are many applications of the binomial distribution. Consider rolling a common die, where one chooses 3 as his "lucky number." The population is then divided into two parts, "3" and "not 3." The probability of success is $\frac{1}{6}$. Ten rolls of the die would be a sample of size 10, and since there is no definable limit on the number of times the die could be rolled, the size of the population in this case is said to be infinite.

Tossing coins is a familiar example, and the probability of obtaining a head is described as $p(\text{head}) = 0.5$, at least approximately. Since there is no limit on the number of times a coin may be tossed, the population is said to be infinite in size. A sample of size 5 can be obtained by tossing the same coin five times or by tossing five different coins.

Drawing for an ace from a deck of cards represents another example, provided that each card drawn is immediately replaced into the deck before the next draw. This is described by the term "with replacement" (wr), and it serves to keep the probability of success constant during successive trials. Opposed to this is the process of sampling "without replacement" (wor), which in the case of the cards can be done by picking out, say, 10 cards all at one time. Particularly where the population being drawn from is limited, the probabilities of success may change rapidly with cumulative withdrawals. Distributions other than the binomial can be used for such cases.

The binomial distribution describes sampling (wor) from a large population or sampling (wr) from a large or small population. In either case the probability of success of the individual trial remains constant. The solution concerns the probability of an exact number of successes in several trials. The basic relation is

$$p(x) = \binom{N}{x} p^x q^{N-x} \tag{4-1}$$

$$= C_x^N p^x q^{N-x}, \tag{4-1a}$$

where

$p(x)$ = probability of obtaining exactly x successes in N trials,

N = size of the sample, or number of trials of an event,

x = number of successes, or favorable outcomes within the N trials,

p = probability of success,

$q = 1 - p$ = probability of failure,

$\binom{N}{x} = C_x^N$ = number of combinations* in which N objects can be displayed as groups of size x, where the order within the individual groups is unimportant.

*Following are ways of evaluating the number of combinations. For example, where $N = 4$ and $x = 2$,

$$C_x^N = \frac{N!}{x!(N-x)!} = \frac{4!}{2!(4-2)!} = \frac{4\cdot3\cdot2\cdot1}{2\cdot1\cdot2\cdot1} = 6.$$

The following is exactly equivalent and often preferable:

$$\binom{N}{x} = \frac{N\cdot(N-1)\cdot(N-2)\ldots(N+1-x)}{x\cdot(x-1)\cdot(x-2)\ldots1} = \frac{4\cdot3}{2\cdot1} = 6.$$

Similarly, where $N = 6$ and $x = 4$,

$$\binom{6}{4} = \frac{6\cdot5\cdot4\cdot3}{4\cdot3\cdot2\cdot1} = 15.$$

$\binom{6}{0}$; $\binom{4}{0}$, etc., are defined as unity. If the denominator exceeds the numerator (group size exceeds population), the value is defined as zero.

The mean, variance, and standard deviation of the binomial are given by

$$m = Np, \tag{4-2}$$

$$\sigma^2 = Npq, \tag{4-3}$$

$$\sigma = \sqrt{Npq}. \tag{4-4}$$

As an example application, the probability of obtaining zero heads when a coin is tossed five times is calculated as follows:

$$p(x) = \binom{N}{x} p^x q^{N-x},$$

$$p(0) = \binom{5}{0}(0.5)^0(1 - 0.5)^{5-0} = (1)(1)(0.5)^5$$

$$= 0.03125.$$

Roughly, the probability is 3 of 100 times. That is, where successive tosses were gathered into groups of size 5 tosses in each group, out of 100 such groups, about 3 would contain no heads.

A second example concerns a case where 10 wells will be drilled. The probability of success is taken to be 0.15. What is the probability of there being more than 2 successful wells? The answer to this can be found in one of two ways: (1) the individual probabilities of 3, 4, 5, 6, 7, 8, 9, and 10 successes can be calculated and added together, or (2) the individual probabilities of 0, 1, and 2 successes can be added together and then subtracted from 1 to obtain the same answer. In the present instance the second method is shorter.

$$p(x) = \binom{N}{x} p^x q^{N-x},$$

$$p(0) = \binom{10}{0}(0.15)^0(0.85)^{10} = 0.1969,$$

$$p(1) = \binom{10}{1}(0.15)^1(0.85)^9 = 0.3474,$$

$$p(2) = \binom{10}{2}(0.15)^2(0.85)^8 = \underline{0.2759}$$

$$0.8202$$

$$p(\text{more than 2 producers}) = 1 - 0.8202 = 0.1798.$$

Additional applications of the binomial will be considered together with the normal distribution, which is often used as a rapid and convenient method for binomial approximation.

4.2-2 Negative Binomial Distribution

This distribution is concerned with the number of failures that will occur before success is attained. As in the binomial, the probability of success is presumed to be constant for each trial. The probability of exactly x failures is given by the relation

$$p(x) = \binom{r + x - 1}{x} p^r q^x, \tag{4-5}$$

where

x = number of failures met before achieving the rth success,

r = required number of successes,

p = probability of success on each trial,

$q = 1 - p$ = probability of failure on each trial.

As an example, let the required number of success be $r = 2$, where the probability of success is $p = 0.1$ for drilling a successful well. The probability of zero failures is

$$\begin{aligned} p(0) &= \binom{2 + 0 - 1}{0} (0.1)^2 (0.9)^0 \\ &= \binom{1}{0} (0.1)^2 (0.9)^0 \\ &= 0.0100. \end{aligned}$$

The probability of exactly one failure is

$$\begin{aligned} p(1) &= \binom{2 + 1 - 1}{1} (0.1)^2 (0.9)^1 \\ &= \binom{2}{1} (0.1)^2 (0.9)^1 \\ &= 0.0180. \end{aligned}$$

At this point the cumulative probability is $p(0) + p(1) = 0.0280$, which is far short of unity, or certainty. Indeed, by subtraction it is evident that there is a 97.2% chance that under the stated conditions, there will be more than one dry hole drilled.

Table 4.1 shows some results obtained by using the negative binomial, which offers a convenient method of calculation for such cases. For $r = 1$ required success and $x = 6$ previous failures, the table indicates a cumulative probability of 0.7903. That is, one would be 79% confident of not having to drill more than 7 wells in order to obtain one producer, where $p = 0.2$. However, if $r = 2$ successes are required, the cumulative probability is

Table 4.1

Probabilities for the Number of Failures preceding the Required Number of Successes*

Number of Failures, x	*Required Number of Successes, r = 1*		*Required Number of Successes, r = 2*	
	p(x)	*Cumulative*	*p(x)*	*Cumulative*
0	0.2000	0.2000	0.0400	0.0400
1	0.1600	0.3600	0.0640	0.1040
2	0.1280	0.4880	0.0768	0.1808
3	0.1024	0.5904	0.0819	0.2627
4	0.0819	0.6723	0.0819	0.3446
5	0.0656	0.7379	0.0787	0.4233
6	0.0524	0.7903	0.0734	0.4967
7	0.0419	0.8322	0.0671	0.5638
8	0.0336	0.8658	0.0604	0.6242
9	0.0268	0.8926	0.0537	0.6779
10	0.0215	0.9141	0.0440	0.7219

*The probability of success, $p = 0.2$.

0.4967, or about a 50% chance that one would need to drill more than 6 dry wells. There is a 1 − 0.7219, or 27.81%, chance that more than 10 dry wells would have to be drilled to obtain two producers.

4.2-3 Normal Distribution

Among all distributions, the normal is generally considered to be most useful. It is a continuous distribution. It approximates random errors involved in measuring a dimension of a machined part. Where balls for bearings are produced by a machine, their diameters, and consequently the cube root of their weights, tend to be normally distributed. Also, many phenomena do not precisely follow the normal, but they approximately follow a normal distribution in the region of interest. An example is the heights of men. In such cases it is common to assume that the normal distribution can be utilized for summarizing statistical properties of many measurements. The normal is a convenient approximation of the binomial distribution, particularly when the sample size is large. Finally, the averages or mean values of sample groups drawn from any one of the many different types of distributions tend to closely follow the normal distribution, so that it is a powerful tool for statistical estimation.

Following are some of the features of the normal curve:

1. It is symmetric about the mean.
2. It has a single mode, or peak; and the mean, median, and mode coincide at the center.

3. The total area under the curve is unity; the area is 0.5 on each side of the mean.

4. The area tends to be clustered about the mean. The area between $-\sigma$ and $+\sigma$ is 0.6826; the area between -2σ and $+2\sigma$ is 0.9544; the area between -3σ and $+3\sigma$ is 0.9974. Probabilities beyond $\pm 3\sigma$ limits are usually of little interest.

The normal distribution curve, or density function, is given by

$$y = \frac{1}{\sigma\sqrt{2\pi}} \exp\left[-\frac{1}{2}\left(\frac{x-m}{\sigma}\right)^2\right]. \tag{4-6}$$

This relation is standardized, so one table will fit all cases, as follows:

$$y = \frac{1}{\sigma\sqrt{2\pi}} e^{-z^2/2}, \tag{4-7}$$

$$z = \frac{X-m}{\sigma} \tag{4-8}$$

$$= \text{number of standard deviations.}$$

Of particular interest is the cumulative probability, which is conveniently defined as follows:

$$F = \int_{-\infty}^{z} y\, dz, \tag{4-9}$$

where

F = cumulative probability, or area under the curve, from $-\infty$ up to the particular value of z.

This is illustrated in Fig. 4.3. The probability areas listed in some tables for the normal distribution is indicated in Fig. 4.4. If z is positive, F can be found by adding 0.5 to the probability area listed in such tables. By virtue of the symmetry, if z is negative, F can be found by subtracting the listed prob-

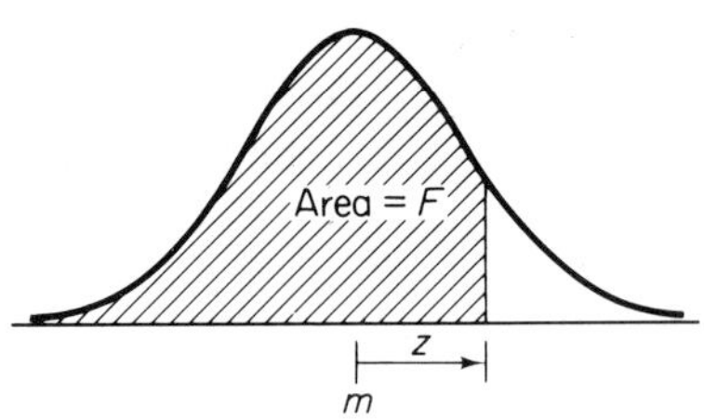

Figure 4.3. Cumulative probability area F under the normal distribution density function.

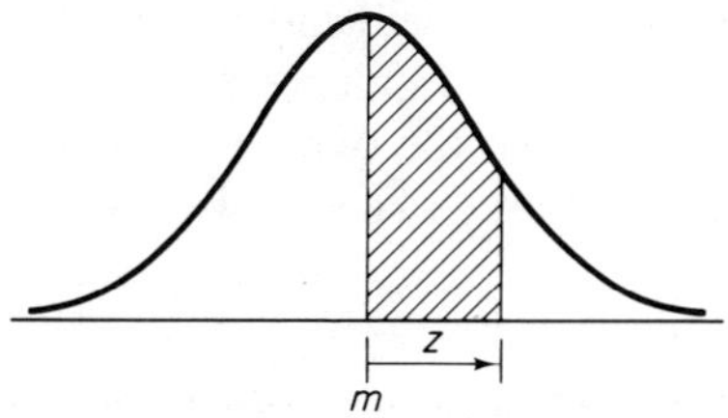

Figure 4.4. Probability area given in many tables for the normal distribution.

ability area from 0.5. In Appendix Table 2, the value of F can be read directly if z is positive; if z is negative, F can be found by subtracting the listed probability area from 1.0.

An example will illustrate the way a normal probability table is used.

Example: A certain type of engine averages 390 days of operation between overhauls, and standard deviation is 100 days. Determine:

1. The probability that an engine will last less than 300 days.
2. The probability that it will run less than 420 days.

Solution: The solution is obtained by calculating the two different z-values and finding the corresponding areas as listed in a normal probability table.

$$1.\ z, \text{ for 300 days} = \frac{X - m}{\sigma}$$

$$= \frac{300 \text{ days} - 390 \text{ days}}{100 \text{ days}}$$

$$= -0.90.$$

Area in table, for $z = 0.90$, is 0.8159.

F, for less than 300 days $= 1.0 - 0.8159 = 0.1841$.

The probability that the engine will fail in less than 300 days is 0.1841. Rephrased, there is 81.59% confidence that the engine will run for more than 300 days.

$$2.\ z, \text{ for 420 days} = \frac{X - m}{\sigma}$$

$$= \frac{420 \text{ days} - 390 \text{ days}}{100 \text{ days}}$$

$$= 0.30.$$

Area in table, for $z = 0.30$, is 0.6179.

F, for less than 420 days $= 0.6179$.

The probability that an engine will fail in less than 420 days is 0.6179. Rephrased, there is only 38.21% confidence that a run will exceed 420 days.

Taking the difference between the two cumulative probabilities gives the probability that the event will take place between the two limits:

$$F_{420} - F_{300} = 0.6179 - 0.1841 = 0.4338.$$

This indicates a probability of approximately 43% that a run will last somewhere between 300 and 420 days. The probability area is shown in Fig. 4.5.

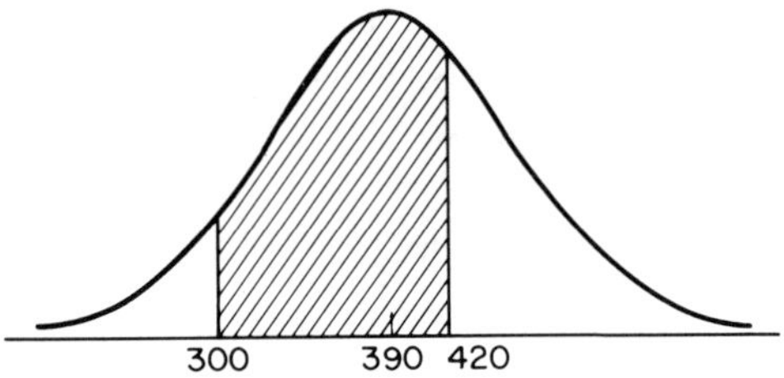

Figure 4.5. Engine operating days between major overhauls.

A graphical method for determining the mean and standard deviation involves the use of probability paper. One scale on this paper is simply linear and unmarked, so any desired values of random variable X can be marked off in arithmetic fashion along that axis. The other scale is linear in z, but it is ruled and labeled in terms of percentage of cumulative probability. The labeled values are closer together near the center and spread farther apart from each other in either direction from the 50% position.

In Table 4.2 are shown the operating days between major overhauls for

Table 4.2
Operating Days Between Major Overhauls for 10 Engines

Operating Days	*% of Cases Less Than*
230	5
280	15
320	25
340	35
380	45
420	55
430	65
450	75
490	85
560	95
3,900	

10 individual engines. These are shown plotted at their assigned "per cent less than" values in Fig. 4.6. If the data corresponded exactly to a normal distribution, they would all fall in a single straight line. As may be observed, there is close correspondence to a normal distribution. At the 50% point the line drawn through the data points reads 390 days on the X-scale, which is the mean of the distribution.

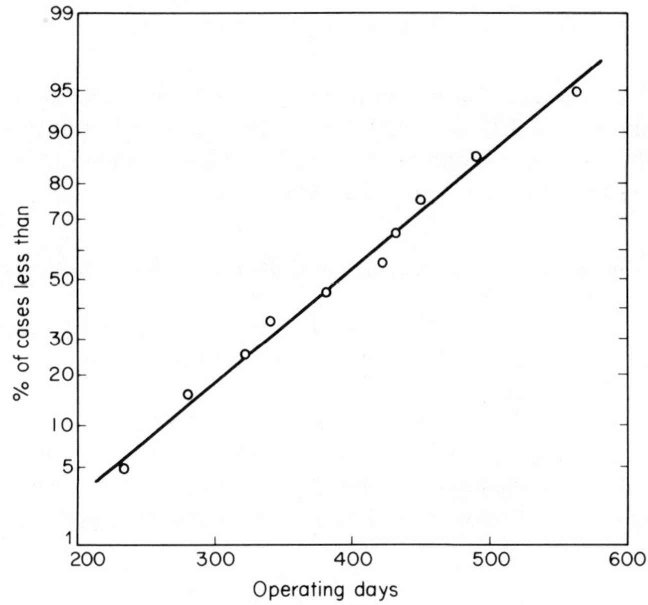

Figure 4.6. Plot on normal probability paper.

Table 4.3 lists the total number of standard deviations, between two tails, for several percentage ranges. These may be used as follows to calculate the

Table 4.3
Number of Standard Deviations Included Between Cumulative Per Cent Probability Values

% Excluded, Each Tail	*z, Each Tail*	*2z, Between Two Tails*	*% Included, Between Two Tails*
1.00	2.327	4.654	1.00–99.00
2.00	2.054	4.108	2.00–98.00
5.00	1.645	3.290	5.00–95.00
10.00	1.282	2.564	10.00–90.00
15.87	1.000	2.000	15.87–84.13
20.00	0.841	1.682	20.00–80.00

standard deviation. The X-values at the percentage points are read directly from the straight line drawn on Fig. 4.6:

$$\sigma = \frac{X_{0.95} - X_{0.95}}{z\text{, between two tails}}$$

$$= \frac{555 \text{ days} - 225 \text{ days}}{3.290}$$

$$= 100 \text{ days.}$$

The value of σ can also be obtained by reading $X = 290$ at the 15.87% line, and since $m = 390$, $\sigma = 390$ days $-$ 290 days $=$ 100 days. Any two points on the line can, of course, be used to calculate σ, and usually less error is introduced by using well-separated points.

4.2-4 Applications of the Normal and Binomial Distributions

It has long been known that the normal curve can be used to approximate the binomial distribution. In 1733 this was stated by deMoivre for the case where $p = 0.5$, and in 1812 Laplace proved that it was true for arbitrary values of p. Figure 4.7 illustrates the case of $p = 0.5$ and $N = 6$. The cross-hatched area represents the probability that the outcome X will be either 1 or 2 or 3 or 4. In this example $a = 1$ and $b = 4$, and these are included in the outcomes of interest. Following is the manner of using the normal approximation:

$$p(a \leq X \leq b) = F(z_b) - F(z_a), \tag{4-10}$$

where $F(z_a)$ and $F(z_b)$ are cumulative probabilities, obtained from a table of normal probabilities, corresponding to z_a and z_b.

The following binomial relations also apply:

$$m = Np,$$

$$\sigma = \sqrt{Npq},$$

$$z_a = \frac{a - 0.5 - m}{\sigma}, \tag{4-11}$$

$$z_b = \frac{b + 0.5 - m}{\sigma}, \tag{4-12}$$

N = sample size,

p = probability of event occurring,

$q = 1 - p$.

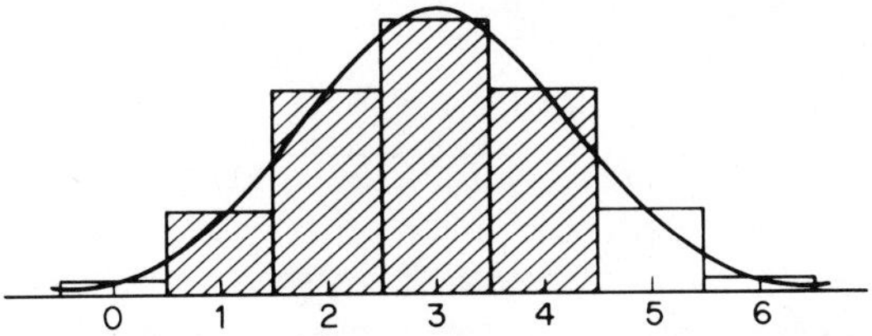

Figure 4.7. Relations between the binomial and normal distributions.

Example: For the case illustrated in Fig. 4.7, calculations are as follows:

$$m = Np = (6)(0.5) = 3,$$

$$\sigma = \sqrt{Npq} = \sqrt{(6)(0.5)(0.5)} = 1.225,$$

$$z_b = \frac{b + 0.5 - m}{\sigma} = \frac{4 + 0.5 - 3}{1.225} = 1.225,$$

$$z_a = \frac{a - 0.5 - m}{\sigma} = \frac{1 - 0.5 - 3}{1.225} = -2.040,$$

$$F(z_b) = F(1.225) = 0.5 + 0.3898 = 0.8898,$$

$$F(z_a) = F(-2.040) = 0.5 - 0.4793 = 0.0207,$$

$$p(1, 2, 3, \text{ or } 4) = 0.8898 - 0.0207 = 0.8691.$$

This solution can be readily checked by using values available in binomial probability tables.

$$p(X = 1) = 0.0938$$

$$p(X = 2) = 0.2344$$

$$p(X = 3) = 0.3125$$

$$p(X = 4) = \underline{0.2344}$$

$$0.8751$$

The two answers check within 1%, which error is small and usually negligible with respect to uncertainties in the input information. When the sample size becomes large, it is much more expedient to use the normal distribution approximation. Extensive binomial tables are not generally available, and the computations are often prohibitive. Fortunately, as the sample size increases, the approximation becomes more accurate.

Example: A check will be made on previous calculations using the binomial. The example concerned the probability of not more than 2 producers in a group of 10 wells, where the probability of success was 0.15. For this case the following calculations are made:

$$m = Np = (10)(0.15) = 1.5,$$

$$\sigma = \sqrt{Npq} = \sqrt{(10)(0.15)(0.85)} = 1.128,$$

$$z_b = \frac{b + 0.5 - m}{\sigma} = \frac{2 + 0.5 - 1.5}{1.128} = 0.887,$$

$$F(z_b) = F(0.887) = 0.5 + 0.3188 = 0.8118,$$

which checks with the 0.8203 obtained by use of the binomial distribution. A calculation will also be made, using the same data, for the probability of obtaining less than one producer.

$$z_a = \frac{a - 0.5 - m}{\sigma} = \frac{1 - 0.5 - 1.5}{1.128} = -0.887,$$

$$F(z_a) = F(-0.887) = 0.5 - 0.3188 = 0.1812,$$

which checks closely with 0.1969, the binomial probability of obtaining zero producers.

Example: The chance of discovering oil is $\frac{1}{8}$. Monetary reserves permit the drilling of 10 wells. Under such conditions, what is the probability of going broke before a discovery is made?
Solution: For these conditions,

$$m = Np = (10)(0.125) = 1.25,$$

$$\sigma = \sqrt{Npq} = \sqrt{(10)(0.125)(0.875)} = 1.045.$$

For the probability of less than one producer,

$$z_a = \frac{1 - 0.5 - m}{\sigma} = \frac{1 - 0.5 - 1.25}{1.045} = -0.718,$$

$$F(z_a) = F(-0.718) = 0.5 - 0.2636 = 0.2364,$$

which is the probability of going broke before an intitial discovery is made. That is, one would be only 76.36% confident of success.

Tests of significance would provide additional examples of these concepts. In such tests the object is to estimate whether or not certain results are consistent with a given distribution. In one-tailed tests, the significant areas are at one end of the distribution. In two-tailed tests, half of the total probability area is at each of the two ends of the distribution. Table 4.4 gives z-values for the normal distribution. If the z-values calculated from the problem exceed the z-value listed in the table, the result is significant and apparently does not fit with the rest of the data. If the calculated z-value is smaller than listed in Table 4.4, the results are accepted as showing no significant differences.

A tool that is widely applicable for making tests of significance is the chi-square distribution. It provides a means for making such tests in cases

Table 4.4
Significance Levels and Corresponding z-Values Based on the Normal Distribution

Significance Level	*One-Tailed-Test z-values (− or +)*	*Two-Tailed-Test z-values (− and +)*
0.10	1.282	1.645
0.05	1.645	1.960
0.02	2.054	2.327
0.01	2.327	2.575
0.005	2.575	2.810
0.002	2.880	3.090

where the normal or other convenient distributions do not apply. The following cases make use of the normal distribution.

Example: The drilling programs of two companies will be compared. The comparison is to be made on wells drilled in the same area and depth range. Over the period of the last 4 years, company C has completed 12 discoveries in 60 wildcats, and company D has completed 8 successes in 30 wildcats. At the 5% level of significance, does it appear that company D has better operating techniques?

Solution: The statistical problem concerns whether or not these samples are being drawn from the same population, and hence whether or not the one company may have better means of delineating the population from which it samples. Three solutions will be provided for this problem.

a. This solution is based on comparing company D to the data established by company C:

$$p = \frac{12}{60} = 0.2,$$

$$m = Np = (30)(0.2) = 6,$$

$$\sigma = \sqrt{Npq} = \sqrt{(30)(0.2)(0.8)} = 2.19,$$

$$z_a = \frac{a - 0.5 - m}{\sigma} = \frac{8 - 0.5 - 6}{2.19} = 0.686,$$

which is less than the $z = 1.645$ required for the data to be significant at the 5% level.

b. This solution is based on comparing company C to the data established by company D:

$$p = \frac{8}{30} = 0.267,$$

$$m = Np = (60)(0.267) = 16,$$

$$\sigma = \sqrt{Npq} = \sqrt{(60)(0.267)(0.733)} = 3.428,$$

$$z_b = \frac{b + 0.5 - m}{\sigma} = \frac{12 + 0.5 - 16}{3.428} = -1.022,$$

which again is less than the $z = \pm 1.645$ required for significance at the 5% level.

c. This solution is arrived at by use of proportions, where the standard deviation is based on pooling all the data. First, the weighted average probability is calculated:

$$p = \frac{N_1P_1 + N_2P_2}{N_1 + N_2} \tag{4-13}$$

$$= \frac{(60)\frac{12}{60} + (30)\frac{8}{30}}{60 + 30} = \frac{20}{90} = 0.222,$$

$$q = 1 - p = 0.778,$$

$$\sigma_{P_1 - P_2} = \sqrt{pq\left(\frac{1}{N_1} + \frac{1}{N_2}\right)} \tag{4-14}$$

$$= \sqrt{(0.222)(0.778)\left(\frac{1}{60} + \frac{1}{30}\right)}$$

$$= 0.0928,$$

$$z = \frac{P_1 - P_2}{\sigma_{P_1 - P_2}} \tag{4-15}$$

$$= \frac{\frac{12}{60} - \frac{8}{30}}{0.0928} = \frac{0.2 - 0.267}{0.0928}$$

$$= -0.723,$$

which again is not significant at the 5% level.

Example: In the evaluation of a new geophysical method, exploratory efforts have established a success ratio of 1 in 8 in a certain region. After adopting a new geophysical method, 6 of the last 30 wells have been successful. Are these results significant?

Solution: The solution will be based on the established probability of success, $p = 0.125$. For the sample of 30 wells,

$$m = Np = (30)(0.125) = 3.75,$$

$$\sigma = \sqrt{Npq} = \sqrt{(30)(0.125)(0.875)} = 1.81,$$

$$z_a = \frac{X - 0.5 - m}{\sigma} = \frac{6 - 0.5 - 3.75}{1.81} = 0.965.$$

To be significant at even the 10% level, z would have to be greater than 1.282. Therefore, the conclusion is that the results to date are encouraging, but further testing (a larger sample) will be required for final evaluation.

4.3 ARRIVALS, RARE OCCURRENCES, AND SERVICING REQUIREMENTS

4.3-1 Poisson Distribution

Certain physical processes, including emissions of radioactive particles, have been observed to follow the Poisson distribution. Appointments to the U.S. Supreme Court per 4-year presidential term of office have been following the same distribution. The same pattern is followed by the number of cars per minute passing through a turnpike toll booth and by the number of ships that require docking facilities. Demands for services or repairs generally follow a Poisson distribution. Bases other than the number of occurrences per unit time can be used. For example, occurrences per unit length of product or the number of flaws per pound of material can also serve as bases of measurements.[3]

The Poisson distribution is given by the following relation:

$$p(x) = \frac{\lambda^x}{x!} e^{-\lambda} \tag{4-16}$$

where

λ = mean,
= variance,
= Np,
x = number of occurrences per basic unit of measure,
N = number of repeated independent trials,
p = probability of success on each trial.

In addition to being a useful distribution in its own right the Poisson is a good approximation for the binomial, particularly when N is very large and p is small, when Np is no larger than 5, and it is a very good approximation when Np is less than unity.

An example of the usefulness of the Poisson probability law is furnished by an oil company's well-servicing requirements, which average two per day in a certain district. On how many days per month will no breakdowns be reported?

The following calculations apply:

$$\lambda = 2,$$

$$p(x = 0) = \frac{\lambda^x}{x!}e^{-\lambda} = \frac{(2)^0}{0!}e^{-2} = e^{-2}$$

$$= 0.135.$$

$$\text{days per month} = (0.135)(30.4) = 4.1.$$

There are 4 days each month when no breakdowns will be reported.

The calculation is readily extended for the cases where 1, 2, and 3 breakdowns will be reported.

$$p(x = 1) = \frac{(2)^1}{1!}e^{-2} = 0.270,$$

$$\text{days per month} = (0.270)(30.4) = 8.2.$$

$$p(x = 2) = \frac{(2)^2}{2!}e^{-2} = 0.270,$$

$$\text{days per month} = (0.270)(30.4) = 8.2.$$

$$p(x = 3) = \frac{(2)^3}{3!}e^{-2} = 0.180,$$

$$\text{days per month} = (0.180)(30.4) = 5.47.$$

The discovery of giant oil fields by a group of companies averages one every 2 years. Over a 10-year period, how would the discoveries be distributed?

For this case, $\lambda = 0.5$ occurrence per year. The probabilities are available in the table of Poisson probabilities. Table 4.5 gives the expected distributions by years.

Table 4.5
Distribution of Number of Discoveries per Year, for a 10-Year Period

Number of Fields Discovered During a 1-Year Period, x	*p(x)*	*p(x) × 10 years*	*Number of Years*	*Fields Discovered*
0	0.6065	6.065	6	0
1	0.3033	3.033	3	3
2	0.0758	0.758	1	2
3	0.0126	0.126	—	—
4	0.0016	0.016	—	—
5	0.0002	0.002	—	—
	1.0000	10.000	10	5

4.3-2 Exponential Distribution

The lengths of time required for repairs to automotive equipment have been observed to follow approximately an exponential distribution. The same is true for the servicing required in many types of operations. The lifetimes of electric light bulbs have been described in terms of this distribution.

Mathematical relations for the exponential distribution are as follows:

$$f(x) = \frac{1}{a} e^{-x/a}, \tag{4-17}$$

$$\begin{aligned} F &= \text{probability of a value less than } x \\ &= \int_0^x \frac{1}{a} e^{-x/a}\, dx \\ &= 1 - e^{-x/a}, \end{aligned} \tag{4-18}$$

$$\text{mean} = m = a,$$

$$\text{variance} = \sigma^2 = a^2,$$

$$\text{standard deviation} = \sigma = a.$$

Figure 4.8 shows an exponential distribution of the repair times required for maintenance of a fleet of trucks. The mean is 5 hours. As an example, a calculation will be made for the percentage of cases when less than 5 hours is required for repairs.

$$\begin{aligned} F &= 1 - \exp\left(-\frac{x}{a}\right) \\ &= 1 - \exp\left(-\frac{5 \text{ hr}}{5 \text{ hr}}\right) \\ &= 0.632, \qquad \text{or approximately } 63.2\% \text{ of repairs,} \end{aligned}$$

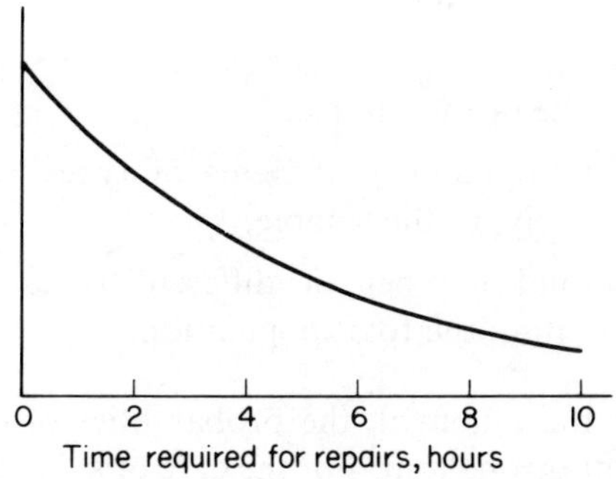

Figure 4.8. An exponential distribution.

require less than 5 hours. As another example, calculation is made for the percentage of cases that are finished in less than 8 hours:

$$F = 1 - \exp\left(-\frac{8\text{ hr}}{5\text{ hr}}\right)$$
$$= 0.798, \text{ or approximately } 80\%,$$

can be finished within 8 hours.

4.4 MULTIPLE KINDS OF OBJECTS AND ECONOMIC OUTCOMES

4.4-1 Multinomial Distribution

In some cases the binomial distribution is not adequate. Whenever the population of interest must be divided into more than two categories, recourse must be had to other probability distributions. One such distribution is the multinomial. Suppose that there are 20 red, 30 white, and 50 blue marbles in a box and that one draws a sample of 4 marbles. The sample could consist of all red, all white, all blue, or any one of numerous combinations of these. Solution of this problem involves listing all possible samples and calculating the probability of each sample.

Following is the multinomial probability relation:

$$p(S) = \frac{N!}{k_1!k_2!\ldots k_m!} P_1^{k_1} P_2^{k_2} \ldots P_m^{k_m}, \tag{4-19}$$

where

$p(S)$ = probability of the particular sample,

$P_1, P_2, \ldots, P_m$ = probabilities of drawing types $1, 2, \ldots, m$ from the population,

$N = k_1 + k_2 + \cdots + k_m$,
= size of the sample,

$k_1, k_2, \ldots, k_m$ = total number of items of types $1, 2, \ldots, m$, respectively, in the sample,

m = total number of different types (or classifications) within the total population.

As in the case of the binomial, the probabilities of occurrence are presumed to stay constant throughout. For the case of $m = 2$, it can be seen that the multinomial reduces immediately to the binomial.

Example: In a certain region, an oil company has grouped the possible outcomes of an exploratory well into three general classes: (1) zero reserves, a dry well; (2) discovery of 15 MM bbl of reserves; (3) discovery of 60 MM bbl of reserves. The corresponding probabilities are 0.5, 0.3, and 0.2. The company plans to drill three wells. What are the probabilities of discovering various total reserves with the three wildcat wells?

Solution: First, the data will be recapitulated:

m = three classes or three types of outcomes,

$N = 3$ wells to be drilled, the sample size,

k_1 = number of wells discovering 0 bbl of reserves,

k_2 = number of wells discovering 15 MM bbl of reserves,

k_3 = number of wells discovering 60 MM bbl of reserves,

$P_1 = 0.5$, probability of a well discovering 0 bbl of reserves,

$P_2 = 0.3$, probability of a well discovering 15 MM bbl of reserves,

$P_3 = 0.2$, probability of a well discovering 60 MM bbl of reserves.

One possible sample consists of $k_1 = 2$, $k_2 = 1$, and $k_3 = 0$. The probability of this particular sample is calculated as follows:

$$\begin{aligned} p(S) &= \frac{3!}{2!1!0!}(0.5)^2(0.3)^1(0.2)^0 \\ &= \frac{(3)(2)(1)}{(2)(1)(1)(1)}(0.25)(0.3)(1) \\ &= 0.225, \end{aligned}$$

$$\begin{aligned} \text{corresponding reserves} &= (2)(0 \text{ bbl}) + (1)(15 \text{ MM bbl}) \\ &\quad + (0)(60 \text{ MM bbl}) \\ &= 15 \text{ MM bbl}. \end{aligned}$$

Table 4.6 is organized according to a list of the values of k_1, k_2, and k_3 which can appear in a sample. Each line in the table represents one possible sample, and the probability of that particular sample has been calculated and listed. Another objective is to calculate the total reserves that correspond to each sample and list its value. Finally, the average or expected value can be calculated by summing up the products of the value of each sample times its probability.

Considerably more insight can be gained by rearranging the calculated results. This has been done in Table 4.7, which is arranged according to the total reserves of each individual sample of three wells. This permits calculation of the cumulative probability so that it can be interpreted as a "prob-

Table 4.6
Probabilities and Values of Possible Outcomes of Drilling Three Exploratory Wells

Sample k_1	Sample k_2	Sample k_3	*Probability, p (sample)*	*Reserves (MM bbl)*	*Expected Reserves (MM bbl)*
3	0	0	0.125	0	0.000
2	1	0	0.225	15	3.375
2	0	1	0.150	60	9.000
1	2	0	0.135	30	4.050
1	1	1	0.180	75	13.500
1	0	2	0.060	120	7.200
0	3	0	0.027	45	1.215
0	2	1	0.054	90	4.860
0	1	2	0.036	135	4.860
0	0	3	0.008	180	1.440
			1.000		49.500

ability of not less than" various amounts of discovered crude oil reserves. The results are plotted in Fig. 4.9, where one starts with 100% probability of at least 0 bbl discovered, progresses stepwise to a 65% chance of discovering at least 30 MM bbls, and finally to an 0.8% chance of discovering 180 MM bbl of reserve crude oil.

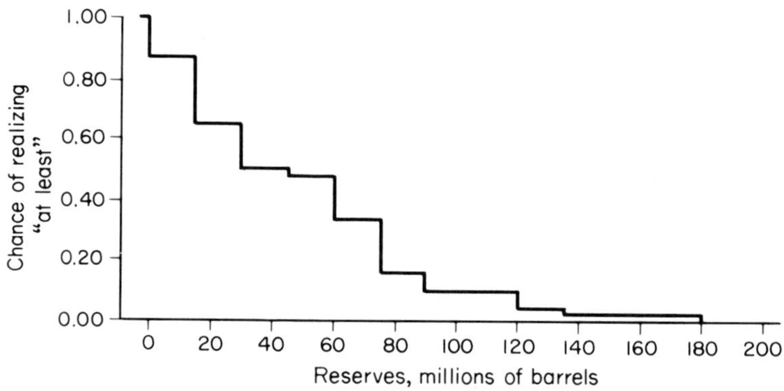

Figure 4.9. Probability versus "at least" reserves discovered by two exploratory wells.

Attention should be directed to the character of information given in Table 4.7 and Fig. 4.9, as contrasted to that in Table 4.6, which is unordered with respect of economic outcomes. The only economic figure that can be extracted from it is the "expected value" of 49.5 MM bbl of reserves.

Table 4.7
Probabilities Arranged According to Total Values of Reserves Discovered by Three Exploratory Wells

Reserves (MM bbl)	*Probability, p (sample)*	*Probability of "At Least These Reserves"*
0	0.125	1.000
15	0.225	0.875
30	0.135	0.650
45	0.027	0.515
60	0.150	0.488
75	0.180	0.338
90	0.054	0.158
120	0.060	0.104
135	0.036	0.044
180	0.008	0.008

This would be the average of very many trials, some of which would be better and some of which would be worse than the average figure. It may be observed that the "expected value," or "expectancy," is not even one of the recognized possible outcomes. If the present three-well venture is one of very many equivalent ventures, then the use of the expected value, or average value, can be justified. In perhaps most cases, however, it is better to make judgments based upon the full spectrum of possible outcomes.

4.4-2 Hypergeometric Distributions

Hypergeometric distributions describe the case of sampling without replacement (wor) from a small population. Consequently, they are most useful. Take, for example, the case where there are 10 geophysical prospects, 4 of which should prove to be productive, and where three exploratory wells can be financed. The probabilities change with each well drilled, so the binomial no longer applies. Or, a problem might be equivalent to that in which a box contains 10 red, 4 white, and 6 blue marbles. What are the chances of various combinations when a sample of 5 marbles is drawn and held, without replacing any until after all 5 are drawn? Another application concerns the problems of distribution of playing cards among various players. The probabilities change as each card is dealt among the players.

Just as the binomial is a special case of the multinomial distribution, so the hypergeometric is a special case of the multiple hypergeometric distribution. It will, therefore, be convenient to consider them together and recognize the hypergeometric as the special case where there are only two ($m = 2$) types of objects in the population. As in the case of the multinomial, a possible sample is described, and then the probability of occurrence of that sample is calculated:

$$p(S) = \frac{\binom{M_1}{k_1}\binom{M_2}{k_2}\cdots\binom{M_m}{k_m}}{\binom{M}{N}}, \tag{4-20}$$

where

$p(S)$ = probability of obtaining the exact specified sample,

$M = M_1 + M_2 + \cdots + M_m$,
= size of the total population,

$M_1, M_2, \ldots, M_m$ = total numbers of objects of types $1, 2, \ldots, m$ in the total population,

m = total number of different types (or classifications) within the total population,

$N = k_1 + k_2 + \cdots + k_m$,
= sample size, total number of objects in the sample,

$k_1, k_2, \ldots, k_m$ = total number of items, of types $1, 2, \ldots, m$, respectively, in the sample.

Example: An example of using the hypergeometric, $m = 2$, is taken from the case of 10 geophysical prospects, 4 of which will be productive. The probability of obtaining one productive well out of a total of three will furnish a sample calculation.
For this case, $M = 10$ total population, $M_1 = 4$ good locations, $M_2 = M - M_1 = 6$ dry locations, $N = 3$ wells in the sample, $k_1 = 1$ good well, $k_2 = N - k_1 = 2$ dry wells.

$$p(k_1 = 1) = \frac{\binom{M_1}{k_1}\binom{M_2}{k_2}}{\binom{M}{N}}$$

$$= \frac{\binom{4}{1}\binom{6}{2}}{\binom{10}{3}} = \frac{\left(\frac{4}{1}\right)\left(\frac{6}{2}\right)\left(\frac{5}{1}\right)}{\left(\frac{10}{3}\right)\left(\frac{9}{2}\right)\left(\frac{8}{1}\right)}$$

$$= 0.5.$$

Example: *Use of the Multiple Hypergeometric Distribution*

A company has 7 geophysical prospects in a certain area. From past experience it is estimated that 4 will prove to be dry, 2 will discover reserves of about 20 MM bbl, and one will discover reserves of about 80 MM bbl. What are the possible outcomes of the next 2 wells?

Solution: First the data will be recapitulated:

$m = 3$ types of outcomes,

$M = 7$ total number of prospects, population size,

$M_1 = 4$ dry locations,

$M_2 = 2$ locations with reserves of 20 MM bbl,

$M_3 = 1$ location with reserves of 80 MM bbl,

$N = 2$ wells to be drilled, sample size,

$k_1 =$ number of wells that will be dry,

$k_2 =$ number of wells with 20 MM bbl of reserves,

$k_3 =$ number of wells with 80 MM bbl of reserves.

Table 4.8 shows the calculations organized according to the composition of each sample. The total of the k-values in each line must add up to 2, the sample size. Also, no k-value can exceed its corresponding M-value, and since $M_3 = 1$, k_3 is limited to values of 0 or 1 in this example. This tabulation also permits listing the reserves corresponding to each sample, which are calculated as the sum of the products of the k-values times their corresponding individual discovery sizes.

Table 4.8
Probabilities and Values of Possible Outcomes of Drilling Two Exploratory Wells

Sample					
k_1	k_2	k_3	*Probability, p (sample)*	*Reserves (MM bbl)*	*Expected Reserves (MM bbl)*
2	0	0	0.286	0	0.00
1	1	0	0.381	20	7.62
1	0	1	0.190	80	15.20
0	2	0	0.048	40	1.92
0	1	1	0.095	100	9.50
			1.000		34.24

Table 4.9 is formed by rearranging the samples according to the total reserves. This permits calculating the probabiltiy of "at least" the reserves listed in each line. One starts, of course, with a 100% chance of at least 0 bbl of reserves, and progresses stepwise to a 33.3% chance of at least 40 MM bbl,

Table 4.9
Probabilities Arranged According to Total Values of Reserves Discovered by Two Exploratory Wells

Reserves (MM bbl)	*Probability, p (sample)*	*Probability of "At Least These Reserves"*
0	0.286	1.000
20	0.381	0.714
40	0.048	0.333
80	0.190	0.285
100	0.095	0.095

and finally to a 9.5% chance of discovering 100 MM bbl of reserve crude oil. Figure 4.10 is a graphical presentation of the results.

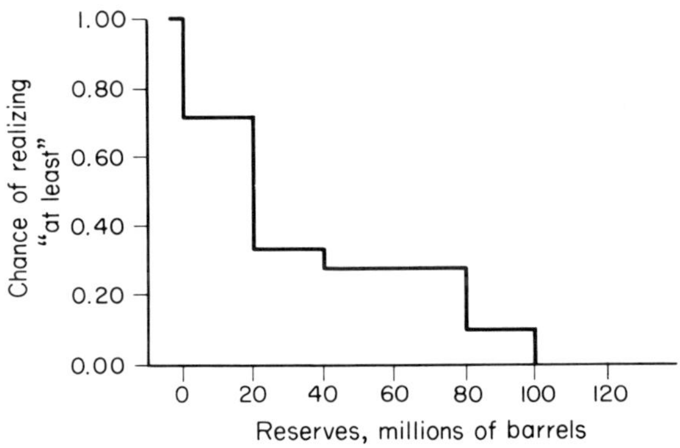

Figure 4.10. Probability versus "at least" reserves discovered by two exploratory wells.

4.5 MINERAL DEPOSITS AND RESOURCES

4.5-1 Log-Normal Distribution

From consideration of the distribution of natural resources, the log normal is an important probability distribution. Sizes of individual deposits of minerals have been observed generally to follow this relationship. Included among these are the sizes of petroleum reservoirs as they occur within geologic provinces. Sizes of pebbles on beaches and other natural occurrences have been observed to follow this distribution. A book has been written that describes its mathematical properties and applications.[1] Its use has been described for decision purposes regarding whether to accept or reject particular financial deals as an annual tax period draws toward its close.[2] It is among the methods currently in use to estimate prospects in developing exploratory operations.

As its name implies, the log normal is a distribution in which the logarithms are normally distributed.[3] Let X be the variable of interest. Then, if the frequencies are plotted versus the logarithmic values, the result is a normal distribution, such as is portrayed in Fig. 4.2. The following definitions will be used: α is the mean of the logarithmic distribution, and β is the standard deviation of the logarithmic distribution.

In working with the log normal, it is customary to use log-probability paper. One of the scales is logarithmic; the other is linear in z, the number of standard deviations. Cumulative probability is plotted versus the quantity of interest. Probabilities of the "less than" type usually slope upward to the right, and those of the "greater than" type generally slope downward to the right. Both work equally well, and the mathematical methods described herein can be used for both cases. Figure 4.11 is a plot of the percentage of

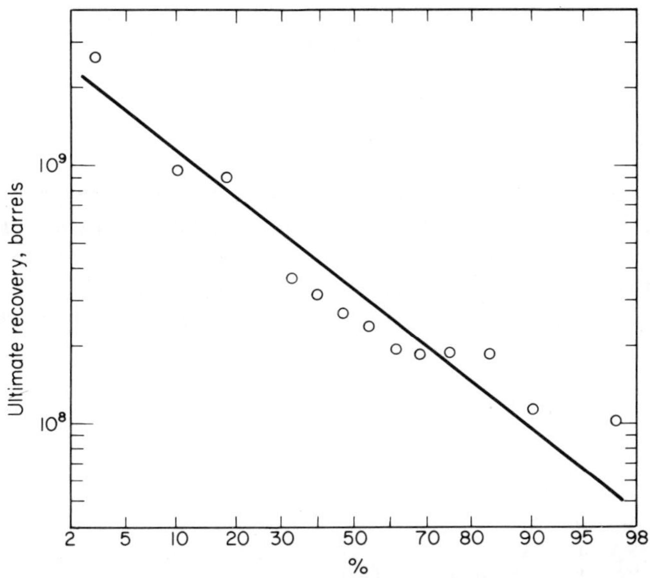

Figure 4.11. Percentage of fields "greater than" for the Los Angeles Basin. Field sizes are estimated barrels of ultimate oil recovery. Log-probability plot.

oil fields "greater than" for the Los Angeles Basin. For example, one reads from the line drawn on the graph that 10% of the fields are greater than 1.1×10^9, or 1,100 MM, barrels in size, and that 80% are greater than 1.45×10^8, or 145 MM, barrels in size. The data for the field sizes plotted are shown in Table 4.10.

Table 4.10
Estimated Ultimate Recovery for Fields in the Los Angeles Basin

Case	*Ultimate Recovery (M bbl)*	*% Plot Point on Graph*
1	2,600,813	3
2	970,197	10
3	896,821	17
4	611,928	25
5	367,800	32
6	316,803	39
7	268,497	46
8	241,466	54
9	194,826	61
10	189,909	68
11	187,607	75
12	183,217	83
13	111,647	90
14	102,470	97

Source: Calculated from data in *Oil and Gas Journal*, Vol. 69, No. 4 (Jan. 25, 1971), p. 124.

Actual values of X are the values of interest, not the logarithms themselves. The highly skewed nature of the log normal is shown in Fig. 4.12, which is a probability density plot calculated from the straight line drawn in Fig. 4.11. However, it appears that the log-probability plot may be the more convenient for calculating statistical properties of interest. The following

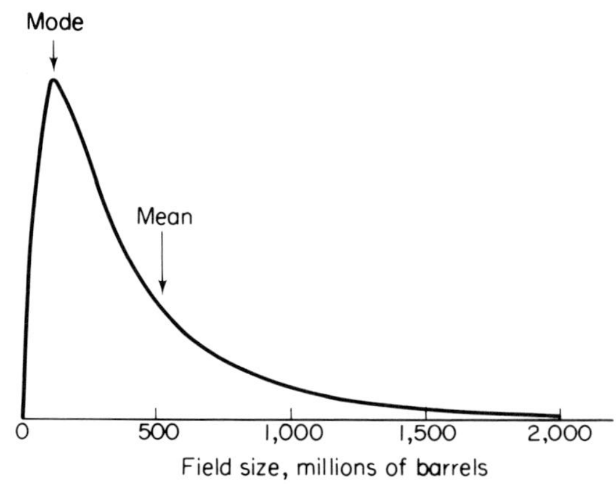

Figure 4.12. Lognormal distribution of field sizes in the Los Angeles Basin.

relations apply:

$$\text{mean of } X = m = e^{\alpha + 0.5\beta^2} = \exp(\alpha + 0.5\beta^2), \tag{4-21}$$

$$\text{variance of } X = \sigma^2 = e^{2\alpha + 2\beta^2} - e^{2\alpha + \beta^2} = \exp(2\alpha + 2\beta^2) - \exp(2\alpha + \beta^2), \tag{4-22}$$

$$\text{mode of } X = e^{\alpha - \beta^2} = \exp(\alpha - \beta^2). \tag{4-23}$$

By algebraic manipulation, the following can be obtained:

$$\sigma^2 = m^2(e^{\beta^2} - 1), \tag{4-24}$$

$$\text{mode of } X = \frac{m}{e^{1.5\beta^2}}. \tag{4-25}$$

An example will be taken from Fig. 4.11. The number found at the 50% probability line is 3.26×10^8 bbl. Since the numbers are plotted in the positions of their logarithms,

$$\alpha = \ln(3.26 \times 10^8) = 19.604$$

The standard deviation can be found from any two points. It will be convenient in this case to use the 2% and 98% points, where values of 2.34×10^9 bbl and 4.55×10^7 bbl are read, respectively. Referring to the normal distribution, or to Table 4.3, one finds that there are a total of 4.108 standard deviations between these percentage points. The standard deviation of the logarithmic distribution is then calculated as follows:

$$\beta = \frac{\ln(2.34 \times 10^9) - \ln(4.55 \times 10^7)}{4.108} = 0.960$$

and

$$\beta^2 = 0.921.$$

Statistical properties of the actual physical distribution, size of the fields in this case, can then be calculated:

$$\begin{aligned}\text{mean} = m &= \exp(\alpha + 0.5\beta^2) \\ &= \exp[19.604 + (0.5)(0.921)] \\ &= \exp(20.064) \\ &= 517 \text{ MM bbl},\end{aligned}$$

$$\begin{aligned}
\text{variance} = \sigma^2 &= m^2(\exp \beta^2 - 1)\\
&= (517 \times 10^6)^2(\exp 0.921 - 1)\\
&= (517 \times 10^6)^2(2.512 - 1)\\
&= 40.481 \times 10^{16},
\end{aligned}$$

$$\text{standard deviation} = \sigma = 636 \text{ MM bbl},$$

$$\begin{aligned}
\text{mode} &= \frac{m}{\exp(1.5\beta^2)}\\
&= \frac{517 \times 10^6}{\exp(1.5 \times 0.921)}\\
&= \frac{517 \times 10^6}{3.98}\\
&= 130 \text{ MM bbl}.
\end{aligned}$$

A check on two of these properties, the mean and standard deviation, can be achieved by conventional calculations performed on the tabulated values of field sizes. The sum of field sizes, and the sum of their squares, are readily calculated to permit the following calculations:

$$\begin{aligned}
m &= \frac{\Sigma X}{n} = \frac{7{,}244 \times 10^6}{14}\\
&= 517 \times 10^6, \text{ or } 517 \text{ MM, bbl},\\
\sigma^2 &= \frac{\Sigma X^2 - [(\Sigma X)^2/n]}{n}\\
&= \frac{9.416 \times 10^{18} - [(7.244 \times 10^9)^2/14]}{14}\\
&= 40.48 \times 10^{16}\\
\sigma &= 636 \text{ MM bbl}.
\end{aligned}$$

A back calculation can be made for α and β. From Eq. (4-21) and (4-22) it follows that

$$\begin{aligned}
\beta^2 &= \ln\left(1 + \frac{\sigma^2}{m^2}\right)\\
&= \ln\left[1 + \frac{40.48 \times 10^{16}}{(5.17 \times 10^8)^2}\right]\\
&= \ln 2.512\\
&= 0.921\\
\beta &= 0.960 \qquad (4\text{-}26)\\
\alpha &= \ln m - 0.5\beta^2\\
&= \ln(5.17 \times 10^8) - (0.5)(0.921) \qquad (4\text{-}27)\\
&= 19.604.
\end{aligned}$$

Finally, the intercepts on the log-probability graph can be calculated. For the left margin, the number on the log scale can be found from

$$\begin{aligned}2\%\text{ intercept} &= \exp(\alpha + z\beta)\\ &= \exp(19.604 + 2.054 \times 0.960)\\ &= \exp(21.575)\\ &= 2.34 \times 10^9.\end{aligned}$$

The number on the log scale, at the right margin, can be found from

$$\begin{aligned}98\%\text{ intercept} &= \exp(\alpha - z\beta)\\ &= \exp(19.604 - 2.054 \times 0.960)\\ &= \exp(17.633)\\ &= 4.55 \times 10^7.\end{aligned}$$

It is evident that the data can be transformed into whichever form is easiest to work with and interpret. Again it should be mentioned that the application here is to an individual mineral deposit, an individual oil reservoir. When interest centers on the average of a group of three, four, or more separate deposits, then statistical evaluation methods should be used. These are presented in the following section.

4.5-2 Several Fields in a Geologic Trend (Sampling from a Log-normal Distribution)

An interesting, as well as very important, problem is that of estimating the combined reserves of several fields which may be found in a geologic trend. Figure 4.13 is a log-normal plot of sizes of such related fields which have already been discovered. The chance that a single field, to be discovered in the future within this trend area, will contain a certain amount of reserves can be read directly from the graph. For example, the probability that a single field will contain 2.5 MM or more barrels is about 0.36, or 36%.

The problem that is to be solved, however, concerns the combined size of several fields. Let it be assumed that an oil company might discover three fields in this trend area. What, then, would the probability be of the three amounting to a total of 7.5 MM barrels, or averaging 2.5 MM barrels in size? This becomes a sampling problem where the sample size is $N = 3$. For samples of this small size, and where sampling is from the highly skewed log-normal distribution, it turns out that the distribution of sample means is not quite a normal distribution. However, by assuming a normal distribution, it is possible to make an estimate and certainly possible to gain insight into the problem.

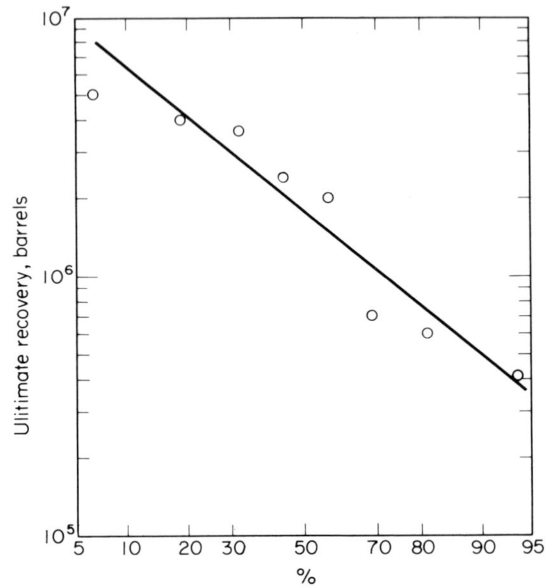

Figure 4.13. Percentage of fields "greater than" for a Wyoming oil play. Field sizes are estimated barrels of ultimate recovery of oil. Log-probability plot.

Properties of the log-normal distribution are calculated as discussed in the preceding section. Units used in the following are in thousands of barrels.

By use of the 50% point one obtains

$$\begin{aligned}\alpha &= \ln (1{,}630 \text{ M bbl}) \\ &= 7.40.\end{aligned}$$

By use of the 10% and 90% points one obtains

$$\begin{aligned}\beta &= \frac{\ln (7.70 \times 10^6) - \ln (3.35 \times 10^5)}{(2)(1.282)} \\ &= 1.22 \\ \beta^2 &= 1.49.\end{aligned}$$

The mean of the log normal then is

$$\begin{aligned}m &= \exp (\alpha + 0.5\beta^2) \\ &= \exp [7.40 + (0.5)(1.49)] \\ &= 3{,}440 \text{ M bbl}.\end{aligned}$$

The standard deviation of the log normal is then calculated.

$$\begin{aligned}\sigma &= m\sqrt{\exp(\beta^2 - 1)} \\ &= (3{,}440 \text{ M bbl})\sqrt{\exp(1.49) - 1} \\ &= 6{,}400 \text{ M bbl.}\end{aligned}$$

The standard deviation of the sampling distribution, for three in each sample, is

$$\sigma_{\bar{X}} = \frac{\sigma}{\sqrt{N}} = \frac{6{,}400 \text{ bbl}}{\sqrt{3}} = 3{,}700 \text{ bbl.}$$

The mean of the sampling distribution is taken equal to the mean of the original parent distribution.

The probability of obtaining an average field size of 2,500 M bbl is calculated as follows:

$$\begin{aligned}z &= \frac{\bar{X} - m}{\sigma_{\bar{X}}} = \frac{2{,}500 \text{ bbl} - 3{,}440 \text{ bbl}}{3{,}700 \text{ bbl}} \\ &= -0.254.\end{aligned}$$

In a table of areas under the standard normal curve, for $z = -0.254$ one finds a cumulative area of $1.0 - 0.6003$. The probability of values greater than this is given by the complementary area, so that the following applies:

$$\Pr(\bar{X} > 2{,}500 \text{ bbl}) = 0.6003.$$

Thus, by assuming that the central limit theorem can be applied in this case, it is calculated that there is a 60% chance that three fields together would average more than 2,500 M bbl each. This compares to the 36% chance of one field by itself being greater than the same size of 2,500 M bbl of reserves.

An estimate of the stability associated with the calculation can be made by investigating the sensitivity of reserve estimates to changes in probability. For example, let the probability be both increased and decreased by 10% and calculate the corresponding changes in reserve estimates. If the probability is increased by 10%, to $60 + 10 = 70\%$, the reserves estimated would be as follows:

$$\begin{aligned}z &= -0.524, \text{ corresponding to cumulative probability of } 30\% \\ \bar{X} &= m + z\sigma \\ &= 3{,}440 \text{ M bbl} - (0.524)(3{,}700 \text{ M bbl}) \\ &= 1{,}500 \text{ M bbl.}\end{aligned}$$

If the probability is decreased by 10%, to 60 − 10 = 50%, the reserves estimated would be calculated as follows:

$$z = 0, \text{ corresponding to 0.5 cumulative probability}$$
$$\bar{X} = m + z\sigma$$
$$= 3{,}440 \text{ M bbl.}$$

These calculations indicate approximately a 40% change in reserve estimates, with a 10% change in probability. Under such conditions the results of a relatively few trials can fluctuate considerably. Allowances for such variations, particularly the unfavorable ones, should be included in financial planning.

Since this is a rather important type of problem, it was solved by simulation using the Monte Carlo method. This was carried out by setting computer-generated decimal random numbers ranging between 0 and 1 equal to the cumulative probability under the normal distribution of logarithms of field sizes. Mean and standard deviation of this distribution are α and β, whose numerical values for the problem at hand have already been determined. Thus a particular decimal random number corresponded to a logarithm which was then converted to a field size in barrels of oil. A total of 200 cases were computed; the results are shown in Table 4.11 and Fig. 4.14. This is a numerical method of obtaining mathematical solutions for cases where there are no exact theoretical or algebraic solution methods available. The particular solution presented here could be improved upon by running two or

Table 4.11
Results of Monte Carlo Simulation of Combined Size of Several Oil Fields (Total M bbl Reserves in Each Group)

% of Cases Greater Than	*Number of Fields in Each Group*			
	3	10	20	30
5	24,450	56,280	103,770	154,150
10	18,500	50,090	93,630	145,540
20	13,380	43,780	82,740	121,350
30	10,580	37,880	73,430	108,120
40	8,580	33,920	67,660	98,066
50	6,970	28,860	63,560	92,710
60	5,550	26,210	56,900	87,760
70	4,590	23,470	51,920	83,763
80	3,760	20,400	46,830	77,070
90	2,610	15,710	42,560	66,030
95	2,070	13,140	36,720	60,190

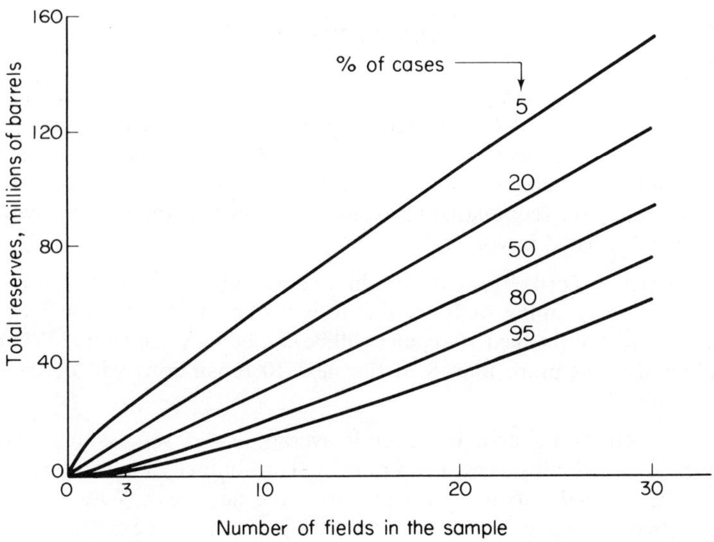

Figure 4.14. Results of Monte Carlo simulation of sampling from a lognormal distribution of oil field sizes.

three times as many trial cases. However, present results are sufficient for practical purposes.

By interpolation in Table 4.11 it can be found that approximately 47% of the groups of three fields have combined reserves greater than 7,500 M barrels. This answer is intermediate between that of 36% for individual fields and the answer of 60% obtained by assuming a normal distribution for the sample means. Stated another way, it is observed not to be safe to assume that the central limit theorem applies exactly for small samples drawn from a highly skewed distribution. This is, of course, a recognized rule. Sample sizes of at least 30 are often recommended, and of course the shape of the parent distribution influences the required sample size. The answers obtained by the different methods have values relative to each other, about as might be expected. For this particular case, the answer obtained by Monte Carlo simulation is to be preferred.

REFERENCES

1. Aitchison, J., and J. A. C. Brown, *The Lognormal Distribution.* New York: Cambridge University Press, 1957.

2. Kaufman, G. M., *Statistical Decisions and Related Techniques in Oil and Gas Exploration.* Englewood Cliffs, N.J.: Prentice-Hall, Inc., 1963.

3. Parzen, E., *Modern Probability Theory and Its Applications.* New York: John Wiley & Sons, Inc., 1960.

PROBLEMS

4-1. For the case of a sample size of 4 ($N = 4$) and where the probability of occurrence is $p = 0.2$, calculate the following probabilities: $p(x = 0)$, $p(x = 1)$, $p(x = 2)$, $p(x = 3)$, and $p(x = 4)$. Calculate also the (cumulative) probability of x being equal to or less than each x-value listed in your table. Also calculate the (cumulative) probability of x being equal to or greater than each x-value listed in your table.

4-2. Well repair records from a certain region indicate that 60% of repairs necessary on pumping wells are due to worn liners. What is the probability that exactly 8 of the next 10 repairs will be caused by worn liners? What is the probability that more than 8 of the next 10 repair jobs will be caused by worn liners?

4-3. In the Shelf Trend area the overall average is 1.5 successful wells of 10 wildcats, an overall success ratio of 0.15. The company plans an exploration program of 8 wildcats in that area during the next year. What is the probability that all these will be dry? What is the probability of exactly one success? What is the probability of exactly two successes? What is the probability of more that two successes?

4-4. A new company has been formed for the purpose of developing oil production in an area where the overall success ratio (probability of success) is 0.25. This new company has limited resources. The owners feel that they would average out all right in the long run. However, they are concerned about the possibilities of a run of bad luck, which might use up all their capital before sufficient discoveries were made. (This is commonly referred to as *gambler's ruin.*) Use the negative binomîal to calculate the probabilities of 0, 1, 2, 3, 4, 5, and 6 failures before one discovery is made. Repeat the calculations for 0, 1, 2, 3, 4, 5, and 6 failures before 2 discoveries are made. In both cases the cumulative probabilities are of interest.

4-5. The table shows the number of well repair jobs that require pulling sucker rods during each month over the past 50 months. Plot the data on probability paper. Do the data form a normal distribution? Determine the mean and standard deviation from the plot. Use the frequency distribution methods decribed in Chapter 3 to calculate the mean and standard deviation. Do the two sets of answers agree?

No. Repair Jobs During Month	*No. Cases*	*No. Repair Jobs During Month*	*No. Cases*
20	1	13	7
19	0	12	4
18	3	11	6
17	2	10	5
16	5	9	2
15	6	8	1
14	8		

4-6. In a particular area the probability of a successful wildcat well is 0.15. A company plans to drill 10 such exploratory wells during the next year.
(a) Use binimial probability methods to calculate the probabilities of zero successes, not more than one success, not more than two successes, and not more than three successes.
(b) Calculate the mean and standard deviation for sample size $N = 10$. Use the normal approximation of the binomial, including of course, the 0.5 correction, to calculate the same quantities required in part a. Compare the two sets of answers.

4-7. Over a long period of time a company has had 1 successful wildcat of 8 in a particular region. A new geophysical method has been adopted and used in locating the past 30 wells, 6 of which were successful. Use the normal approximation of the binomial to calculate if these results are significant at the 1, 5, or 10% levels.

4-8. Two companies have been operating in the same area and depth range. Over the past 2 years company *A* has completed 12 discoveries of 50 wildcats. During the same time company *B* has completed 9 of 25 wildcats. At the 5% level of significance, does it appear that company *B* had better operating techniques? Propose three solutions, first using the data of *A* as a basis, second using the data of *B* as a basis, and third using proportions and pooled data for a basis.

4-9. The chance of discovery is taken to be 0.125. Monetary reserves permit drilling 10 wells. Under such conditions, what is the probability of going broke? Solve for $p(0)$ by both binomial and normal approximation methods.

4-10. The company has averaged one successful wildcat of 8 in district 5C. But of the last 27 wells, only 2 have resulted in discoveries. Is this statistically significant? What is your interpretation and recommendation?

4-11. At a certain turnpike entrance, there is an average of 2 cars entering per minute during the 9 to 10 o'clock period. Use the Poisson distribution to estimate the number of 1-minute periods, from 9 to 10 o'clock, when 0 cars will pass through, when 1 car will pass through, 2 cars, 3 cars, etc., until the tabulation accounts for all the data.

4-12. Between 11:30 and 12:30 a total of 240 people enter the cafeteria. Use the Poisson distribution to estimate the number of people entering per period for periods 1 minute long.

4-13. (a) A material contains flaws that average 1% of the total volume. The volume of each flaw is 0.0025 cubic inch. On the average, how many flaws are there per cubic inch of the material?
(b) Of a total of 50 cubic inches of the material, use the Poisson distribution to estimate how many individual cubes, of 1 cubic inch each, would contain 0, 1, 2, 3, etc., flaws per cube.

4-14. A certain metal contains flaws that constitute 0.1% of the total volume of the material. Each flaw is $\frac{1}{32}$ inch in diameter. The material has been made into tubes, and threads are to be cut on the ends of the tubes. On the average, how many flaws will be exposed per square inch in the machined surface? A complete thread will have 30 square inches of machined surface. If this is divided into 30 squares of 1 square inch each, use the Poisson distribution to estimate how many squares would show 0 flaws, how many would show 1 flaw, 2 flaws, etc., up to the maximum indicated by the distribution.

4-15. If on the average there are 3 giant fields discovered each year, make a table for a 10-year period, showing the number of years in which there would be 0, the number of years in which there would be 1, 2, . . . such discoveries. Use the Poisson distribution for making the estimate.

4-16. Where arrivals per period are distributed according to the Poisson distribution, it can be shown that the time between arrivals is distributed according to the exponential distribution. The mean of the exponential is the reciprocal of the mean of the Poisson distribution. The variable number of arrivals per period of the Poisson becomes the variable length of time between arrivals in the exponential. Referring to Problem 4-15, what is the probability that there will be less than $\frac{1}{3}$ year between discoveries of two giant fields? What is the probability that there will be more than $\frac{1}{2}$ year between two such discoveries?

4-17. A department has been averaging 18 service calls per month. (a) What is the probability that there will be more than 1 call on any one day? (b) What is the probability that there will be more than 2 calls on any one day?

4-18. In the Central Trend area there have been 2 discoveries per year over the past 5 years. However, there have been no discoveries during the past 14 months. Exploratory activity has remained at approximately the same level. Provided that the probability of discovery has remained constant, what is the probability of 14 months or greater between discoveries?

4-19. In a bin there are 30 items, 6 of which are defective. What is the probability that 1 of these will be found in a sample of size 5? What is the probability that more than 1 defective item will be found in a sample of size 5? Solve by using (a) the binomial and (b) the hypergeometric distributions. Which method is more accurate in this case?

4-20. A bin was described as containing 30 items, with not more than 5 being defective. However, a sample of size 4 contained 2 defective items.

(a) Use the hypergeometric distribution to calculate the probability of obtaining this particular sample.

(b) At what significance level could the description of the bin be rejected, based on this particular sample?

Solve by using the hypergeometric distribution.

4-21. In an area, there are 7 prospective well locations, and it is believed that 3 of these will prove to be successful. The company owns 4 of these locations. If the company drills 3 of these locations, what are the probabilities of exactly 0, exactly 1, exactly 2, and exactly 3 successes? Solve using the hypergeometric distribution.

4-22. The table shows representative sizes of possible discoveries by exploratory wells.

Size of Discovery (bbl)	*Probability*
0	0.60
2×10^6	0.20
20×10^6	0.15
50×10^6	0.05

A company drills 3 wells. What is the probability that this group (sample) contains one dry well, one discovery of 2×10^6 bbl, and one discovery of 20×10^6 bbl?

4-23. An exploration area contained 18 geophysical prospects originally. It is believed that half of these would prove to be dry, that 3 fields of 5×10^7 bbl size, and 6 fields of 5×10^6 bbl size would be found. Eight of the prospects have been tested, 2 larger and 2 smaller fields have been found. If the original estimates are correct, what are the reserves (and their probabilities) that could be discovered by the next 3 wells? Plot a cumulative probability diagram of the "greater than" type. Solve by mutiple hypergeometric probability.

4-24. Plot the following oil field sizes on log-probability paper. Use the plot to determine the mean and standard deviation of the log-normal distribution.

Field	*Ultimate Recovery (MM bbl)*
A	170
B	94
C	46
D	30
E	12

5

APPRAISAL OF UNCERTAIN VENTURES

5.1 INTRODUCTION

5.1-1 Statistical Decision Principles

An appraisal is essentially the settling of a price. Competition, or the market, usually plays a part in this, even in cases of uncertain ventures. It becomes essential, then, to determine whether or not a particular company or economic unit should enter into a particular business deal or arrangement. The worth to a company may be either more or less than the going price. For the parties concerned, a judgment must be made in weighing the advantages of possible gains versus the disadvantages of possible losses. Both the objectives and resources of the company must be taken into account in the evaluation, for failure of a single project would hurt a large company much less than it would hurt a small one. In this chapter the role taken is generally that of a going concern which customarily purchases additional ventures in its usual line of business. However, the principles used are quite general and could be applied to other appraisal situations.

Any potential investment is usually judged by several standards. It must coincide with the general purpose of the company. Ventures must be sought out that will increase the stature and future opportunities and capabilities of the company. Required investments and return cash flow must fit into the financial schedule. Expected rate of return or expected profit/investment ratio must be satisfactory. Some of these and perhaps other criteria must be met. In addition, the risk must not be too great.

Where risk is involved, a high chance of success is achieved by participation in a large number of ventures. In essentially all cases, capital and resources are limited, so that a limit exists on how much can be invested in any one project. Not all risks can be eliminated, and for many people affairs would become too dull if they were. However, the common measure of the efficiency of any type of organization is its continued survival.[1] In this chapter evaluation methods are developed that are based on statistical sampling theory. Of special interest are the relations between possible rewards, the degree of risk, and the price paid for a particular business venture. A decision-making group should know the relations between the amount of initial investment paid for a venture and the attendant degree of jeopardy to the group and to the company.

5.1-2 Portraying Individual Ventures

There are several ways to describe the outcomes of ventures. One way is to use only the expected value. This, by definition, is the average value that would be reached after participating in a large number of similar ventures. Use of the expected value alone may be sufficient if indeed the company will engage in a large number of the ventures. However, in this chapter it will be presumed that use of only the expected or average value will not be sufficient. Interest is centered on the many cases where allowances must be made for variations in the outcome.

Estimation of the physical quantities is of primary concern. These can be converted to dollars and final appraisal calculations made in terms of dollars. For evaluating a drilling venture where an exploratory oil well will be drilled, preliminary estimates must be made regarding the possible amounts of oil that may be discovered. The range of the estimates, and the probabilities associated with various possible outcomes, can be expressed in any of the following ways:

1. Frequency distribution tables.
2. Discrete distributions.
3. Continuous distributions of many kinds.
4. Mixed discrete and continuous distributions, as a triangle combined with a spike.

Relative frequency distributions and discrete distributions are equivalent for calculation purposes. In these cases it is generally to be presumed that the size given acts as a class mark. For example, the estimator might mentally group the possibilities of obtaining 40,000 to 60,000 bbl of oil into one figure and say there is, for example, a 20% chance of a 50,000-bbl well. Possibilities of greater and lesser amounts would be similarly listed until the full 100% probability range was described. Another estimator might prefer to use the continuous form and say there is a 20% chance of the well recovering somewhere between 40,000 and 60,000 bbl. Both of these forms can be used directly in the necessary calculations. Examples are included using both methods. It will be presumed that the form in which the data are given most accurately describes the physical situation within the understanding of the person making the estimate.

Sources of information include geologic data on formation thickness and structures. Experience with the same or similar formations often permits descriptions of ultimate oil recovery in terms of barrels per acre-foot. When available, well logs yield information on reservoir rock thickness, porosity, and fluid content. Where distribution estimates can be made best on several individual quantities, such as rock formation thickness, porosity, water saturation, and other important quantities, these can be molded into a composite oil recovery distribution by the Monte Carlo process. The computer output of this process is generally in the form of a frequency distribution. For cases where the venture may include a series of steps such as obtaining lease options, seismic exploration, and exploratory drilling, the economic model may be in the form of a decision tree. Analysis of decision trees yields discrete distributions. These forms of information can be used directly in the calculation methods developed in the following sections.

5.1-3 Review of Previously Developed Methods

Two methods that have been used for the appraisal of exploratory drilling ventures are discussed in this section. The first is fascinating in its own right and of historical interest. The second is a modern method and applicable wherever the binomial distribution describes the probability of obtaining a discovery well.

1. *Hayward's Application of Whitworth's Equation.* An early appraisal equation, based on a lottery, was developed by Whitworth.[2] Where there is a total of m tickets sold in the lottery, only one of which can win, the following average conditions prevail. The winning ticket multiplies the player's funds by the factor

$$\frac{M + A - C}{M},$$

and each losing ticket multiplies his funds by

$$\frac{M - C}{M}.$$

If the funds of the player are to remain the same through one winning ticket and $m - 1$ losing tickets, the following must hold:

$$\frac{M + A - C}{M}\left(\frac{M - C}{M}\right)^{m-1} = 1. \tag{5-1}$$

Taking the $1/m$ root gives the following relation:

$$\left(1 + \frac{A - C}{M}\right)^p \left(1 - \frac{C}{M}\right)^{1-p} = 1, \tag{5-2}$$

where

M = total speculative funds,

C = cost of each ticket,

A = reward,

$p = \frac{1}{m}$ = probability of success.

Where there exists the possibility of more than one reward, the general equation becomes

$$\left(1 + \frac{A_1 - C}{M}\right)^{p_1} \left(1 + \frac{A_2 - C}{M}\right)^{p_2} \left(1 + \frac{A_3 - C}{M}\right)^{p_3} \dots = 1. \tag{5-3}$$

These equations can be solved for C, and approximate solutions have been offered by Hayward and also by Pirson.[3,4] Another solution, of the trial-by-error variety, will be offered here. Trial values of C are to be used in the right-hand side of the following equation. Let the first trial value be $C = pA$. Then proceed to solve

$$C = M\left[1 - \left(\frac{M - C}{M + A - C}\right)^p\right]. \tag{5-4}$$

Hayword used a form of the Whitworth equation to determine maximum appraisal values that could be paid for drilling ventures and oil royalties. Whitworth reasoned that, since the resources of the world were greater than those of any individual, the individual must pay less than the amount obtained by this relation or he would eventually be ruined. The problem of just how much less was not solved at that time.

2. *Method of W. M. Campbell and F. J. Schuh.* This method combines the binomial probability of a well discovering oil with the log-normal distribution of sizes of oil fields.[5] The required average size of a field to be discovered in a drilling program is calculated as follows:

$$F = \frac{\text{cost of all } N \text{ exploratory wells + profit}}{(x, \text{ number of fields found})(\text{value of oil, dollars/bbl})}, \tag{5-5}$$

where F = average field size, barrels. Setting the profit at zero gives the average field size required to break even. The profit can also be set higher, corresponding to various desired rates of return.

The confidence level of obtaining at least this average-sized field, and hence at least the corresponding rate of return, can be calculated by summing up the unconditional probabilities as follows:

$$\text{confidence level} = \sum_{x=1}^{x=N} (p(x))(p_F), \tag{5-6}$$

where

$p(x)$ = binomial probability of discovering exactly x fields with a total of N exploratory wells,

p_F = probability of exactly x fields containing an average of at least F barrels of recoverable oil.

The confidence level here is the total probability of discovering at least the stipulated F barrels average per field.

The binomial probabilities were calculated by assuming such values as 0.1 probability of success, in the usual manner.

Values of p_F were obtained from a log-normal plot of oil field sizes.[6] For $x = 1$ the probability of a field containing at least F barrels can be read directly from the log-normal plot. For values of x greater than 1, a sampling distribution needs to be used, as discussed in Chapter 4. Interestingly enough, these authors also preferred values obtained by the Monte Carlo process for $x = 2, 3, 4$, and other small values, and they found that for $x \geq 20$, the normal distribution approximation is sufficiently accurate.

There would appear to be the possibility of using field size distributions other than the log normal in this general solution method. In many cases the company that discovers a field would not end up owning the entire field, and some modifications would be desirable in such cases. In the following sections, the probability of a dry well is combined with possible variations in field size into one single probability distribution. This procedure somewhat simplifies the necessary calculations and is similarly based on sampling theory.

5.2 STATISTICAL APPRAISAL METHOD FOR SEVERAL VENTURES

5.2-1 Basic Relations

An appraisal equation can be developed on the basis that a company will participate in a number of similar ventures.[7] Drilling exploratory oil wells furnishes an example. Let Fig. 5.1 represent the possible financial outcomes of a single venture. For the case illustrated, there is a 60% chance of the well proving to be dry, resulting in a loss of \$150,000. If the well is productive, the outcome could vary from a loss of \$50,000 to a gain of \$1,000,000, as indicated by the probability distribution. These figures are present values of operating profit less the initial investment.

It is necessary next to consider the relations that prevail when a group of

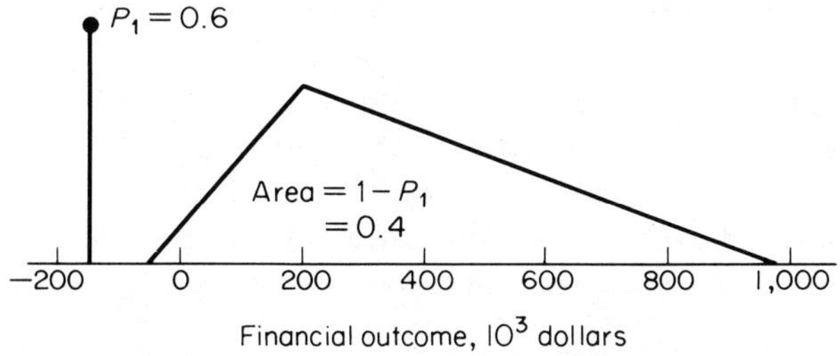

Figure 5.1. Probability distribution of possible outcomes from drilling an exploratory oil well.

wells are drilled, where the wells are located independently of each other. When the rewards are added together, the statistical distribution of results is as shown in Fig. 5.2. Half of the groups of wells turn out better than

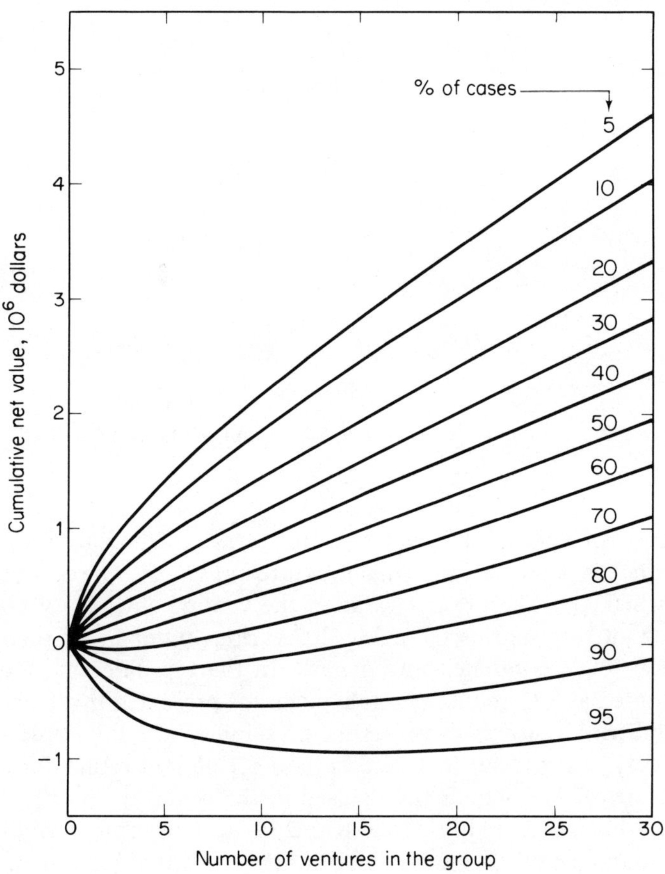

Figure 5.2. Probability distribution of the cumulative values of groups of the individual venture shown in Fig. 5.1.

expected, half turn out worse. The several curved lines show the values that are exceeded by various percentages of the groups. In accordance with the central limit theorem of statistics, the distribution becomes normal as the number of wells in the groups increases. Numerical proof of this is given by virtue of the straight-line relationship shown in Fig. 5.3, which is a plot on probability paper of Monte Carlo simulation of groups of size 30 randomly selected from the individual venture shown in Fig. 5.1.

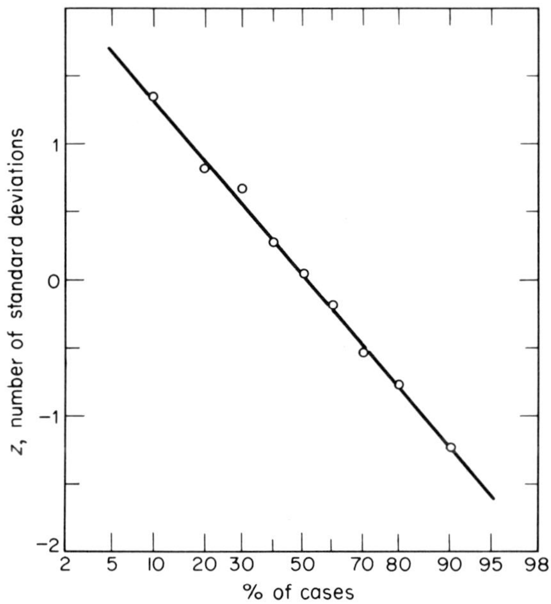

Figure 5.3. Analysis of Monte Carlo simulation of the probability distribution shown in Fig. 5.2.

The effects of various different discount rates are shown with a second example, which is portrayed as a single venture in Fig. 5.4. The cost of drilling this well is \$60,000. The present value of the venture shrinks, of course, as higher discount rates are used. In Fig. 5.5 various numbers of such similar ventures have been combined into groups. In the top diagram values have been discounted at 8 % and a favorable picture is presented in which increasing the number of ventures increases the probability of a favorable outcome (see Table 5.1). It turns out that 23 % is the DCF rate of return for this particular venture. This discount rate is used in the center diagram, where the 50 % of cases line becomes horizontal with zero present value. Actual present values, discounted at 23 % yearly interest, are, nevertheless, dispersed both above and below this average line. It is always possible to use a sufficiently high discount rate which will make a venture appear unfavorable. This is illustrated in the bottom diagram. Finally, in Fig. 5.6, the DCF rate of return

Table 5.1
Present-Value Factors Used for Discounting in Figures 5.4 and 5.5

Interest Rate (%)	*Present-Value Factor*
0	1.000
8	0.748
20	0.549
23	0.518
30	0.453
50	0.345

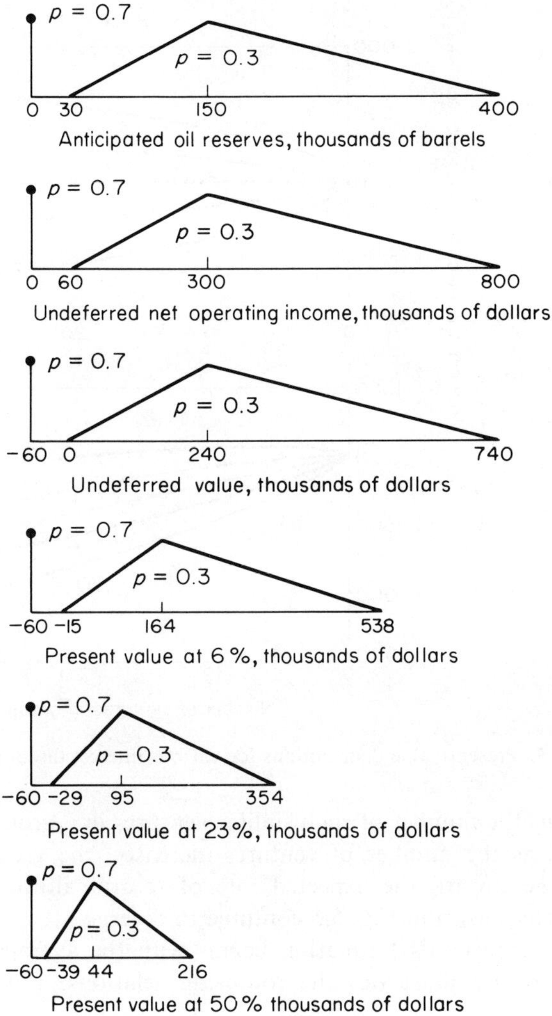

Figure 5.4. Expressions of income distribution.

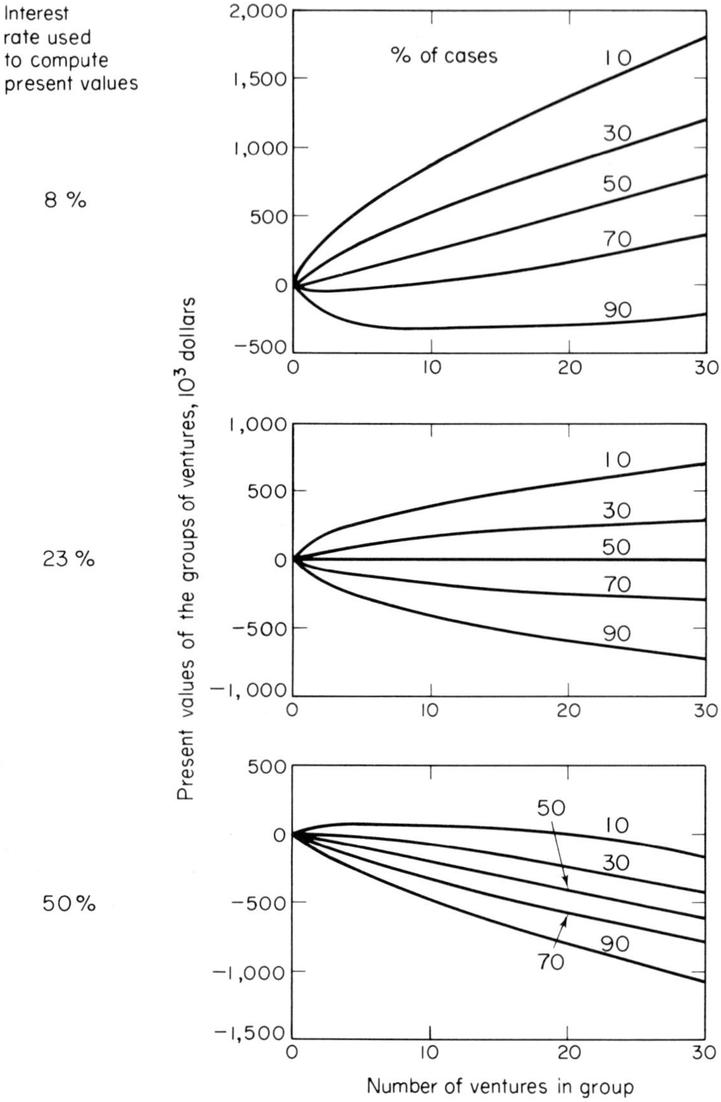

Figure 5.5. Present-value distributions for various interest discount rates.

is plotted versus the number of such similar ventures in a group. Here it can be noted that, as the number of ventures increases, the great majority of groups converge toward the expected rate of return, although their final asset positions as shown in Fig. 5.5 continue to diverge.

Developing an appraisal equation begins with the average or expected outcome. Figure 5.7 illustrates the following relations. For a total of n

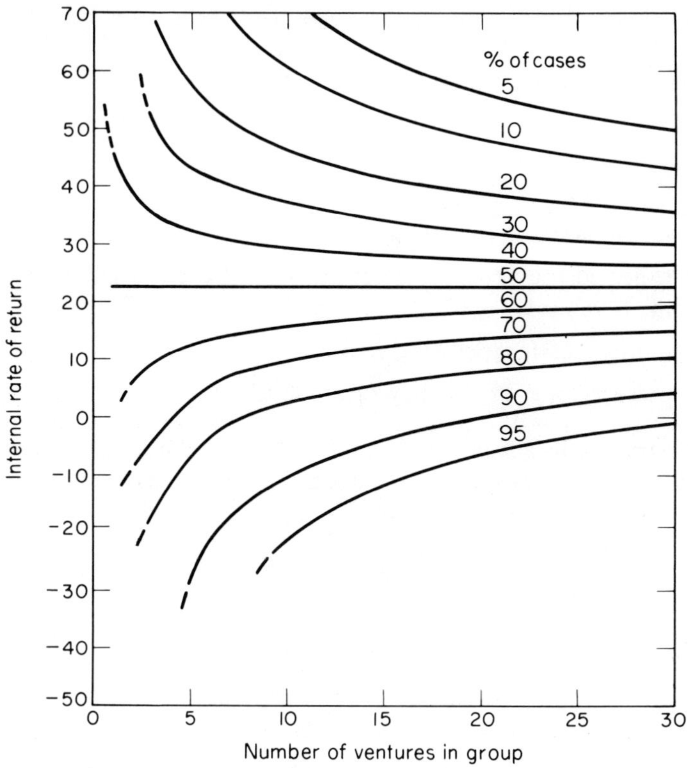

Figure 5.6. Discounted cash flow rate of return for groups of ventures.

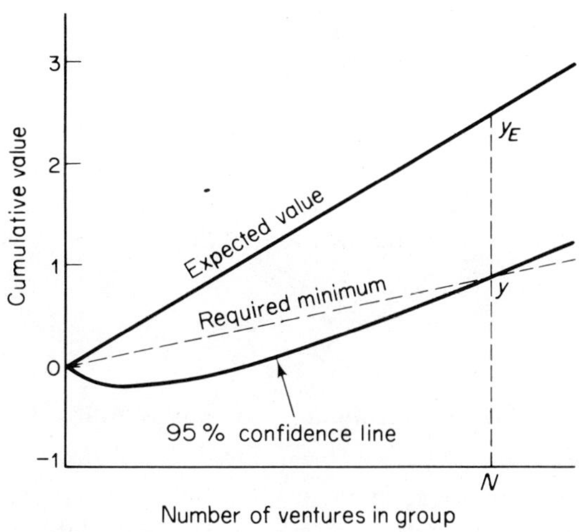

Figure 5.7. Relations among the number of ventures, the required return, and the confidence limit.

similar ventures, the total expectancy y_E can be expressed as

$$y_E - y_0 = nX_E, \tag{5-7}$$

where

y_0 = initial assets of the company,

y_E = expected present-value asset position of the company after resolving n ventures,

n = number of independent ventures,

X_E = present-value expectancy of a single venture.

Total variance in all groups is $n\sigma^2$, so the standard deviation becomes $\sigma\sqrt{n}$. Thus, if one considers the curved line, where 95% of the cases are greater in value than the position of the line, the following relation applies:

$$y - y_E = z\sigma\sqrt{n}, \tag{5-8}$$

where

y = asset position that will be exceeded, with a specified probability,

z = number of standard deviations corresponding to the specified probability; z is negative in cases of interest,

σ = standard deviation of a single venture, reduced to present value.

Combining Eqs. (5-7) and (5-8) gives the following:

$$y - y_0 = nX_E + z\sigma\sqrt{n}. \tag{5-9}$$

This equation forms the basis for various solutions that are developed in following sections.

Quite often a constraint placed on an investment is that there be some stated high probability that no money will be lost, in which cases $y - y_0 = 0$. A somewhat more general solution can be obtained by specifying this condition as some fraction of the initial investment that must be gained per venture. This results in

$$y - y_0 = nwC, \tag{5-10}$$

where

C = initial speculative investment; in drilling ventures C is the cost of a dry hole,

w = minimum fractional gain per venture (minimum present-value profit/investment ratio); if $w = 0$, risk calculations are based on zero loss of money.

Development of an appraisal equation requires that the expectancy be broken up into essentially the present value of operating profits and the speculative cost that must be spent to test the venture.

$$X_E = \bar{X}_A - C, \tag{5-11}$$

where

$\bar{X}_A$ = discounted expected value of possible outcomes, including everything except the initial speculative cost.

Relative variations in the final outcome are independent of the value of C. This relation can be visualized in the simple case presented in Fig. 5.4. The topmost distribution shows the estimated possible reserves of oil that a well may discover. Directly under this is a distribution that shows the corresponding undiscounted operating profits reckoned at \$2 per barrel. Next is a diagram that shows the net undeferred value, which includes the \$60,000 cost of drilling the well. Note that the size and shape of the diagram does not change, and therefore its variance does not change, by making this lateral shift. Discounting, however, does reduce both the size of the diagram and the variance. The same final result can be reached by an alternative pathway. First let the operating income be reduced to a present value. The expectancy of the diagram would be $\bar{X}_A$, a present value. At this point the variance and standard deviation can also be calculated, and they will not be changed by a lateral shift of the entire diagram, due to initial investment.

Completion costs and their inclusion present an interesting variation. Quite often there are appreciable additional expenses that will occur only if the well, or other venture, is successful. These include such things as additional pipe, cementing, acidizing, hydraulic fracturing, and surface equipment. Since these costs are generally a present-value figure, they can simply be subtracted from the present value of operating profits. These procedures are illustrated in examples that follow. In some cases the different implied time patterns might dictate the use of different present-value factors for the upper and lower extremes of reserve estimates. However, in many cases the use of an average discount factor would appear to be adequate. The important point to be made here is that a representative present-value distribution can be constructed. This distribution should include everything except the speculative or dry hole cost. The standard deviation of this distribution will not be altered by a lateral shift of the entire distribution. Inclusion of the speculative cost produces such a lateral shift without altering the variance in the possible outcomes of a venture.

5.2-2 Appraisals for Which the Number of Ventures Is Known

In some instances the number of exploratory wells that will be drilled during an accounting period is known. If the possible returns can be described probabilistically, a solution can be made for the maximum price per venture.

Combining the foregoing relations and solving for C, the initial investment, yields the following appraisal equation:

$$C \leq \frac{1}{1+w}\left(\bar{X}_A + \frac{z\sigma}{\sqrt{n}}\right). \tag{5-12}$$

Example: A company plans to drill 8 development wells in a shallow limestone reservoir. Because of local variations in porosity within the reservoir rock, there are variations in the expected rewards. The prospects have been summarized as illustrated in Fig. 5.8, which shows the present values

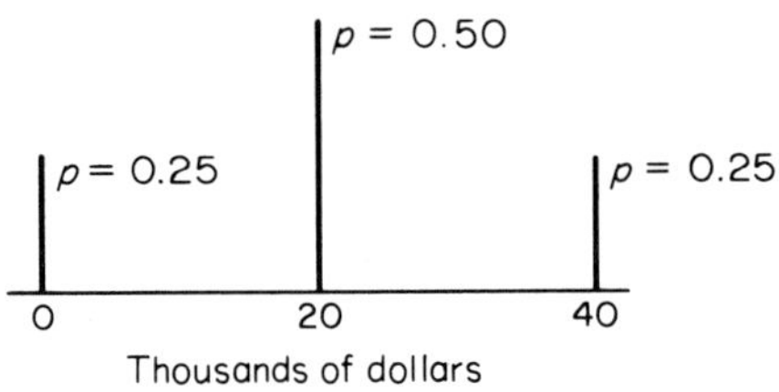

Figure 5.8. Possible present values of operating income. Shallow limestone field.

of net operating income. The present-value calculations were based on 8% annual interest, which rate covers the cost of invested capital. Based on the 8% interest rate, and assuming that the company will accept a 10% risk in this case, how much can the company pay for each well?

Solution: The expectancy and standard deviation are calculated as follows:

$$\begin{aligned}
\bar{X}_A &= \sum p_i X_i \\
&= (0.25)(0) + (0.50)(20{,}000) + (0.25)(40{,}000) \\
&= \$20{,}000, \\
\sigma^2 &= \sum p_i (X_i - \bar{X}_A)^2 \\
&= (0.25)(0 - 20{,}000)^2 + 0.50(20{,}000 - 20{,}000)^2 \\
&\quad + 0.25(40{,}000 - 20{,}000)^2 \\
&= 200{,}000 \\
\sigma &= \$14{,}140.
\end{aligned}$$

The appraisal price can be calculated next.

$z = -1.282$, for 10% risk,

$w = 0$, since capital costs are included in the 8% present-value discount,

$$
\begin{aligned}
C &\leq \frac{1}{1+w}\left(\bar{X}_A + \frac{z\sigma}{\sqrt{n}}\right) \\
&\leq \frac{1}{1+0.0}\left[20{,}000 - \frac{(1.282)(14{,}140)}{\sqrt{8}}\right] \\
&\leq \$13{,}600.
\end{aligned}
$$

This means that if \$13,600 per well were paid for such a group of 8 wells, then in 10% of the cases, or 1 time in 10, such a group of 8 wells will result in a net loss. Of course, if one paid a price equal to the expected present worth, then one would lose money 50% of the time.

5.2-3 Fractional Participation in Large Ventures

Two concepts that should be brought into any general analysis relate to the possibility of fractional participation and to the constraint placed on the number of ventures by limited capital. Fractional participation in speculative drilling deals has been a way of life for smaller and independent operators. By sharing the risk they have been able to survive. Two rather spectacular occasions when major companies joined forces to overcome specific obstacles were in the offshore drilling in the Gulf of Mexico in the mid-1940s and again in Alaska North Slope drilling in the late 1960s.

The number of speculative ventures in which a company can participate is limited by the total amount of speculative capital available. This produces the following relationship:

$$nfC \leq M, \tag{5-13}$$

where

M = total venture capital available to the management group; this is the amount that is, or can be, invested in unresolved ventures,

f = fractional participation in a venture,

fC = amount invested in a single venture.

The value of M will not in general be the total resources of a company nor even the total speculative capital. It will be the portion allotted within an accounting period to a management or decision-making group, which must stand on its own performance. In some company situations, M would be the exploration budget. Equation (5-13) can be used to eliminate the quanity n from the preceding relations. In turn, this permits judging ventures as they relate to the total capital available in terms of their inherent risks. Combining with Eq. (5-12) yields the following:

$$f \leq \frac{M}{C}\left[\frac{\bar{X}_A - C(1+w)}{z\sigma}\right]^2. \tag{5-14}$$

Example: A relatively small company is considering developing a structure where the chances of success have been estimated to be approximately as shown in Fig. 5.9. The cost would be 12 MM dollars. The company has

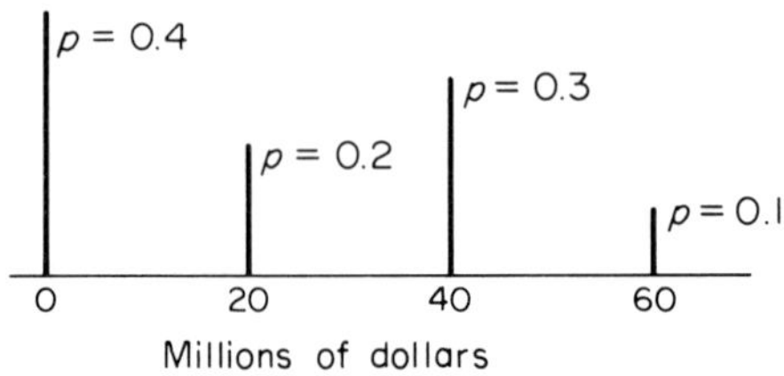

Figure 5.9. Possible outcomes of a development venture, in present-value dollars.

already committed 25 MM dollars to attractive ventures whose outcome is still to be decided, and there remains 15 MM dollars of venture capital available. Should the company make this investment? Management will take a 10% risk.

Solution:

$$\begin{aligned}
\bar{X}_A &= \sum p_i X_i \\
&= (0.4)(0) + (0.2)(20\text{ MM}) + (0.3)(40\text{ MM}) + (0.1)(60\text{ MM}) \\
&= 22 \times 10^6 \text{ dollars, present value of expected net operating income,} \\
\sigma^2 &= \sum p_i (X_i - \bar{X}_A)^2 \\
&= (0.4)(0 - 22)^2 + (0.2)(20 - 22)^2 + (0.3)(40 - 22)^2 \\
&\quad + (0.1)(60 - 22)^2 \\
&= 434.0, \text{ in } 10^{12} \text{ dollars} \\
\sigma &= 20.83 \times 10^6 \text{ dollars,} \\
M &= 25 \times 10^6 \text{ committed} + 15 \times 10^6 \text{ available} \\
&= 40 \times 10^6 \text{ total venture capital,} \\
z &= -1.282, \text{ based on management's willingness to take a 10\% risk,} \\
w &= 0, \text{ assuming that discounting covers cost of using capital,} \\
f &= \frac{M}{C}\left[\frac{\bar{X}_A - C(1 + w)}{z\sigma}\right]^2 \\
&= \frac{40 \times 10^6}{12 \times 10^6}\left[\frac{22 \times 10^6 - 12 \times 10^6(1 + 0.0)}{(-1.282)(20.83 \times 10^6)}\right]^2 \\
&= 0.466, \text{ or } 46\%.
\end{aligned}$$

This indicates that even by assuming a 10% risk, the company should not take more than about a 45% interest and should look for possible partners in this development operation.

5.2-4 Appraisals Limited by Risk

It remains only to solve Eq. (5-14) for the speculative cost in order to obtain a general appraisal equation. The solution may be given as follows:

$$C \leq [B_1 + B_2]\left[1 - \sqrt{1 - \left(\frac{B_1}{B_1 + B_2}\right)^2}\right], \tag{5-15}$$

where

$$B_1 = \frac{\bar{X}_A}{1 + w}, \tag{5-15a}$$

$$B_2 = \frac{f}{2M}\left(\frac{z\sigma}{1 + w}\right)^2, \tag{5-15b}$$

C = appraisal value for 100% interest,

fC = appraisal value for f fractional interest.

Table 5.2 simplifies the solution of this equation.

Example: A management group has an exploration budget of \$1.2 MM for the fiscal year. They are considering drilling a development well whose various possible outcomes are shown in the top diagram of Fig. 5.10. Operating profit can be taken as \$2 per barrel, and a present-value factor of 0.7 may be

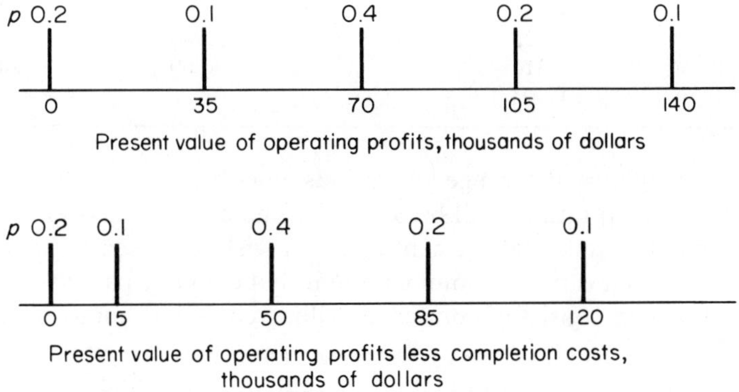

Figure 5.10. Possible rewards from drilling a development well.

Table 5.2
Ratios of Appraisal Value to Expected Value
[Solution of Eq. (5.15)]

B_2/B_1	C/B_1	B_2/B_1	C/B_1
0.000	1.000	0.850	0.294
0.005	0.905	0.900	0.284
0.010	0.868	0.950	0.276
0.015	0.841	1.000	0.268
0.020	0.819	1.100	0.253
0.025	0.800	1.200	0.240
0.030	0.783	1.300	0.229
0.035	0.768	1.400	0.218
0.040	0.754	1.500	0.209
0.045	0.742	1.600	0.200
0.050	0.730	1.700	0.192
0.055	0.719	1.800	0.185
0.060	0.708	1.900	0.178
0.065	0.699	2.000	0.172
0.070	0.689	2.100	0.166
0.075	0.681	2.200	0.160
0.080	0.672	2.300	0.155
0.085	0.664	2.400	0.150
0.090	0.656	2.500	0.146
0.095	0.649	2.600	0.142
0.100	0.642	2.700	0.138
0.150	0.582	2.800	0.134
0.200	0.537	2.900	0.130
0.250	0.500	3.000	0.127
0.300	0.469	3.100	0.124
0.350	0.443	3.200	0.121
0.400	0.420	3.300	0.118
0.450	0.400	3.400	0.115
0.500	0.382	3.500	0.113
0.550	0.366	3.600	0.110
0.600	0.351	3.700	0.108
0.650	0.338	3.800	0.105
0.700	0.325	3.900	0.103
0.750	0.314	4.000	0.101
0.800	0.303		

assumed throughout the range of various possible reserve discoveries. If the well is successful, there will be $20,000 completion costs over and above the costs of a dry well. Management does not wish to exceed 5% risk. Based on risk alone, what is the maximum amount that could be paid as speculative cost, which would consist of combined drilling costs and bonus paid for the lease?

Solution: The first step is to obtain a present-value distribution of the possible rewards, exclusive of the speculative cost. The second diagram in Fig.

5.10 shows the net operating profit reduced to a present value by the 0.7 discount factor. Finally, the completion costs of $20,000 are subtracted from the successful outcomes but not from the unsuccessful case. This produces a present-value distribution that describes the possible rewards, exclusive of the speculative or dry hole cost.

The second step in the solution is to calculate $\bar{X}_A$, the expectancy, and σ, the standard deviation, of the present-value distribution of possible outcomes, exclusive of the speculative cost. For a discrete distribution, which was used in this case for describing the various possible outcomes, these quantities can be calculated as illustrated in Table 5.3.

Table 5.3
Possible Outcomes from Drilling a Development Well

Reserves Discovered (bbl)	*Probability,* p	*Net Operating Profit* ($)	*Present Value of Net Operating Profit* ($)	*Present Value of Reward,* X_i ($)	$p_i X_i$ ($)	$p(X_i - \bar{X}_A)^2$($)
0	0.2	0	0	0	0	510.1×10^6
25,000	0.1	50,000	35,000	15,000	1,500	126.0
50,000	0.4	100,000	70,000	50,000	20,000	0.1
75,000	0.2	150,000	105,000	85,000	17,000	238.1
100,000	0.1	200,000	140,000	120,000	12,000	483.0
					50,500	1357.3×10^6

$$\bar{X}_A = \$50{,}500,$$
$$\sigma = \sqrt{1{,}357.3} \times 10^3$$
$$= \$36{,}800.$$

The third step in the solution is to solve the statistical evaluation, which in this case is Eq. (5-15). Let $w = 0$ with the present-value distribution and $z = -1.645$ for 5% risk.

$$B_1 = \frac{\bar{X}_A}{1+w} = \frac{50{,}500}{1+0} = \$50{,}500,$$

$$B_2 = \frac{f}{2M}\left(\frac{z\sigma}{1+w}\right)^2 = \frac{1}{(2)(1{,}200{,}000)}\left(\frac{-1.645 \times 36{,}800}{1+0}\right)^2$$
$$= \$1{,}530,$$

$$C \leq (B_1 + B_2)\left[1 - \sqrt{1 - \left(\frac{B_1}{B_1 + B_2}\right)^2}\right]$$
$$\leq (52{,}030)\left[1 - \sqrt{1 - \left(\frac{50{,}500}{52{,}030}\right)^2}\right]$$
$$\leq \$39{,}500.$$

Interpretation can be stated as follows. If no more than this amount is paid for the venture, the company can participate in a sufficient number to ensure averaging out 95 times out of 100 groups of ventures.

Table 5.2 can be used to solve Eq. (5-15). For the present example,

$$\frac{B_2}{B_1} = \frac{1{,}530}{50{,}500}$$
$$= 0.030.$$

From Table 5.2 the corresponding value is

$$\frac{C}{B_1} = 0.783.$$

Hence the maximum appraisal value is

$$\begin{aligned} C &= 0.783B_1 \\ &= (0.783)(50{,}500) \\ &= 39{,}500. \end{aligned}$$

This value is, of course, in agreement with the direct arithmetic solution indicated previously.

Example: Determine how much can be paid for combined lease bonus and speculative drilling costs. The prospect is a 40-acre tract, pay thickness is 30 ft. The probability of a dry hole is set at 50%. If the well is successful, recovery in barrels per acre-foot may be as low as 50 to as high as 200, with the most probable recovery being 100 bbl/acre-ft. Operating profit may be taken as $2 per barrel. Completion costs for a successful well will exceed the costs of a dry well by $30,000.

Solution: The first step is to obtain a present-value distribution of the possible rewards (Table 5.4). It will be assumed that a discount rate of 8% will cover costs of capital, and this rate will be used to calculate the present-value distribution. Present-value factors were estimated by using methods described

Table 5.4
Calculation of Present-Value Distribution of Rewards
(Triangle with Spike)

Bbl/ acre-ft	*Total bbl*	*Present-Value Factor*	*Present Value at $2/bbl ($)*	*Completion Costs ($)*	*Net Present Value ($)*
0	0	1.000	0	0	0
50	60,000	0.805	96,500	30,000	66,500
100	120,000	0.720	172,800	30,000	142,800
200	240,000	0.559	268,400	30,000	238,400

in Chapter 11. A maximum production rate of 60 bbl/day was assumed, followed by an exponential decline to an economic limit of 3 bbl/day. Calculation of the present values is shown in Table 5.4 and on Fig. 5.11.

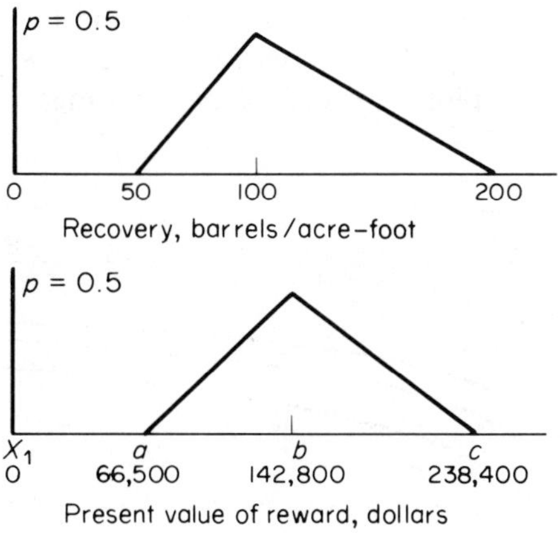

Figure 5.11. Possible rewards of a drilling venture. Triangle with a spike.

The second step in the solution is to calculate $\bar{X}_A$ and σ of the present-value distribution of rewards. For the triangular distribution with a spike, use will be made of equations given in Chapter 3.

$$\bar{X}_A = p_1 X_1 + (1 - p_1)\frac{a + b + c}{3}$$

$$= (0.5)(0) + (1 - 0.5)\frac{66{,}500 + 142{,}800 + 238{,}400}{3} \quad (3\text{-}28)$$

$$= \$74{,}600,$$

$$p_1 = 0.5,$$

$$p_2 = (1 - p_1)\frac{b - a}{c - a} = (0.5)\frac{142{,}800 - 66{,}500}{238{,}400 - 66{,}500}$$

$$= 0.222,$$

$$p_3 = 1 - p_1 - p_2 = 1 - 0.5 - 0.222$$

$$= 0.278,$$

$$\sigma^2 = \frac{p_1}{1 - p_1}(\bar{X}_A - X_1)^2 + \frac{(c - a)^2}{18}\left(1 - p_1 - \frac{p_2 p_3}{1 - p_1}\right)$$

$$= \frac{0.5}{0.5}(74{,}600 - 0)^2 + \frac{(238{,}400 - 66{,}500)^2}{18} \quad (3\text{-}29)$$

$$\times \left[1 - 0.5 - \frac{(0.222)(0.278)}{1 - 0.5}\right]$$

$$= 6{,}183 \times 10^6$$

$$\sigma = \$78{,}600.$$

The third step in the solution is to solve the statistical evaluation equation (5-15). We shall let $f = 1$ for 100% working interest and let $w = 0$ with the 8% discounted present-value distribution. In this problem no acceptable risk factors have been stated nor has any total amount of risk capital been given. We shall let z have values of 2.327, 1.645, and 1.282 for risk values of 1, 5, and 10%, respectively. We shall let M assume values from 0 to 10 MM dollars. The results are plotted as curved lines on Fig. 5.12.

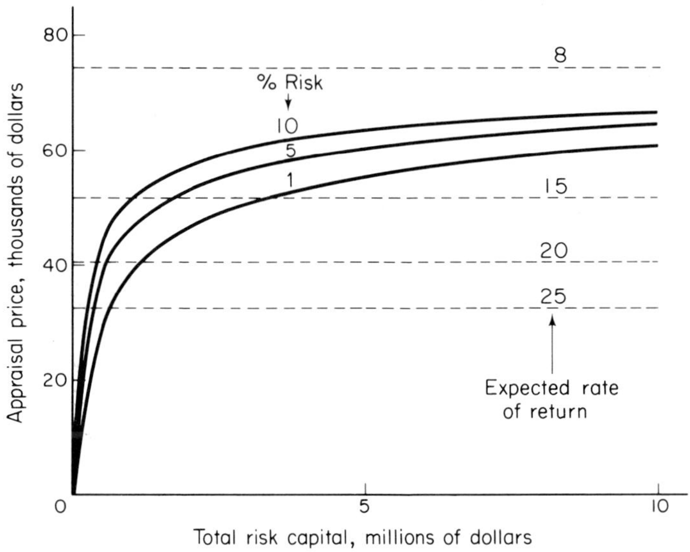

Figure 5.12. Appraisal price versus risk capital for a specific venture.

For comparative purposes, the appraisal values calculated by discounting the expected oil recovery are also shown in Fig. 5.12. The curved risk lines, which are based on 8% interest, approach asymptotically the 8% expected value. This is a rather good prospect, and by inspecting this graph it is clear that any company with more than about $1 million in risk capital would probably set their maximum appraisal value by the discounted-cash-flow method. Companies smaller than this should limit their appraisal value of this prospect according to the risk calculations.

5.3 VALUE OF ADDITIONAL INFORMATION

Additional geophysical efforts or geological investigations can sometimes reduce the regional percentage of failures among wildcat wells. The maximum value of information that decreases the probability of failure can be estimated by the following equation:

$$V_I = C_{\text{new}} - C_{\text{old}} \tag{5-16}$$

where

V_I = value of the information,

C_{new} = appraisal value based on new information,
C_{old} = based on the original information.

There are two bases on which such calculations can be made. The appraisal values can be based on risk or on discounting the expected values. Both methods will be illustrated. In the following example, the original probability of a dry well was 0.7. An added condition arises because of the claim that certain additional geophysical work can reduce the number of dry wells by 20%. New probabilities are calculated in Table 5.5, on the basis of 10 original wells.

Table 5.5
Calculation of New Probabilities

	Original Conditions		*New Conditions*	
	Wells	*Probability*	*Wells*	*Probability*
Dry	7	0.7	5.6	0.651
Producers	3	0.3	3.0	0.349
	10	1.0	8.6	1.000

First the value of the additional information will be calculated from a risk basis. The values that define the triangle with spike distribution, as in Fig. 5.12, are calculated in Table 5.6.

Table 5.6
Values for Calculations Based on Risk
(Based on 8% Discount Factors)

Point	*Undiscounted Net Operating Profit* ($)	*Present-Value Factor*	*Present Value of Net Operating Profit* ($)	*Completion Costs* ($)	*Net Present Value* ($)
X_1	0	1.000	0	0	0
a	100,000	0.804	80,400	40,000	40,400
b	250,000	0.714	178,500	40,000	138,500
c	600,000	0.531	318,600	40,000	278,600

Equation (5-15) was used to calculate the appraisal values, based on 5% risk and total risk capital of \$1 million for both the original and new conditions (see Table 5.7.) Following are the results:

$$\begin{aligned} V_I &= C_{\text{new}} - C_{\text{old}} \\ &= \$30{,}690 - \$25{,}670 \\ &= \$5{,}020. \end{aligned}$$

Table 5.7
Probabilities for Risk Calculations

	Original Conditions	*New Conditions*
p_1	0.700	0.651
p_2	0.124	0.144
p_3	0.176	0.205
	1.000	1.000

Next the value of the additional information will be calculated from an expected value basis. One factor that must be decided upon is the interest discount rate that is to be used for the evaluation. In Table 5.8, 20% will be used. It is necessary to again calculate the defining points in the distribution.

Table 5.8
Values for Calculations Based on Expectancy
(Based on 20% Discount Factors)

Point	*Undiscounted Net Operating Profit* (\$)	*Present-Value Factor*	*Present Value of Net Operating Profit* (\$)	*Completion Cost* (\$)	*Net Present Value* (\$)
X_1	0	1.000	0	0	0
a	100,000	0.602	60,200	40,000	20,200
b	250,000	0.480	120,000	40,000	80,000
c	600,000	0.288	172,800	40,000	132,800

The appraisal values are then taken as the expected values, as follows:

$$C = p_1X_1 + (1 - p_1)\left(\frac{a + b + c}{3}\right).$$

Following are the results:

$$\begin{aligned} V_I &= C_{\text{new}} - C_{\text{old}} \\ &= \$27{,}110 - \$23{,}300 \\ &= \$3{,}805. \end{aligned}$$

At this stage one is confronted with two different answers. The one that is valid would depend upon whether the appraisal price was limited by risk or limited by the necessary discounted rate of return. In this example it is most probable that rate of return would be the limiting factor. Therefore, one would set the value of the additional information at about $3,800 per exploratory well. For a program of 10 wildcat wells, this would be a savings of about $38,000.

5.4 GAMBLER'S RUIN BY SUCCESSIVE LOSSES

The connotations associated with gambler's ruin lend fascination to its study. The idea is that, even though a series of ventures is attractive and eventually will yield a profit, there is nevertheless some possibility that a run of losses could dissipate the resources of a player and force him to leave the game. Possibilities are controlled by the expectations and variance of the ventures as well as by the amount of venture capital resources of the player.

A study of this phenomenon can be based on Fig. 5.7 and Eq. (5-9). Setting the first derivative of the equation equal to zero permits a solution for n where the financial position reaches its minimum value.

$$\frac{d(y - y_0)}{dn} = 0 = X_E + \frac{z\sigma}{2\sqrt{n}},$$

whence

$$n = \left(\frac{-z\sigma}{2X_E}\right)^2. \tag{5-17}$$

Substituting this expression for n back into Eq. (5-9) yields the following relation:

$$y - y_0 = -\frac{(z\sigma)^2}{4X_E}. \tag{5-18}$$

Next replacing $-(y - y_0)$ with M_G permits writing

$$M_G = \frac{(z\sigma)^2}{4X_E}, \tag{5-19}$$

where

M_G = venture capital required to survive the lowest ebb of fortunes.

The significance of this relationship will be illustrated using the same data of the fractional participation example. The venture in that case was quite attractive, but it would cost $12 MM dollars and the company had a total of $40 MM risk capital. The company was advised to limit its participation to about 45% in that venture. By spreading its investments, the company could be 90% confident of not showing a loss within the current series of ventures and accounting period.

Let us presume that a company purchases one after another of such ventures, with full ownership. Equation (5-19) can be used to calculate the cash reserves required to survive an initial run of bad luck. Of course, the first few ventures may turn out to be very favorable. But if the company wishes to be 95% confident of surviving the first few ventures, some initial reserves should be provided for. Using the data of the foregoing problem, solution is made as follows:

$$\begin{aligned} X_E &= \bar{X}_A - C \\ &= 22 \text{ MM} - 12 \text{ MM} \\ &= \$10 \text{ MM}, \\ M_G &= \frac{(z\sigma)^2}{4X_E} \\ &= \frac{(-1.645 \times 20.83 \text{ MM})^2}{4 \times 10 \text{ MM}} \\ &= \$29.5 \text{ MM}. \end{aligned}$$

The corresponding value of n is

$$\begin{aligned} n &= \left(\frac{-z\sigma}{2X_E}\right)^2 \\ &= \left(\frac{-1.645 \times 20.83 \text{ MM}}{2 \times 10 \text{ MM}}\right)^2 \\ &= 2.9 \text{ ventures}. \end{aligned}$$

This means that, for the 95% confidence line, the low point in fortunes occurs after three ventures, and the company might be out of pocket as much as \$29.5 MM at this point. Of course, one hopes that his group of three ventures is among the 95% which are better, and some certainly much better, than this. Remember, in this chapter the chief consideration has been survival. Other objectives are discussed in later chapters.

REFERENCES

1. BARNARD, C. I., *Functions of the Executive*. Cambridge, Mass.: Harvard University Press, 1964.

2. HAYWARD, JOHN, "Probabilities and Wildcats Tested Through Mathematical Manipulations." *Oil and Gas Journal*, Nov. 15, 1934, p. 129.

3. WHITWORTH, W. A., *Choice and Chance*. Cambridge, 1878.

4. PIRSON, S. J., "Probability Theory Applied to Oil Exploration Ventures." *The Petroleum Engineer*, Feb.–May, 1941.

5. Campbell, W. M., and F. J. Schuh, "Risk Analysis: Overall Chance of Success Related to Number of Ventures." Presented at SPE Annual Fall Meeting, Dallas, Oct. 1961.

6. Lahee, F. H., "How Many Fields Really Pay Off?" *Oil and Gas Journal*, Sept. 17, 1956, p. 369.

7. McCray, A. W., "Evaluation of Exploratory Drilling Ventures by Statistical Decision Methods." *Journal of Petroleum Technology*, Sept. 1969, p. 1199.

PROBLEMS

5-1. The possible outcomes from drilling an exploratory well are summarized in the table. Calculate the mean, variance, and standard deviation of this distribution.

Probability	*Present-Value Outcome* ($)
0.3	−120,000
0.3	−75,000
0.1	100,000
0.2	200,000
0.1	300,000

5-2. In the table are summarized representative outcomes for a proposed well.

Ultimate Recovery (bbl)	*Present Value of Operating Income* ($)	*Probability of Outcome*
0	0	0.5
50,000	100,000	0.2
100,000	200,000	0.2
150,000	300,000	0.1

(a) Calculate the corresponding net present-value outcomes where it will cost $20,000 to drill and complete the well. Calculate the mean and standard deviation of these net present-value outcomes.
(b) Repeat the calculations for the case where it will cost $30,000 to drill and complete the well.

5-3. The table summarizes possible outcomes for a proposed well.

Ultimate Recovery (bbl)	*Present Value of Operating Income* ($)	*Probability of Outcome*
0	0	0.6
100,000	200,000	0.1
200,000	400,000	0.2
400,000	800,000	0.1

(a) The cost of drilling the well is \$50,000. If the well discovers oil there will be additional completion costs of \$25,000. Calculate the corresponding net present-value outcomes. Calculate the mean and standard deviation of the net present-value outcomes.

(b) Repeat the calculations for the case where drilling costs are \$60,000 and completion costs remain at \$25,000.

(c) Repeat the calculations for the case where drilling costs are \$50,000 and completion costs are \$35,000.

5-4. In a certain province there is a binomial probability $p = 0.2$ that an exploratory well will discover a new oil field. The undiscovered fields in this province are believed to have a mean size of 800 MM bbl and a standard deviation of 600 MM bbl. Calculate the per cent confidence that the average discovery size will be at least 400 MM bbl for a total of $N = 5$ trys and again for a total of $N = 10$ trys. Use the following joint probability relation:

$$\text{confidence} = \sum_{x=1}^{N} (p(x))(p_F).$$

Calculate $p(x)$ as the binomial probability of exactly x successes in N trys. Calculate p_F, the probability of the average-sized field, being at least the required 400 MM bbl in size, by assuming that the sampling distribution is normally distributed with a mean of 800 MM bbl and a standard deviation of 600 MM bbl$/\sqrt{x}$.

5-5. A workover is being planned in which a zone, presently behind the casing, will be tested. This will require a workover rig, perforating, treating, and testing. Productive portions of the formation will produce 150 bbl/acre-ft. Spacing is 40 acres. There is doubt about the thickness of the possible producing interval, and the prospects have been summarized in the table. The

Probability	*Net Producing Interval (ft)*
0.4	0
0.2	15
0.2	20
0.2	25

operating profit will be about \$2 per barrel and a present-value factor of 0.7 is applicable. The company will engage in approximately 20 such projects during the current program. On this basis, what is the maximum that could be invested in the workover, so that at least a 50% present-value profit/investment ratio could be realized, with 90% confidence?

5-6. A company has the opportunity to participate in an offshore drilling venture. The object formation has 50 ft of pay and when productive will produce 300 bbl/acre-ft. Based on similar developments, there is estimated to be a 50% chance that the venture will find no oil, a 25% chance of a field of 800 acres, and a 25% chance of finding a field of 1,600 acres. The present value of the operating profit can be taken as \$2 per barrel. It will cost \$10 MM to obtain the lease, build a platform, and test the structure. The company has a total of

$30 MM invested in similar unresolved projects. There is available an additional $20 MM venture capital to be invested. Management is willing to assume a 5% risk. What is the maximum fractional participation that would be advisable in this venture?

5-7. An oil lease of 160 acres has four wells already drilled on 40-acre spacing. The net pay is 25 ft. It is believed that the possible ultimate recovery is best estimated by a triangular distribution with a low of 100, a most likely yield of 200, and a high of 350 bbl/acre-ft. The four wells combined have already produced a total of 120,000 bbl. A company is considering buying this lease. The operating profit can be taken as $2.40 per barrel, and a present-value factor of 0.75 is applicable. However, an allowance is to be made for one workover per well, which would average $30,000 when performed, with the same present-value factor acceptable on an average basis. Because of inherent risks in a venture of this type, the company desires to be 95% confident of realizing at least a 60% present-value profit/investment ratio. What is the top figure that the company could pay for this oil production?

5-8. A company has developed information on an offshore structure. The investment necessary to test the structure for oil is approximately $12 MM. There is a 50% chance that the structure might be dry. However, if it is successful, there is a 60% chance of a net present-value profit of $30 MM and a 40% chance of a $50 MM net present-value profit. The company has $80 MM venture capital. Should the company develop alone or look for partners?

5-9. An exploratory well is given the following payoff table:

Probability	*Present Value of Operating Profit* ($)
0.60	(dry)
0.10	60,000
0.20	120,000
0.10	180,000

If the well is a producer, completion costs will amount to $25,000. What is the maximum dry hole cost that can be tolerated? The management group has $1 MM venture capital. Base the cutoffs at 5% risk or 15% return, whichever applies.

5-10. An exploratory well has the prospects given in the table. The cost of drilling a dry hole is $20,000. Completion costs require an additional $30,000. Management has $1 MM venture capital and will accept a 5% risk. How much can be paid for the lease (right to drill)?

Probability	*Present Value of Operating Profits* ($)
0.65	(dry)
0.10	50,000
0.10	100,000
0.10	150,000
0.05	200,000

5-11. If an exploratory well proves to be dry, it will cost \$80,000. If it is completed as a producer, there will be an additional \$50,000 cost. The present value of the operating profits for a successful well can be represented by a uniform distribution from \$200,000 to \$500,000. Calculate the mean, variance, and standard deviation of the distribution of possible net present-value outcomes.

5-12. In a certain region wells are drilled on 80-acre spacing. The producing formation is 20 ft thick, and total recovery may be anywhere in the range from 100 to 300 bbl/acre-ft. However, there is only 40% probability of obtaining production. Operating profit is about \$2 per barrel and the present-value factor can be taken as 0.6. A dry well costs \$80,000 and completing a producer costs an additional \$50,000. How much venture capital is required to drill such wells at the 10, 5, and 1% risk levels?

5-13. If a wildcat well is successful, the reserves discovered will be in the range 200,000 to 600,000 bbl. The well completion costs will be \$50,000 greater than the costs of a dry well. Operating profits will be \$2 per barrel, and a present-value factor of 0.65 is applicable. Under present operating techniques, the company's experience is that 60% of such prospects prove to be dry. It is claimed that certain improved geophysical techniques would eliminate 20% of the prospects that later prove to be dry. The management group is restricted to \$1.5 MM venture capital. On the basis of 5% risk, what is the value of the improved geophysical techniques per prospect?

5-14. There is 0.75 probability that a wildcat in a certain region will be dry and cost \$60,000. If the well is completed, the cost will be \$80,000. The undeferred value of the operating profits can be represented by a triangular distribution where \$70,000, \$300,000, and \$800,000 represent the lower limit, mode, and upper limit, respectively. Deferment factors can be estimated by assuming that 40% will be produced on allowable over 5 years and the remainder on exponential decline, where the ratio of initial rate to final rate is 10 : 1 over a 15-year period. Plot the appraisal value versus the total risk capital. Show the 1, 5, and 10% risk lines, based on an 8% cost of capital. Also show the appraisal values for 10, 20, and 30% expected DCF rates of return.

6

DECISION TREES AND ECONOMIC MODELS

6.1 INTRODUCTION

6.1-1 Purposes Served by Decision Trees

A plan of action is desirable and quite often essential. This applies in political, governmental, industrial, and personal instances. Most commonly the path followed depends only partially upon free or deliberate choice. It depends also upon external forces or outcomes which in their nature tend to be probabilistic. Portraying in some fashion both the actions that may be taken and the results that may follow will help in the anticipation of both dangers and opportunities. *Decision trees* serve as a format for drawing up and evaluating plans of action.

One way of forming an economic plan, which can serve as a decision model, is shown in Fig. 6.1.[1] Here an oil company has made preliminary geologic investigations which indicate some possibility of locating an oil field. The company is now faced with making the decision either to buy or not to buy options on leases. These would permit drilling for and producing oil on certain properties. If the options are bought, additional investments would be made to conduct seismic surveys. These would serve to determine with a high degree of certainty whether or not oil would be found. Thus, further action would await the outcome of the seismic survey, and this could be said to be tomorrow's decision. Based upon past experience, there is considered to be only a 25% chance that the prospect will prove to be successful.

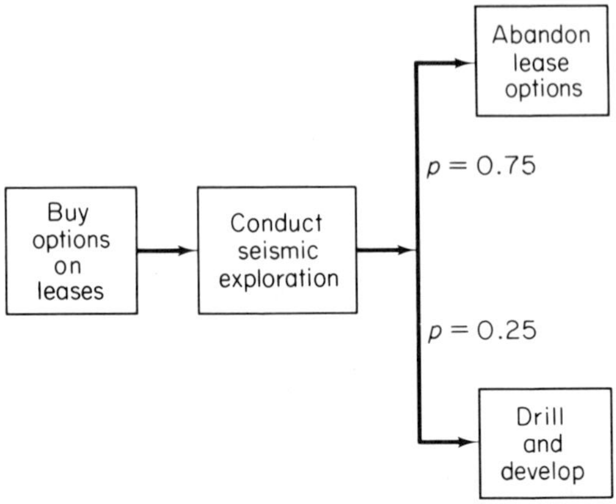

Figure 6.1. Economic model for oil exploration in the Mid-Continent region of the United States. (After Louis F. Davis[1])

Today's decision must allow for the chance of unfavorable outcomes as well as provide for achieving the goals of the company.

One path only will be followed through a decision-tree network. The number of branches should adequately describe the number of choices and outcomes. A desirable solution would lead toward accomplishing the objectives of the company, consistent with the risks that can be taken. Such objectives could include profit from an individual venture, increased reserves of petroleum resources, growth of the company, the service and public image of the company, or other objectives. Information should be distilled into a form where the difference between available choices becomes understandable and meaningful. Indeed, if there is no discernable difference between two alternatives, this should be pointed out, for then the choice itself is trivial and no time should be wasted on it. One should proceed to make decisions where the choices are meaningful and important.

Most of the discussion in this chapter is concerned with investment decisions. The choice of investments determines the future course of a company, its profits or losses, its growth—in fact, the very nature of its business. Figure 6.2 shows the arrival of decision opportunities. In large companies, one of the functions of the staff is to search out opportunities. Men who control great wealth are assailed by opportunities brought in by others, by promotors of many kinds. In general, however, the opportunities must be sought in the economic environment. Figure 6.2 indicates that the chance of participating in venture *A* has been lost, presumably by purposeful default. Project *B* was accepted and now awaits the march of time until its outcome is known and any benefits are returned from it. Projects *C* and *D* are in the current decision set. Choices must be made whether or not to buy or participate in one or both of these. One may prefer to wait for better opportunities in the future, and this

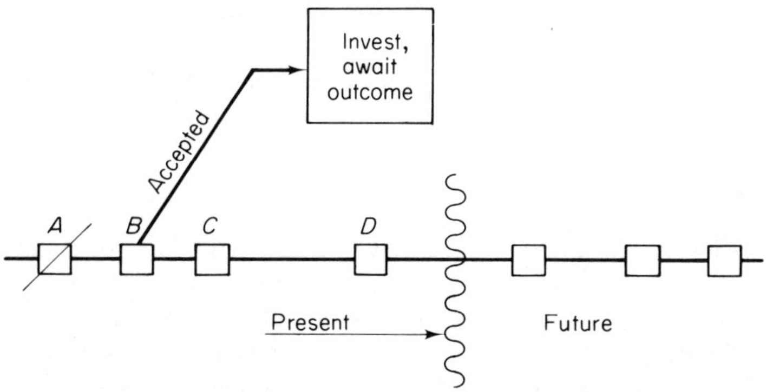

Figure 6.2. Arrival of investment opportunities.

could be considered as one of the alternatives in the current decision set. Decision trees present a method for helping to make such decisions.

6.1-2 Variations in Degree and Kind

Decision trees can be described according to their complexity. The following groups, except possibly for the last, differ more in degree than in kind.

1. Probability trees, the ends of whose branches represent possible final outcomes of a venture, which will be determined by chance. These may be isolated or combined as parts of larger trees.
2. A concise map which indicates the chronological order in which a series of actions will be performed. Major choices or decision that must be made are indicated. The few branches present are generally reduced to their expected values.
3. A more detailed map whose branches indicate the range of possible outcomes and whose decision nodes are emphasized according to their future position values.
4. An expanded tree with many branches for indicating various degrees of success. This becomes unwieldy unless programmed for a computer, as for calculating expected values.
5. A tree fabricated with selected probability distributions at the probability nodes and with decision rules built into the decision nodes. This becomes operable as a computer program whose output is a frequency distribution describing present-value variation. These, because of the high degree of interaction between successive outcomes and decisions, are referred to as *stochastic decision trees.*[2]

Variations occur in the degree of probabilistic treatment incorporated into decision trees. Insofar as evaluating risk is concerned, the chief requirement is a means of indicating the expectancy and variance within each branch. As with any means of expression, the form depends upon the degree of detail desired by those who will use it as well as the general level of the purpose to be served.

6.1-3 Symbols and Layout

A decision tree is a network, somewhat equivalent to a road map, which describes several possible courses, one of which will be followed. Most often it is drawn chronologically beginning at the left. Moving toward the right indicates the passage of time. It is constructed with two kinds of nodes, decision nodes and chance nodes. Preferably, although not always, these are drawn with identifying symbols. At the present time the symbol most consistently used to mark decision nodes is the square. Chance nodes are commonly marked by a circle (but may, as a matter of convenience, be left unmarked). Stochastic trees present a special case that will be discussed later.

In Fig. 6.3, starting at the left, there is indicated to be the choice of drilling or of not drilling a particular well. This is indeed a decision node. If the well is drilled, the various possible outcomes can be grouped into three representa-

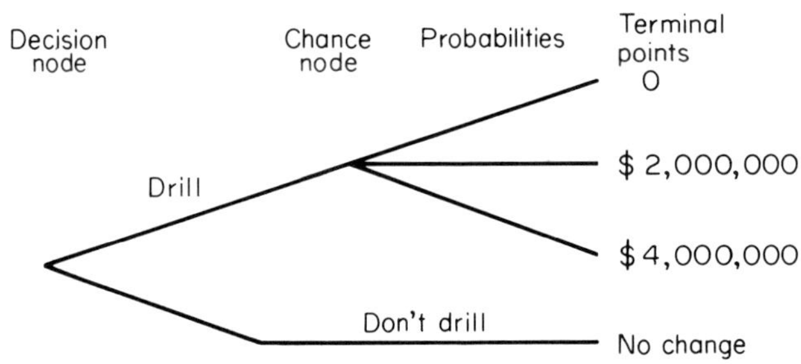

Figure 6.3. Decision tree with unidentified nodes and probabilities missing.

tive cases, which are shown on the diagram. These emanate from a chance node. But essential information is missing. No evaluation can be made without estimates of the probabilities of each of these representative cases.

In Fig. 6.4 the customary symbols have been included for the decision node and for the chance node. Also, the probabilities assigned to the various terminal events have been noted on the individual branches. Immediately one should observe that the information present is sufficient for the estimation of both the expectancy and the variance associated with the chance node. Another observation is that the sum is unity for probabilities of branches emanating from a chance node. This convention is commonly used, so prob-

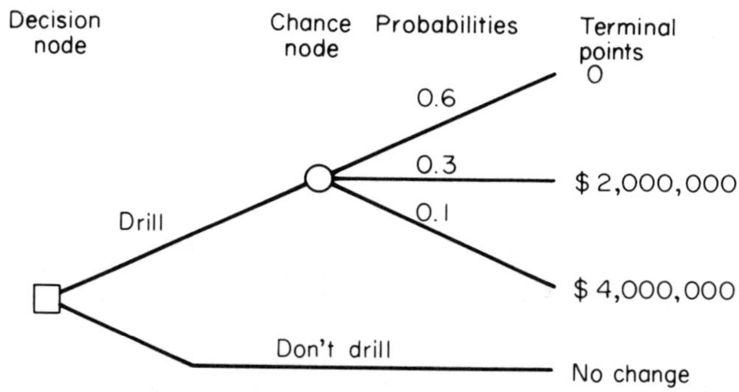

Figure 6.4. Decision tree with identified nodes and conditional probabilities.

abilities listed on a tree are usually conditional probabilities, dependent upon the particular node actually being reached in the course of future events in the real world.

6.2 ANALYSIS OF A PROBABILITY TREE

6.2-1 Evaluating a Development Drilling Venture

A probability tree is a special case of a decision tree. In its typical form the only provision made for a decision is at its beginning: either the whole is accepted or the whole is rejected. There are implied decisions, but they are presumed to be automatic, depending upon the outcomes of prior chance nodes. This is not always a good assumption, and the omission of decision nodes may in some cases produce a faulty analysis. In other cases it is an acceptable procedure. In the following example it is assumed that, if the oil operator accepts the venture, he will drill until he hits a dry well or until he completes four oil-producing wells.

The example concerns an oil operator who has the opportunity of buying a lease containing four locations where wells may be drilled. Cost of the lease would be $20,000. Figure 6.5 is a map showing the present best estimate of the direction of the structural contours that control oil accumulation. There is an offset well which justifies a probability of 60% that well *A* would be successful. If well *A* is successful, there is a probability of 80% that well *B* would be successful. If well *B* is successful, there is a 90% chance that well *C* will be successful. But even if well *C* is successful, there is only a 70% chance that well *D* would be successful. A dry well would cost $40,000, and a successful well would incur an additional $15,000 in completion costs.

The probability tree for this problem is shown in Fig. 6.6. Probability values noted on the branches are conditional probabilities. The unconditional

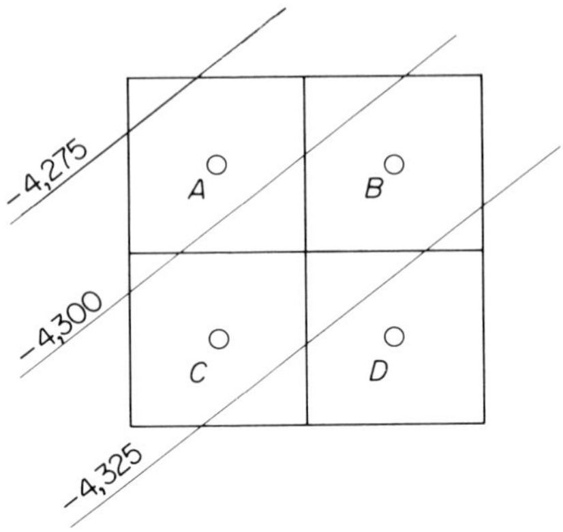

Figure 6.5. Map of well locations and structural elevation contours.

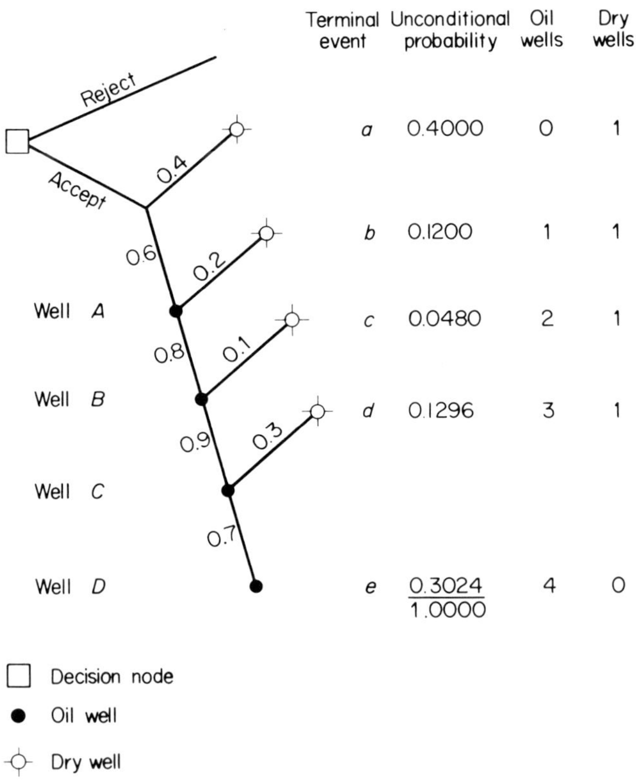

Figure 6.6. Probability tree for a development drilling venture.

probabilities can be calculated by multiplying together all conditional probabilities along the path leading to each terminal event. Table 6.1 shows the calculation of the net value of each terminal event. Next the tree will be evaluated by two methods: (1) rollback, and (2) conventional statistical methods.

6.2-2 Rollback Evaluation of a Probability Tree

The *rollback* procedure starts with the terminal events of a decision tree. The method is based fundamentally upon evaluating each branch leading from a decision node.[3] Thus it helps to predict which way a future decision will be made, and in the conventional procedure, only the one branch with the greatest expected value is retained at the decision nodes. In turn, this makes it possible to make a prior or present decision in a more intelligent way. Projected future conditions and opportunities are carried stepwise back to the present so that maximum opportunity may be realized at each step. The rollback method has also been used to sum up a probability tree that describes a drilling venture.[4] In both cases the method has been used only to obtain expected values.

Table 6.1
Calculation of Net Values of Terminal Events in the Probability Tree

Terminal Event	*Total Net Operating Profits ($)*	*Costs ($)*			*Net Value ($)*
		Lease Bonus	*Producing Wells*	*Dry Wells*	
a	0	20,000	0	40,000	−60,000
b	140,000	20,000	55,000	40,000	25,000
c	240,000	20,000	110,000	40,000	70,000
d	340,000	20,000	165,000	40,000	115,000
e	420,000	20,000	220,000	0	180,000

One of the attributes of the rollback method is that it places an emphasis on the concept of *position value.* This is a combination of the immediate reward, to which is added the expected value of the events or decisions that, normally, should follow. For a single node, this can be expressed as follows:

$$\begin{pmatrix}\text{expected}\\ \text{position}\\ \text{value}\end{pmatrix} = \sum p_b \cdot \begin{pmatrix}\text{immediate} & & \text{expected value}\\ \text{reward} & + & \text{of subsequent}\\ \text{on branch} & & \text{events on branch}\end{pmatrix}, \tag{6-1}$$

where p_b is the conditional probability of following each branch that emanates from the node. For a probability node, all branches would be included in the summation, and the sum of the conditional p_b values would equal unity.

For a decision node, a *decision rule* is needed. One such rule is to use only one path from a decision node, the path with the highest expected value.[3] This, as long as one chooses by expectancies alone, is indeed the expected rule. However, it would be remiss not to mention another rule, applicable where a more complete analysis is being made. This rule is to eliminate from among alternative choices those branches which have both a lower expectancy and a greater variance.[2] The remaining decision paths would then be considered individually or successively combined with other parts of a decision tree.

A rollback evaluation of a decision tree consists of successive applications of Eq. (6-1). The process starts at the extreme ends of the branches, at the terminal events, and works backward to the present. A solution for the development drilling venture is given in Table 6.2. No decision nodes as such

Table 6.2
Rollback Evaluation of Probability Tree for Development Drilling Venture

Position: Prior to Drilling Well	*Outcome Event*	*Conditional Probability, p_b*	*Immediate Reward ($)*	*Expected Value of Subsequent Decisions ($)*	*Net Present Value ($)*	*Expected Value of the Position ($)*
D	Dry	0.3	−40,000	0	−40,000	−12,000
	Producer	0.7	25,000	0	25,000	17,500
						5,500
C	Dry	0.1	−40,000	0	−40,000	−4,000
	Producer	0.9	45,000	5,500	50,500	45,450
						41,450
B	Dry	0.2	−40,000	0	−40,000	−8,000
	Producer	0.8	45,000	41,450	86,450	69,160
						61,160
A	Dry	0.4	−60,000	0	−60,000	−24,000
	Producer	0.6	65,000	61,160	126,160	75,696
						51,696

occur in this example, and the choices are rather obvious. The table begins at the point in time just before drilling the last well, well *D*, and the position value calculates to be simply the expected value of well *D*, which is $5,500. The next calculation is for the point in time just before drilling well *C*. Only two possibilities are considered here. First, there is the 0.1 probability of a dry hole, which means a $40,000 loss, and in which case the value of subsequent wells becomes zero. Second, there is the 0.9 possibility of a good well worth a net present value of $45,000, and in which case the expected value of subsequent developments is $5,500. These considerations permit calculating the position value prior to drilling well *C*. The same process is continued, until one works back to the point in time prior to drilling well A, which completes the solution for the expected present value.

If the knowledge of the reservoir conditions warranted, more than two possible outcomes could have been listed for each well. This would have made both the decision tree and the rollback table more complicated in structure. Also, after the expected value has been found, there is no reason why a second pass could not be made through the tree for determining the total variance about the expected value. However, determination of both the mean and the variance appears to be more readily performed by the customary statistical methods, which are illustrated in the following section.

6.2-3 Statistical Evaluation of a Probability Tree

Evaluation of a decision tree by statistical methods means primarily the determination of its mean, or expected value, and its variance, or standard deviation about the mean. The particular example at hand reduces to a simple discrete distribution, which is shown in Fig. 6.7. Unconditional probabilities are of course used in the calculations, which are shown in Table 6.3.

Table 6.3
Calculation of Mean and Variance of the Probability Tree

Terminal Event	*Unconditional Probability,* p_i	*Net Present Value,* X_i *($)*	*Expectancy,* p_iX_i *($)*	*Variance,* $p_i(X_i - \bar{X})^2$ *(MM $)*
a	0.4000	−60,000	−24,000	4,990.4
b	0.1200	25,000	3,000	85.5
c	0.0480	70,000	3,360	16.1
d	0.1296	115,000	14,904	519.4
e	0.3024	180,000	54,432	4,978.1
	1.000		51,696	10,589.5

$$\sigma = \sqrt{10{,}589.5 \text{ MM}} = \$102{,}900$$

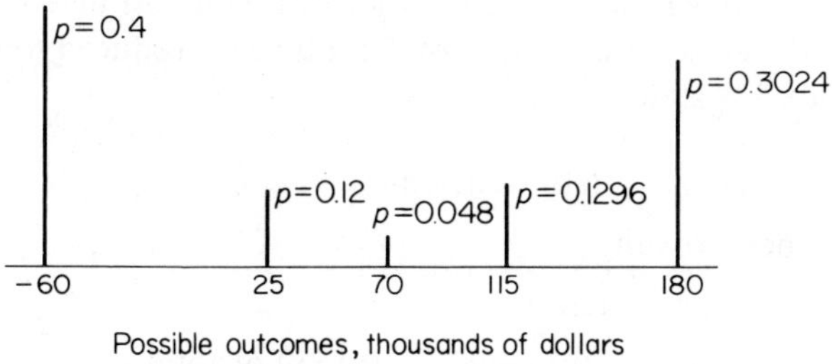

Figure 6.7. Decision tree of Fig. 6.6 expressed as a probability diagram.

This problem can now be further evaluated in the light of statistical decisions. As an example, Eq. (5-19) will be used to calculate the financial resources needed to be 95% sure of avoiding gambler's ruin if one accepted a sequence of such ventures.

$$\begin{aligned} M_G &= \frac{(z\sigma)^2}{4X_E} \\ &= \frac{(-1.645 \times 102.9 \text{ M})^2}{(4)(51{,}696)} \qquad (5\text{-}19) \\ &= \$1.38 \text{ million.} \end{aligned}$$

This answer can be explained in the following way. The venture is attractive from the standpoint of its expectancy. That is, over a number of such ventures the overall average would be positive. But this is an average of high, mediocre, and negative outcomes, so considerable fluctuation can occur. Only if a company is able to survive such fluctuations should it invest in this particular venture.

6.3 COMPARING ALTERNATIVES: RETAINING PARTIAL WORKING INTEREST VERSUS OVERRIDING ROYALTY INTEREST

6.3-1 Economic Description

In this example a company must choose between two alternative courses.[5] The decision-tree format will be used to clarify the choices. The problem arises because the company now owns the oil and gas lease on 120 acres in the SW/4 Section 5-73 N-26 W, Adams County, Oklahoma. The owner of leases on the remaining 520 acres in Section 5 has proposed a 6,300-ft Jones Limestone test. Well spacing is 640 acres and the combined leases make up one drilling unit, in which one well can be drilled. The company has its choice of either participating in the working interest on an acreage basis, or it can farm out its interest for $\frac{1}{16}$ of $\frac{7}{8}$ overriding royalty interest under its 120 acres.

Ownership interests as a fraction of the total gas produced were calculated on the bases of 5% state tax, as follows:

120 acres/640 acres = 3/16 of acreage.

1/8 landowners' royalty.

For partial working interest:

$$\left(\tfrac{3}{16}\right)\left(\tfrac{7}{8}\text{ WI}\right)(0.95) = 0.1559 \text{ of total.}$$

For overriding royalty interest:

$$(\tfrac{7}{8}\text{WI})(\tfrac{3}{16})(\tfrac{1}{16}\text{ORRI})(0.95) = 0.009741 \text{ of total.}$$

Probable reserves of a successful well are estimated as follows:

Maximum:	5,300 MM ft^3 (30% of wells)
Average:	3,750 MM ft^3 (50% of wells)
Minimum:	2,500 MM ft^3 (20% of wells)
Regional wildcat success ratio:	35%
Regional development success ratio:	85%
Spacing:	640 acres/well
Dry hole costs:	\$40,000
Completion costs:	\$32,000
Operating costs:	\$150 per well per month
Gas price:	0.15 per M ft^3 or \$150 per MM ft^3
Producing life:	18 years

6.3-2 Constructing the Decision Tree

Management decisions in this particular company are based on present values calculated at a 15% yearly discount rate. Present values are to be used throughout the decision tree. One of the first things to be done is to estimate a present-value factor. In this area the estimated life of the gas wells is about 18 years, and since no better information is available, the production will be assumed to be uniformly distributed over this period of time. For sums received at the midpoint of each year for 18 years, discounted at 15%, the factor is 0.36587. However, there will be a delay from the date of well completion until gas sales commence, which averages 2 years, an additional factor of 0.75614. Hence

$$\text{present-value factor} = (0.75614)(0.36587)$$
$$= 0.277.$$

The operating costs are estimated as follows:

$$\text{total operating costs} = (\$150 \text{ per month})(12 \text{ months/yr})(18 \text{ yr})$$
$$= \$32,400,$$
$$\text{net share of operating costs} = (\tfrac{3}{16})(\$32,400)$$
$$= \$6,080.$$

The net additional investment required for the $\frac{3}{16}$ working interest is: For a completed well:

$$(\tfrac{3}{16})(\$40{,}000 + \$32{,}000) = \$13{,}500.$$

For a dry hole:

$$(\tfrac{3}{16})(\$40{,}000) = \$7{,}500.$$

Table 6.4 shows the calculation of the values of the terminal events for the case where the company would elect to participate in the working interest. The operating income is calculated for the net interest of a 0.1559 fraction of the total, at a selling price of \$150 per million cubic feet of gas. In Table 6.5, the overriding royalty interest income is calculated as a 0.009741 fraction of the total.

Table 6.4
Calculation of Net Present Value of Possible Terminal Events for Partial Working Interest

Total Reserves ($MM\,ft^3$)	*Operating Income* (\$)	*Operating Costs* (\$)	*Undis-counted Value* (\$)	*Present Value at 15%* (\$)	*Invest-ment* (\$)	*Net Present Value* (\$)
5,300	123,940	6,080	117,860	32,600	13,500	19,100
3,750	87,690	6,080	81,610	22,610	13,500	9,110
2,500	58,460	6,080	52,380	14,510	13,500	1,010
0	0	0	0	0	7,500	−7,500

Table 6.5
Calculation of Net Present Value of Possible Terminal Events for Overriding Royalty Interest

Total Reserves ($MM\,ft^3$)	*Overriding Royalty Income* (\$)	*Net Present Value at 15%* (\$)
5,300	7,710	2,140
3,750	5,450	1,520
2,500	3,640	1,010
0	0	0

6.3-3 Calculating Position Values by Rollback

A decision tree for evaluating the alternative choices is shown in Fig. 6.8. The values of the possible terminal events are shown at the ends of the branches. The probability values on each branch are used to calculate the position values at the probability nodes. The details of the calculations are shown in Table 6.6. If it were certain that the well would be a producer,

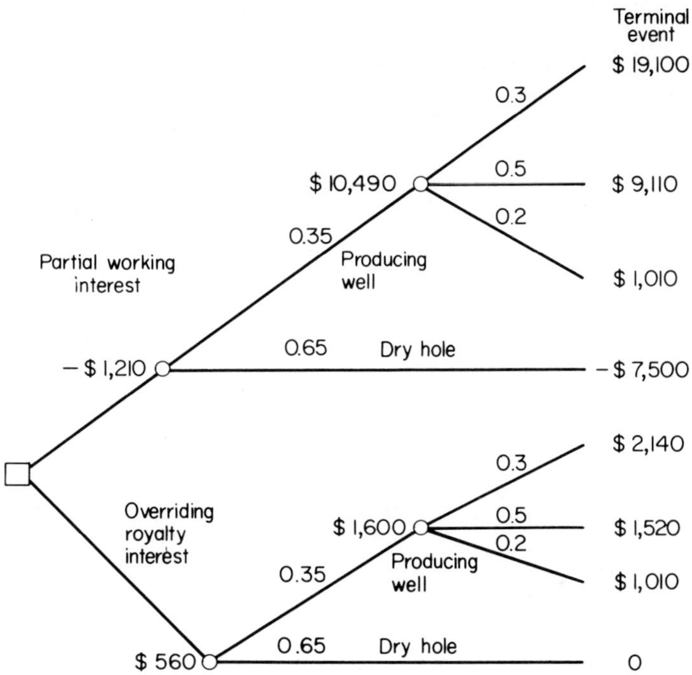

Figure 6.8. Decision tree for choosing between a partial working interest and an overriding royalty interest.

by comparing the position value of $10,490 to $1,600, it becomes obvious that participating in the working interest would be the preferred choice. However, after allowing for the 65% chance of a dry hole, the preferred choice becomes the overriding royalty. Rather obviously the other company, who own the 120 acres in the drilling unit, rates the prospects something greater than 35% for obtaining a good well. Such differences in estimates can, of course, be settled by the drill.

Table 6.6
Rollback Evaluations of
Working Interest Versus Overriding Royalty Interest

Outcome Event	*Conditional Probability*	*Reward, Present Value ($)*	*Expected Present Value ($)*
	For partial working interest, producing well		
5300 MM ft^3	0.30	19,100	5,730
3750 MM ft^3	0.50	9,110	4,560
2500 MM ft^3	0.20	1,010	200
			10,490
	For partial working interest, prior to drilling		
Producing well	0.35	10,490	3,670
Dry hole	0.65	−7,500	−4,880
			−1,210
	For overriding royalty interest, producing well		
5300 MM ft^3	0.30	2,140	640
3750 MM ft^3	0.50	1,520	760
2500 MM ft^3	0.20	1,010	200
			1,600
	For overriding royalty interest, prior to drilling		
Producing well	0.35	1,600	560
Dry well	0.65	0	0
			560

6.4 EVALUATING ACCEPTANCE OF A FARM-OUT

6.4-1 Economic Description

A company must decide whether or not to accept a farm-out of 8,000 acres in Smith County, Kansas, subject to royalties and overriding royalties amounting to a total of 20%.[5] Acceptance of the farm-out would incur the obligation to drill a minimum of three wells. The objective horizon would be a gas sand at a depth of approximately 2,800 ft. This particular management judges its ventures according to present value calculated at a 15% annual discount, neglecting income tax.

The property is north of and adjoins a producing gas field. The prospects of discovering gas on the property are considered to be 25%, which is the exploratory success average for the area. The regional average for development wells is 70% successful. The spacing is 640 acres, which provides for 12 possible drilling locations on the property.

The data are as follows:

Wildcat success ratio:	25%
Development success ratio:	70%
Spacing:	640 acres
Dry hole cost:	\$15,000
Completion cost:	\$7,500
Gas price:	\$0.15 per M ft^3 or \$150 per MM ft^3
State tax:	5%
Probable reserves of a successful well:	
Maximum:	2,000 MM ft^3 (20% of wells)
Average:	1,500 MM ft^3 (60% of wells)
Minimum:	750 MM ft^3 (20% of wells)
Operating costs:	\$150 per well per month
Producing life:	10 years

6.4-2 Present Values of Individual Wells

A present-value factor will be estimated on the basis of a 10-year producing life. It will be assumed that production is evenly distributed over that period of time. In this case the management desires to use 15% per year. For the midyear convention, which assumes all monies to be received at the middle of the year, the corresponding discount factor is 0.53874. It will be assumed further that there will be a 2-year delay between well completion and gas sales, a further discount of 0.75614. This gives the following:

$$\text{discount factor} = (0.75614)(0.53874) = 0.407.$$

For an average producing well:

(0.2)(2000 MM ft^3)	400 MM ft^3
(0.6)(1500 MM ft^3)	900 MM ft^3
(0.2)(750 MM ft^3)	150 MM ft^3
Expected production	1,450 MM ft^3

Income of an average producing well:

$$(1{,}450 \text{ MM ft}^3)(\$150/\text{MM ft}^3)(0.95)(0.80) = \$165{,}300$$

Operating expense:

$$(10 \text{ yr})(12 \text{ months/yr})(\$150/\text{month}) = 18{,}000$$

Net operating income: \$147,300

Present-value net operating income	= (0.407)($147,300)
	= 59,950
Drilling and completion	= 22,750
Net present value	= $37,200 per well

For a dry hole:

net loss = $15,000 per well.

6.4-3 Constructing a Decision Tree

Both the economic values and the probable drilling development program must be incorporated into the decision tree. A minimum of three test wells must be drilled. The outcomes of the test wells will govern the fill-in drilling of adjacent wells. Of the three test wells, either 0, 1, 2, or 3 could be successful. A probability of 75% has been assigned to the 0 case. In the absence of more specific knowledge, a conditional probability of $\frac{1}{3}$ each will be assigned to the cases of 1, 2, and 3 successful wells, which are all within the 25% chance of discovering gas on the property. Finally, there will be the regional tendency of only 70% of the development wells proving successful. Table 6.7 attempts to incorporate these ideas into a representative likelihood plan.

Table 6.7
Likelihood Development Plan
Clustered Around Three Test Wells

Outcome Case	*Test Wells*		*Fill-in Wells*		*Total Wells*		*Present Value ($)*
	Dry	*Good*	*Dry*	*Good*	*Dry*	*Good*	
a	0	3	3	6	3	9	289,800
b	1	2	2	4	3	6	178,200
c	2	1	1	2	3	3	66,600
d	3	0	0	0	3	0	−45,000

The corresponding decision-tree diagram is shown in Fig. 6.9. Both the table and the tree could be described as already trimmed, because they do not show all possible ultimate variations in results. Specifically, in outcome *a*, where there are three successful test wells, a final expected value of three dry and nine producers is listed, which reflects the overall development success ratio of 70%. The possibilities would include 12 good wells, on the one hand, and perhaps an end result of 6 good wells and 6 dry wells; and each well has various possibilities. As long as the concern is chiefly with expected values, it does not make any difference whatsoever that the tree is trimmed in this way. Even in the case of making a risk analysis, it can be

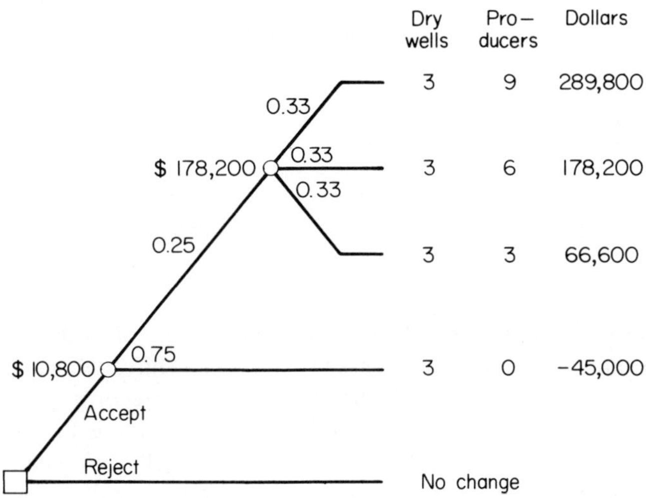

Figure 6.9. Decision tree for evaluating the acceptance of a farm-out.

argued that the major portion of the variance is sufficiently described by the tree as it is shown. That is, the variance within each part, about the expected value of the whole tree, is approximately equaled by the variance of the lumped average value of the part, also taken about the expected value of the whole tree. This of course, depends upon the individual case, but it is often sufficiently accurate.

6.4-4 Evaluating the Decision Tree by Rollback

The expected value, provided gas production is obtained, is calculated in Table 6.8. This turns out to be \$178,200. There is, however, only a 25% chance of this, and a 75% chance of drilling three dry holes with a total loss of \$45,000. The latter are combined in Table 6.9, where the value of the "decision to accept" is calculated to be \$10,800.

Table 6.8
Expected Value if Gas Production is Obtained

Case	*Conditional Probability*	*Event Value (\$)*	*Expected Value (\$)*
a	0.333	289,800	96,600
b	0.333	178,200	59,400
c	0.333	66,600	22,200
			178,200

Table 6.9
Expected Value of the Decision To Accept the Farm-Out

Event	*Probability*	*Event Expected Value (\$)*	*Expected Value (\$)*
Gas discovered	0.25	178,200	44,550
No discovery	0.75	−45,000	−33,750
			10,800

The answer obtained by the rollback method is, of course, simply the expected value. The desired answer is usually a present value, which is dependent upon the discount rate, in the present case 15%. According to the criteria mentioned, the recommendation would be to accept the farm-out, which includes an obligation to drill at least three test wells. We shall, however, take an additional look at this, which includes a risk analysis.

6.4-5 A Statistical Risk Evaluation

For making a risk analysis, a 10% per year cost of capital will be assumed. The factor for uniform amounts, received at midyear, over a 10-year period, is 0.64445. The 2-year delay factor is 0.82645. Accordingly, the present-value factor is

$$D = (0.82645)(0.64445) = 0.533.$$

The expected present value of a completed well is

present value of expected net operating income	= (0.533)(\$147,300)
	= \$78,510
drilling and completion costs	22,750
net present value per good well	55,760

For 100% working interest, $f = 1$, and let $w = 0$ for capital cost based on 10% interest. For 95% confidence of success, let $z = -1.645$. Equation (5-15) and Table 6.10 can now be used to calculate the maximum appraisal value.

$$B_1 = \frac{\bar{X}_A}{1 + w} = \frac{83{,}640}{1 + 0} = \$83{,}640.$$

For purposes of making an analysis, it will be assumed that the company has a total of \$1 million in risk capital.

Table 6.10
Calculation of $\bar{X}_A$ and Variance for Statistical Risk Evaluation

Outcome Case	Unconditional Probability, p_i	Value ($)	Shifted Value, X_i ($)	$p_i X_i$ ($)	$p_i(X_i - \bar{X}_A)^2$ (MM $)
a	0.0833	456,840	501,840	41,820	14,574
b	0.0833	289,560	334,560	27,880	5,247
c	0.0834	122,280	167,280	13,940	575
d	0.7500	−45,000	0	0	5,175
	1.0000			83,640	25,571

$\bar{X}_A = \$83,640$
$\sigma_A = \sqrt{25,571 \text{ MM}} = \$159,910$

$$B_2 = \frac{f}{2M}\left(\frac{z\sigma}{1+w}\right)^2 = \frac{1}{2 \times 1 \times 10^6}\left(\frac{1.645 \times 159,910}{1+0}\right)^2$$
$$= \$69,200.$$
$$B_1 + B_2 = \$152,840.$$

The maximum appraisal value can be calculated next.

$$C \leq (B_1 + B_2)\left[1 - \sqrt{1 - \left(\frac{B_1}{B_1 + B_2}\right)^2}\right]$$
$$\leq 152,840\left[1 - \sqrt{1 - \left(\frac{83,640}{152,840}\right)^2}\right]$$
$$\leq \$24,920.$$

This answer is considerably less than the $45,000 that would be obligated for the drilling of three test wells. It approaches the $30,000 that would be required for the drilling of two test wells and suggests trying to negotiate a deal on that basis.

Another approach would be to consider how the prospects would look after the initial two test wells had proved to be dry. Presumably about two-thirds of the original possible locations would then have been eliminated as favorable prospects, and only about four possible locations would remain under consideration. Figure 6.10 probably represents a charitable view of the prospects of the third test well, just after paying for two dry holes. This again suggests limiting the obligations to the drilling of two test wells. Another course that should be considered is the possibility of obtaining additional information. This could either increase the assigned probabilities of a discovery, or show the prospect to be quite unattractive and thus clarify the decision.

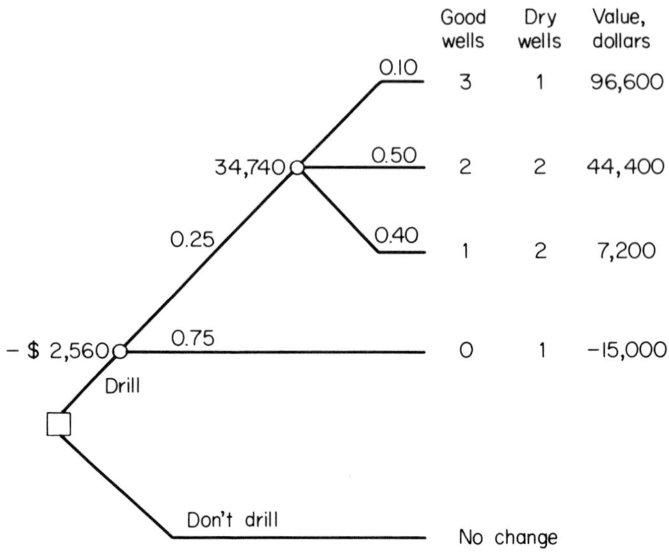

Figure 6.10. Decision tree for evaluating the third test well, following two failures.

6.5 STOCHASTIC DECISION TREES

6.5-1 Functional Design

One of the chief goals of statistical decision methods is to present the available information in its most usable form. Many times it is sufficient to know the expected value and variance for a proposed venture. In other cases, and particularly if the venture is relatively large, it is desirable to consider the form of the entire distribution. Figure 6.11 shows two distributions, which have approximately the same expectancy. In project *A* there is about a 30% chance of losing money, while in project *B* there is less than a 10% chance of losing money. Such characteristics are important from the view-

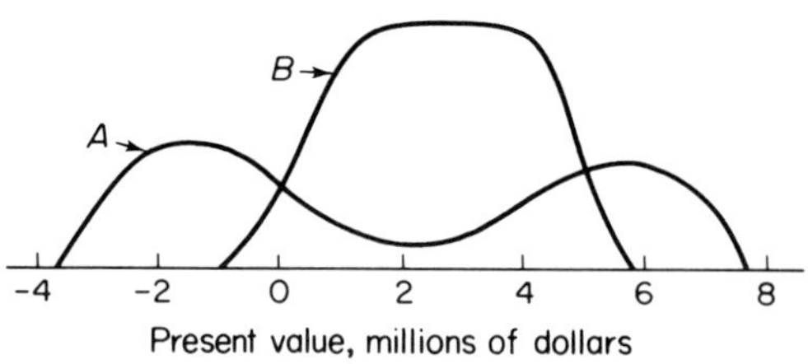

Figure 6.11. Probability diagrams for possible outcomes of two different ventures.

point of management. This type of probability diagram is readily constructed from a frequency distribution, wherein the entire range of possible outcomes is divided into 10, 20, or more classes and the percentage of outcomes occurring within each class is recorded. This section discusses ways of converting decision trees into frequency distributions that could be used to construct detailed probability distributions of present worth.

Stochastic decision trees consist first of adapting the Monte Carlo simulation technique to the decision tree format.[2] Repeated trials, in which chance events are picked according to random numbers, are used to produce the required distribution of present values. This method of many repeated trials is practical only when using a computer. The inclusion of additional features, which make possible the interactions desired in the stochastic tree models, is enhanced by computer simulation. Interactions are provided for in the following ways:

1. Random numbers are used to individually select several chance events which occur within an individual trial.

2. Dependent distributions can be used to properly conform to any tendencies toward relationships between the quantities, such as indiciated in Fig. 6.12.

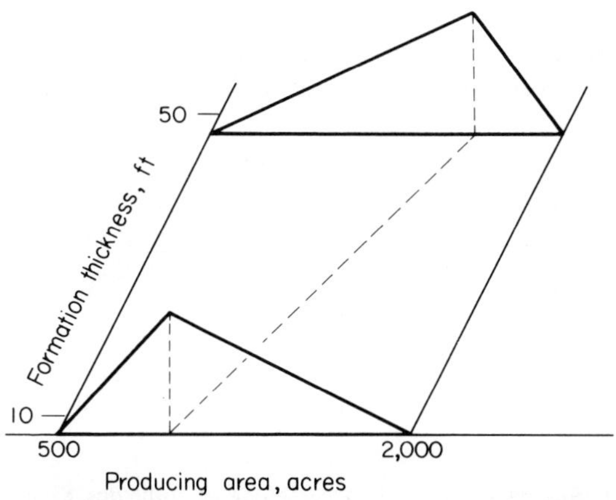

Figure 6.12. Dependent probability distribution. The figure indicates that the most likely area and the formation thickness are interrelated.

3. Decision rules can be built into the program:

a. Where the nature of previous chance outcomes would provide a clear choice, control within the program would be directed to follow one specific course toward the remainder of the trial.

b. In unclear areas, where the choice would not be clear, control within the program would be directed to split and follow two (or more) paths. Each of these paths would be weighted according to the most likely fractional degree of preference which would be given it at that future point in time when the decision would have to be made.

Surprisingly, it will be seen that the diagramming of the decision tree becomes simpler, an unexpected benefit. Consider first Fig. 6.13, which shows part of a conventional approach. Here chance node *A* is shown with three

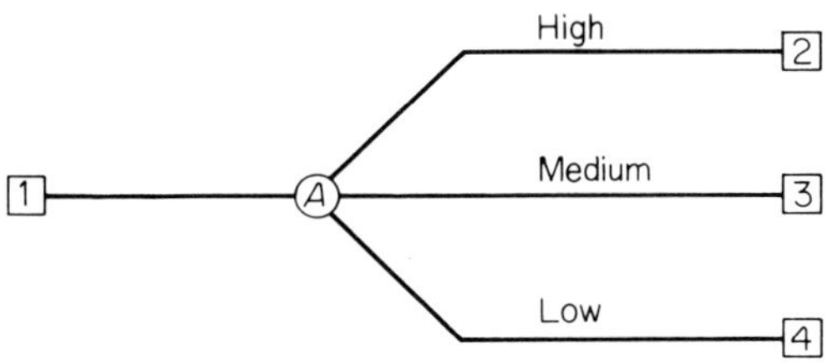

Figure 6.13. Conventional decision tree using high, medium, and low estimates.

representative possible outcomes, consisting of high, medium, and low cases. These can be quite satisfactory at node *A*, but the decision nodes 2, 3, and 4 each have several branches, which lead to subsequent probability nodes, each of these having several branches, so the final tree ends up with very many branches. It often turns out that many of the terminal event branches are really the same final outcomes. Some simpler means of presentation should then be possible.

Consider next Fig. 6.14, where chance node *A* is designated by the sector of a circle and has only one path leading from it to a decision node. This symbol, the sector of a circle, is used here to designate that random-number

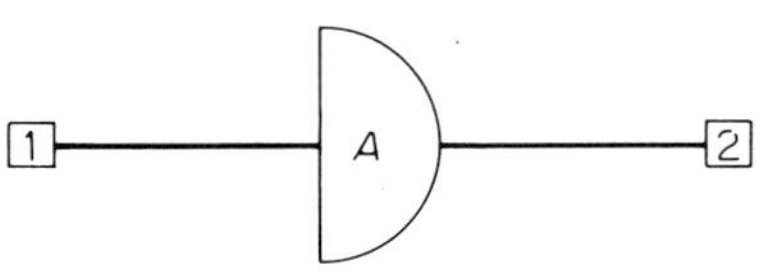

Figure 6.14. Stochastic decision tree. Node *A* indicates a probabilistic choice.

choices of one or more probabilistic events are made at that point in the network. This is, of course, an integral part of the Monte Carlo process. The appropriate type of distribution(s) should be used, such as uniform, triangular, normal, log-normal, or a discrete distribution if that type is preferred. The symbol designates only that the necessary simulation is done before proceeding to the next step in the decision tree.

Decision nodes commonly follow chance nodes. For example, in an oil-exploration venture, if the results of a test well were rather poor, we would prefer to farm out the entire venture, and the income from that simulation trial would be based on the revenue to a $\frac{1}{16}$ overriding royalty or other possible way out. On the other hand, if the test results were good, we would prefer to drill and develop the entire venture ourselves, and the income from that simulation trial would be based on the net revenue of the working or operating interest. The exact point, in terms of the outcomes of the test well, where the paths will divide, must be predetermined and ordinarily entered as data input into the computer program. As explained previously, if a gray area existed where it is not clear which way a decision would go, both subsequent paths could be followed and each weighted by, say, 0.5 or other considered preference weighting fraction. The decision node indicates that the necessary decision rules must be established and built into the program at that point in the decision-tree network.

As far as drawing the decision-tree network is concerned, the decision nodes can be omitted. The rule, then, is that whenever more than one path emanates from a chance node, a decision node and program mechanism is implied as following the random selection process associated with the chance node. However, these two processes are basically different, and the use of the decision node will be retained here to emphasize that point.

At the completion of a pass, or trial, through the decision tree, the final outcome is usually expressed in dollars at their present worth. Each individual trial will fall into a particular class, since its value will be greater than the lower class limit and smaller than the upper class limit. The result, then, is expressed by adding one to the number of outcome cases that fall into the particular class. An exception to this would be where a trial pass had been split at a decision node. If the pass had been split into two equal halves, 0.5 would be added to the number of cases in each of the two classes that contain the respective trial values. Values other than 0.5 could be used as readily. At the conclusion of N total trials, the number of cases tallied into each class is divided by N to form the relative frequency distribution.

6.5-2 A Petroleum Development Example

In this section a venture is described which involves exploratory drilling. In the following sections the manner of solution will be discussed, first by the conventional decision-tree approach and later by the stochastic decision-tree approach.

The exploration staff of a company has located a prospective producing area where leases can be obtained on 20 locations of 40 acres each. The leases will cost \$50 per acre. A decision must be made whether to lease all 20 locations, or to "checkerboard" leases on 10 locations, or to pass up the opportunity.

One well would be sufficient for testing the area, for obtaining information on formation thickness, porosity, oil saturation, and other characteristics. Dry hole cost is about \$60,000 and completion costs amount to \$20,000. Present estimates are that the net pay would average between 10 and 30 ft, the porosity somewhere between 12 and 20%, and the connate water somewhere between 25 and 40%.

Depending upon the outcome of the exploratory well, a later decision must be made whether to continue the development of the property or to farm it out for $\frac{1}{16}$ of $\frac{7}{8}$ overriding royalty. If development is continued by the company, it is quite possible that some dry holes will be encountered later, and some allowance should be made for this contingency.

In the event that the governing factors were very favorable, secondary recovery operations would be conducted in the future which would further increase the recovery. Primary production can be assumed to have a present value of \$1 per barrel. Additional secondary recovery oil can be assumed, for purposes of the present decisions, to have a present value of \$0.50 per barrel.

In order to make the present decision it is necessary to produce two probability diagrams. One diagram is required for the case of leasing all 20 locations. Another is required for the case of checkerboarding and leasing only 10 locations interspersed in the area. The final solution involves the calculation of the means and variances and the financial base required for each case.

6.5-3 A Tree Using Low, Medium, and High Estimates

Just about the simplest, but still fairly satisfactory way of drawing a conventional decision tree is to use representative values for low, medium, and high possible outcomes from each chance event node. This helps to arrive at a better estimate of the expected value and it also permits an estimate of the variance involved in the possible outcomes. The beginnings of such a tree, for the petroleum development example, are shown in Fig. 6.15. Even though more than 20 terminal events are indicated, the tree is incomplete as shown. For example, there are two decisions to farm out, after drilling the test well, which are not explicitly provided with the probabilistic low, medium, and high estimates of the value of subsequent developments.

Such a tree as in Fig. 6.15, although it is cumbersome and in some places redundant, can of course be evaluated. Particularly, the expected value is not extremely difficult to determine. The rollback method can be used, and as one progresses backward, all but the path with the highest expected value can be loped off at each decision node, until only the one preferred path remains and is clearly evident. Or so it seems. But by neglecting the variance, we may also have omitted common sense from our answer.

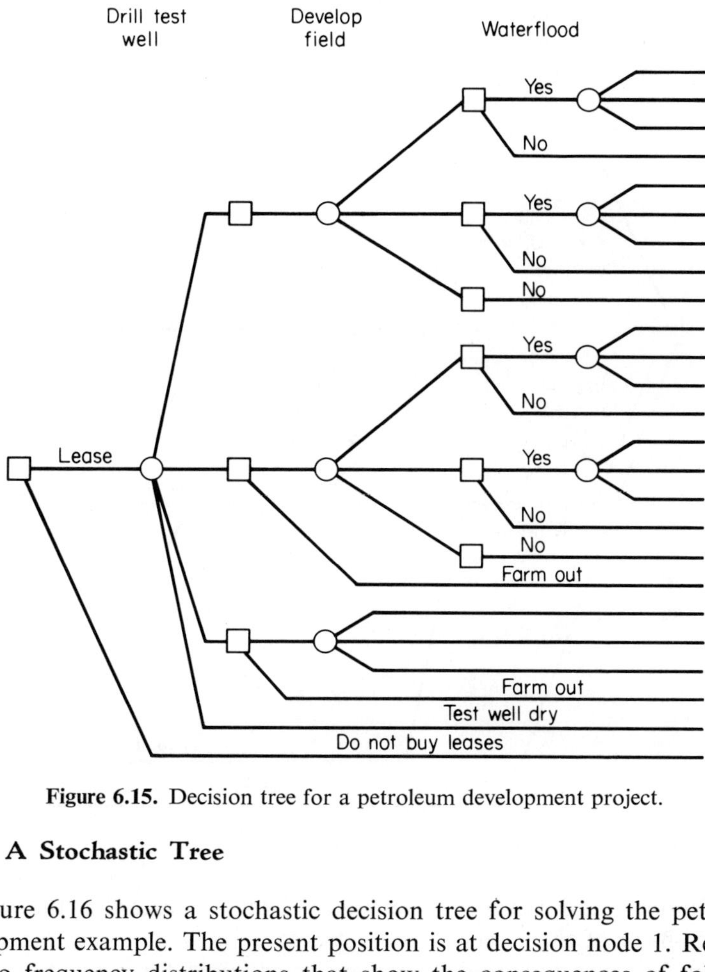

Figure 6.15. Decision tree for a petroleum development project.

6.5-4 A Stochastic Tree

Figure 6.16 shows a stochastic decision tree for solving the petroleum development example. The present position is at decision node 1. Required are two frequency distributions that show the consequences of following paths, one of which consists of leasing 20 locations and the other of leasing 10 locations. When these are obtained, the present decision can be made.

A trial starts by using random numbers in the one or more distributions at chance node *A*, where the outcome of the test well is simulated. Control then goes to decision node 2, where the built-in decision rules convert the outcome of *A* into the decision to develop or farm out the leases and retain, say, $\frac{1}{16}$ overriding royalty; or, if the test well is very poor, to drop the leases. In the latter event, the money spent for the leases and for drilling the test well is simply lost.

Chance nodes *B* and *C* have been shown separately here, not so much to show that two successive random numbers are used but to indicate that two

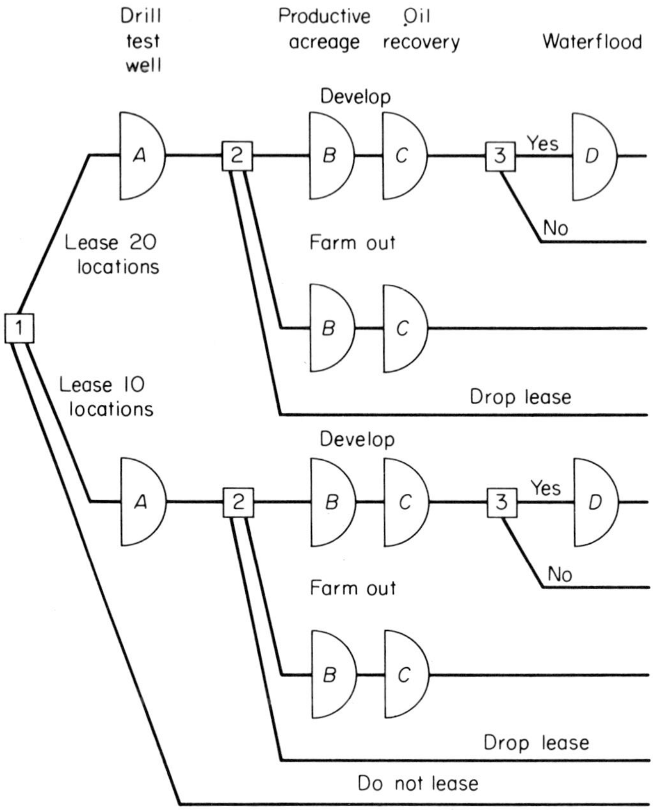

Figure 6.16. Stochastic decision tree for a petroleum development project.

types of information are needed. Geologic information regarding the probable size and shape of the reservoir rock structure, as well as the thickness of the formation and the producing zone, is required for estimation of the probable size of the reservoir. This in turn will influence the number of good wells and the number of dry wells which will probably be drilled. These can be described much better after information is available on the outcome of the test well. This means that the distributions used in chance node *B* will, for the most part, be dependent distributions, whose limits are adjusted according to the values obtained in node *A*.

Chance node *C* is used here to designate the simulation which has to do with the oil recovered per acre-foot of reservoir rock. Again, the limits of the distribution used here would be adjusted according to the outcome obtained at node *A*. So, again the distribution would be dependent, for the most part.

Control within the simulated trial passes to decision node 3, where built-in rules convert previous results into the decision to water-flood or not to water-flood. If the decision is to water-flood, then continued simulation is used to randomly distribute the probable results.

6.5-5 A Streamlined Stochastic Tree

In Fig. 6.16, chance nodes *B* and *C* are reproduced four times. They are exactly the same nodes, containing the same probability distributions, because they deal with the same subject and only the best available information would be built into them. It appears, therefore, that the diagramming can be made simpler by showing more than one path between decision node 1 and chance node *A*. This has been done in Fig. 6.17. The diagram becomes greatly simplified.

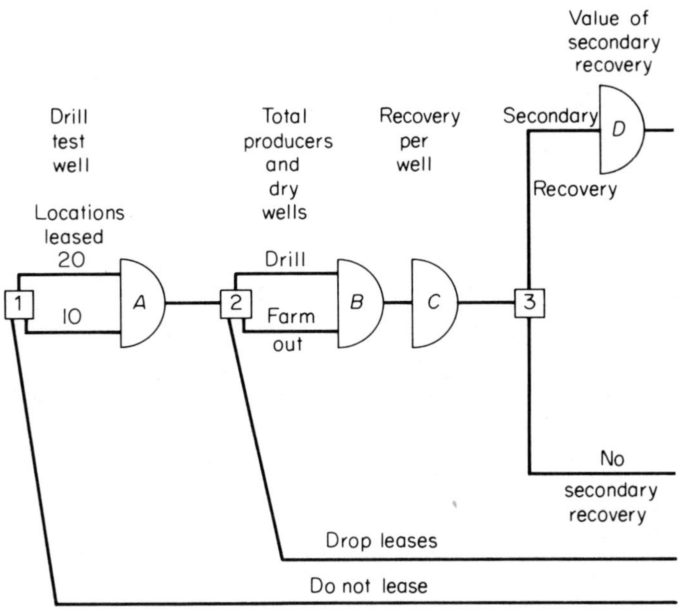

Figure 6.17. Streamlined stochastic decision tree.

A rule of network presentation has been broken in Fig. 6.17, as constructed. The rule is that there should be only one path going directly from one node to another. The usual way of complying with this is to use dummy segments or "activities." That is, present (1, *A*) "Lease 20 locations" could be broken into (1, 14) "Lease 20 locations," followed by (14, *A*) blank. The present parallel line (1, *A*) "Lease 10 locations" could be broken into (1, 15) "Lease 10 locations," followed by (15, *A*) blank. Thus all ambiguity is removed.

For purposes of the present discussion, it will be kept in mind that the path labeled "Lease 20 locations" implies one initial investment and the path labeled "Lease 10 locations" implies a smaller initial investment. Similarly, the line labeled "Drill" implies being the operator, paying drilling and operating costs, and receiving the revenue due to the working interest. The line labeled "Farm Out" implies signing over the major portion of the interest to some other potential operator and receiving a small fraction, such as 1/16, which would, however, be free of drilling and operating expenses.

6.6 FORECASTING AND PLANNING

6.6-1 Planning and Selection

More than technique is needed to provide guidance. Vision, philosophy, luck, and good or bad fortune all play their parts. To a considerable extent these depend on the relations among present positions, capabilities, and future developments and events. Before leaving the discussion of economic decisions and models, it seems appropriate to consider briefly two other facets, forecasting and planning.

Both planning and forecasting are done either purposely or by default; there is no escaping the consequence. The degree of reliance placed on trend analysis needs to be monitored. Certain strategic factors may change and result in changing the complexion of an industry and requiring new kinds of capabilities. Sometimes decisions are needed which relate to anticipated new developments, which will outmode present technology and therefore make obsolete any actions designed on present knowledge.

Changes that affect values are inevitable. Forecasting techniques are often used to narrow the ranges and better define the probabilities of future trends and events. Some of these changes occur within an industry but some occur outside the industry. Substitutions, changing demands, changing emphasis, and changing regulations are to be expected. These are the reasons top management is often most concerned about the things a company is not doing.

Worthwhile risks should be sought—those with the desirable payoffs. Risks cannot be avoided, but many are not worth taking. Risks may be classified as to consequences. For one type, the question to consider is: Can we afford to take it? This is generally answered in terms of the present rules of the game. For another type of risk, the question to consider is: Can we afford not to take it? The rules of the game may be changed and we should be ready. There may be major changes in supply or demand, outmoded processes or products, and diminishing resources. As one era ends, a new era begins.

6.6-2 The Uncertainty Spectrum

Any provision for the future requires estimates of events and conditions that will determine future needs and values. These are not known, certainly not precisely known. Different degrees of uncertainty have been described.

The following distinctions are due to Knight.[6] Risk is characterized by situations where the individual outcome is not known with certainty but where the probabilities of alternative outcomes are known. Uncertainty refers to situations where the probabilities of outcomes cannot be predicted in probabilistic terms. Most current writers feel that, in regard to real-life or real-world situations, the distinction is one of degree and not of kind. Most real-world situations fall between the two extremes. There is serious doubt that the distinction applies in the organizational context. Decision makers must act from presumed probabilities of future events, which range from near certainty to ill-defined feelings that approach mysticism. Rapport uses the words *uncertainty* and *risk* interchangeably.[7]

6.6-3 Forecasting Techniques

Following are some of the commonly used types of predictions.[8]

Persistence predictions depend upon a phenomenon that continues over a long period of time. Many geologic processes would fit into this category. For short-term predictions, the preferred modes of transportation could be treated as persistent.

Trajectory predictions follow a trend. Curve fitting, in which the constants in a mathematical equation are adjusted so that one physical quantity is apparently related to another, is an example of this. Relations between vapor pressure and temperature for a mixture of hydrocarbons is an example.

Cyclic predictions are well founded. A good example is the fall and winter seasonal demand for natural gas and fuel oil for heating purposes.

Associative predictions are another powerful means of forecasting. Exploration geologists make estimates of oil recovery in new areas based upon the number of cubic miles of sediments present. Estimates of total future energy demand have been based upon population estimates made by the Census Bureau, combined with the trend of changes in per capita consumption of energy.

Analogue predictions utilize the observed correspondence between sets of events. If *A* is present and *B* occurs, *C* occurs also. Mineral prospectors learn that mineral *A* is found where mineral *B* or *C* occurs.

Hindsight predictions are recognized as the most accurate of all the methods. They are said to be unavailable when needed the most. However, a review of past estimates of an individual may reveal a bias, a tendency to be too high or too low. Where consistent bias is evident, it can be used to adjust current estimates.

6.6-4 Long-Range Planning

Planning is associated with decision making, not with forecasting. Long-range planning deals with relations between present and future decisions. Ideally it should include a review of past decisions, policy decisions, to see how they have fared in the light of real-world developments. This is akin to the feedback and control loop used in industrial production. Unfortunately, judgment sometimes goes more against those who have made wrong decisions than against those who have contributed nothing. Any planning done must be in the light of available material and human resources. Many times it has been pointed out that lower and middle management is concerned chiefly with current problems related to production. On the other hand, higher management becomes more concerned with the personnel who constitute the organization and concerned with those things which the company is not doing.

The complexity of organizations requires common understanding and vision for effective performance. In many cases, the organization of present knowledge will lead toward enlightened decisions. A changing technology and economy require the assimilation of new knowledge. It does not do to withdraw from reality and use only preconceived ideas. On the other hand, one cannot make the presumption of an irrational world wherein one decision is as good as another. Decisions that affect the future of a company are made at all levels. Such decisions are commonly made by groups or teams which include experts and those with special knowledge. The cumulative effect of such decisions provides for the continued operating success of an organization.

The time span associated with long-range planning varies from one industry to another. "Short range" refers usually to something less than the gestation period, which in farming and forestry would be the time required to grow a crop. In manufacturing it would be the time required to put a new product on the market. Time spans up to 20 years are not uncommon in planning the development of an effective organization. In laying out decision trees, it has been suggested that the ensuing events be traced only to the point where the effect of the initial decision becomes insignificant.[3] In this context the present tree can be visualized as part of a continuing pattern of interwoven events.

All planning decisions should have objectives, desired results. They are generally based on certain assumptions. Both the objectives and assumptions need periodic review, for changes are inevitable. As companies grow in size, the moral stature and obligations become much more important. Decisions should also be reviewed from the stand point of feasibility, whether or not they could be made effective, also whether or not they would be premature. If a decision causes an impact of emphasis in one area, quite possibly it would cause neglect in other areas. Alternative courses of action can be outlined and evaluated. The Bayes methods are useful for this purpose. The general concept of position value is useful, and sometimes a slight advantage in economy is sacrificed in order to retain flexibility in operating capabilities. The relation to the entire decision structure needs to be considered, for sometimes a present decision will obligate future commitments to a continuing program.

REFERENCES

1. DAVIS, L. F., "Economic Judgment and Planning in North America Petroleum Exploration." *Journal of Petroleum Technology*, May 1968, p. 467.
2. HESPOS, R. F., and P. A. STRASSMAN, "Stochastic Decision Trees for the Analysis of Investment Decisions." *Management Science*, Vol. 11, Aug. 1965, p. 244.
3. MAGEE, J. F., "Decision Trees for Decision Making." *Harvard Business Review*, July–Aug. 1964.
4. TOWNES, H. L., "Using Economics in Exploration Decisions." Presented at Oklahoma City Section of SPE of AIME, Dec. 5, 1967.
5. GULLATT, E. M., "Petroleum Economics and Evaluation—the State of the Art." Presented at Oklahoma City Section of SPE of AIME, Dec. 12, 1967. Examples are reconstituted from those presented in the lecture.
6. KNIGHT, F. H., *Risk, Uncertainty and Profit*. Boston: Houghton Mifflin Company, 1921.
7. RAPPORT, ALFRED, *Information for Decision Making*. Englewood Cliffs, N.J.: Prentice-Hall, Inc., 1970, p. 174.
8. BROSS, I. D. F., *Design for Decision*. New York: Macmillan Publishing Co., Inc., 1953.

PROBLEMS

6-1. Calculate the mean, variance and standard deviation of each branch that leaves the decision node. Branches that have a lower expected value and greater variance are often trimmed from decision trees. On this basis, can the tree be simplified?

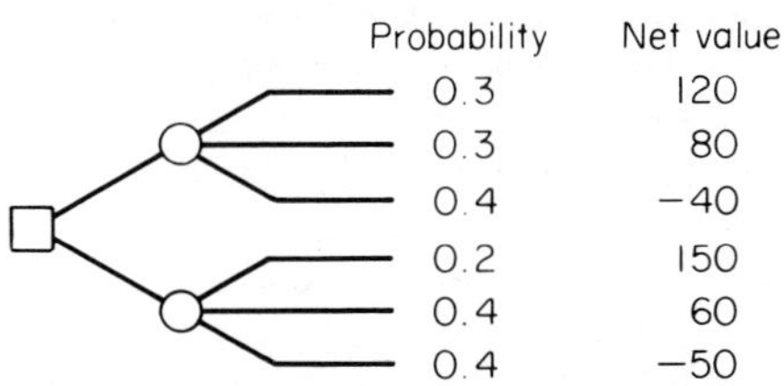

6-2. Calculate the mean, variance, and standard deviation of each branch that leaves the decision node. Can any branch be eliminated?

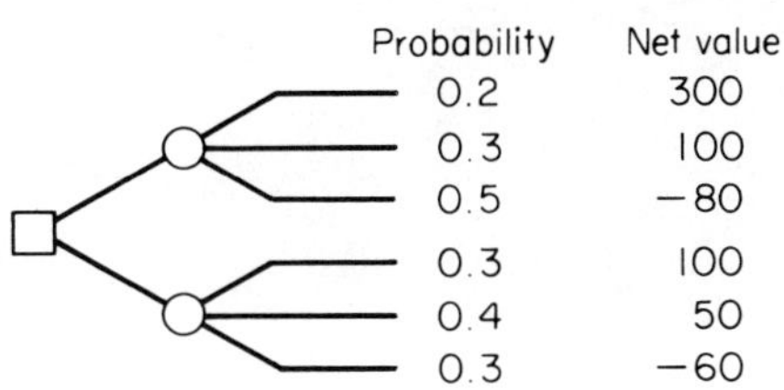

6-3. Operator W is contemplating drilling well A on his lease. A dry hole will cost \$90,000, and a completed well will cost \$150,000. Another operator has an adjacent lease on which he may drill well B. This is contingent upon operator W contributing \$15,000 toward the expenses of well B in the event that it turns out to be dry. However, if well B turns out to be good, the prospects for well A would be much brighter, as indicated on the diagram. According to expected values, should operator W proceed immediately to drill well A, or should he encourage the other operator to drill well B by guaranteeing \$15,000 dry hole money?

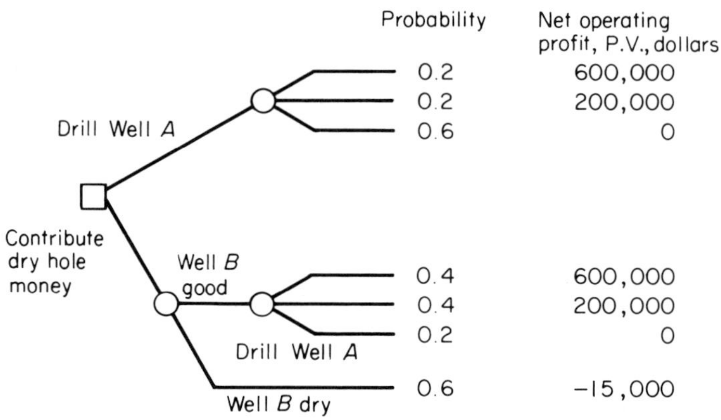

6-4. Calculate the minimum amount of total venture capital required for each course of action contemplated in Problem 6-3. Base the answers on 5% risk. Use the methods of Chapter 5.

6-5. An independent operator has paid \$6,000 for a lease on an 80-acre drilling unit. The landowner retained 1/8 royalty on all oil and gas produced. Drilling and completing a well will cost an additional \$130,000. After state severance taxes, it is estimated that \$3.95 per barrel will remain to be divided among all interests. Operating costs will amount to \$2,400 per year for an estimated 14-year producing life. A present-value factor of 0.68, based on 10% annual interest, may be applied to both future income from production and operating expenses.

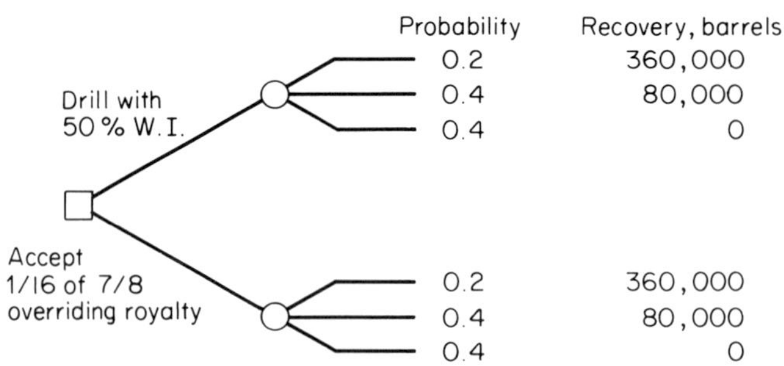

(a) A second operator has offered to go in as a partner and pay the first $10,000 drilling costs, with the remaining costs to be equally divided. Each would then have a 50% working interest, as shown on the diagram. Probability is 0.2 of 360,000 bbl recovery, probability is 0.4 of 80,000 bbl recovery, and probability is 0.4 of zero recovery. The first operator has a total of $400,000 venture capital. If he accepted this offer, what would be his expected present-value profit and what per cent risk would he be assuming on the venture? *Ans.:* $51,500 present-value profit, 20% risk.

(b) A larger operator has offered to drill and produce the well with the first operator retaining 1/16 of 7/8 overriding royalty. What would be the present-value profit under this arrangement, and what per cent risk would he be assuming? *Ans.:* $16,500 present-value profit, approximately 0% risk.

(c) At 5% risk, what is the maximum percentage of a 7/8 total working interest that the first operator should own? What would be the corresponding present-value profit? *Ans.:* 10% working interest, $9,500 present-value profit.

6-6. The decision tree concerns the purchase of a lease and the drilling of a test well. The lease may be purchased and drilling proceed immediately. Or, after the lease is purchased, a geophysical survey may be made to better delineate the prospects of discovering oil, before drilling the expensive test well. The data are as follows:

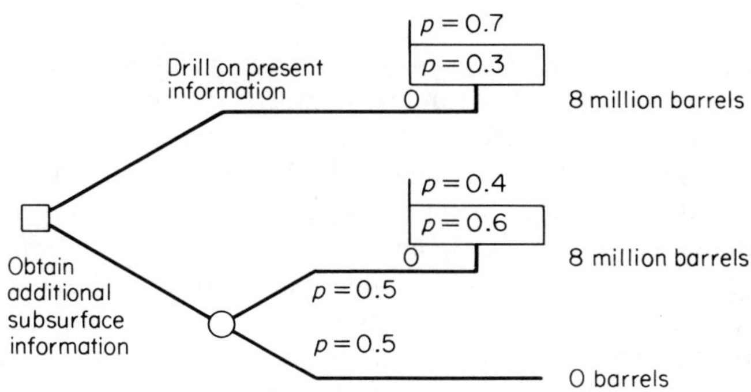

Maximum ultimate discovery size	8 MM bbl
Present value of operating profit, per barrel	$1.40
Cost of additional wells, if successful	$1.35 MM
Cost of test well	$500,000
Cost of lease	$150,000
Cost of additional subsurface information	$200,000

Calculate the expected value of each branch that leaves the decision node. Also calculate the per cent risk associated with each branch, based on $10 MM venture capital available to the management group. *Ans.:* Drill on present information, expected value = $827,500, with 11.9% risk. Obtain additional information, expected value = $877,500, with 3.7% risk.

7

SIMULATION—THE MONTE CARLO METHOD

7.1 INTRODUCTION

Simulation is a powerful tool for obtaining solutions by numerical methods. Because of the use of random numbers, it is often called the Monte Carlo method. *Simulation* refers to the group of numerical methods characterized by the process of making repeated solutions. Values entering into the calculations are repeatedly selected by random numbers taken from an appropriate range of values. From several hundred to several thousand trials are generally used to obtain suitable results, which makes the technique less suitable to hand calculations than to the electronic computer. Essentially, it is a method in which the answers can be arranged in a form that gives the fractions of the total results which fall within certain ranges. Thus it is readily adapted to solving probability problems and constructing probability distributions.

Some of the types of solutions that have been obtained by simulation are given in the following list:

1. Estimates such as probability distributions of rates of return that might be obtained in a business venture.
2. Probability distributions of the possible "state of nature," such as the probable ultimate recovery from an oil reservoir, which reflect the variations permitted or undefined by present incomplete information and knowledge about a subject.

3. Determination of the most profitable number of parallel facilities, such as ship berths or toll booths, where the arrivals and required service per unit time are not uniform but are probabilistically grouped.
4. Determination of probability diagrams formed by combining quantities whose variations can be expressed algebraically with those whose variations can be expressed only empirically or graphically.
5. Illustrations of the validity of theorems, such as the central limit theorem, in the study of probability. Also, determinations of approximate values of π and other quantities which can be defined geometrically.
6. Sensitivity analysis. The sensitivity of the results to changes in any one input factor can be determined by changing the range or probability distribution of that factor. The changes in the final results can then be observed.

The foregoing list is far from complete. The method is very flexible and the extent of its uses depends largely upon the ingenuity of the person designing the solution. The process is also widely used in the social sciences and in solving problems in economic theory.

Several reasons may be suggested for using simulation. A most striking reason is that it offers a means of combining the knowledge of many individuals into a single expression. It is adapted to the team concept, the project-oriented group. As an example, the sales manager can give his estimates of the market, the plant superintendent his estimates of manufacturing costs. Combined with other estimates, these can be transformed by simulation into a probability distribution of rates of return from a prospective undertaking.

Another type of reason for using simulation concerns the organization of knowledge. Information is often easiest and best presented in segments. Indeed, there is a continuous effort to gather information into classifications where it can be stored and relocated. Experts are usually honored the most within their own fields. The individual can work best on one thing at a time, record his findings, and pass on to the next phase. In all such endeavors where a diversity of quantities must be considered in their joint effect, some method of combining them is necessary. In some limited cases, this combining can be done within commonly achieved mental processes. However, in most complex cases the combining should be done mechanically when numerical interpretation is possible. It is faster and easier in the long run. Also, it introduces no personal bias during the combining process.

7.2 CASES SOLVED BY SIMULATION

Some problems and solutions that have been published are given here as illustrations. Both the procedural methods and the results obtained give an indication of what can be done toward solving other problems that may arise.

Example 1: *Calculating Reservoir Porosity (φ) and Water Saturation (S_w) from Well Logs*[1]

The logs available were:

a. The compensated density log, which gave values of ρ_B, the bulk density, of 2.36 to 2.38 g/cm³.

b. The induction log 6FF40, from which it was estimated that the true resistivity R_t of the formation varied from 19 to 21 Ω-m.

c. The gamma-ray log, which was used only to interpret the lithology.

Other variables that had to be estimated were:

R_w, connate water resistivity,
n, exponent in the Archie equation,
m, exponent in the Archie equation,
a, coefficient in the Archie equation,
ρ_{MA}, rock mineral density,
ρ_F, reservoir fluid density.

Uniform distributions were assumed for all the foregoing quantities, except parameters a and ρ_F, which were assumed to be constants. The limits are shown in Table 7.1.

Table 7.1
Ranges of Parameters Used in a Log-Interpretation Example

Parameter	*Lower Limit*	*Upper Limit*
R_t	19.000	21.000
R_w	0.055	0.075
n	1.800	2.200
a	0.620	0.620
m	2.000	2.300
ρ_B	2.360	2.380
ρ_{MA}	2.580	2.630
ρ_F	0.900	0.900

The calculations proceeded as follows. When a particular quantity was used in the calculations, a random number was used to select a value for that quantity from within the prescribed range. Random numbers customarily used vary between 0 and 1. Thus, if a value for R_t were needed and the random number turned out to be 0.5, a value of 20.0 would be used for R_t in that particular calculation. And, of course, other random numbers, one for each quantity involved, would be used to select values for the other quantities. The calculation steps were as follows:

1. Determine φ from the relation

$$\rho_B = \rho_F \varphi + \rho_{MA}(1 - \varphi).$$

2. Use the value of φ to determine F from the relation

$$F = a\varphi^{-m}.$$

3. Use the value of F, and randomly chosen values of the other quantities, to determine S_w from the relation

$$R_t = FR_w S_w^{-n}.$$

The results of the simulation are shown in Figs. 7.1, 7.2, and 7.3, which reflect the results of several hundred cases, wherein each case was processed

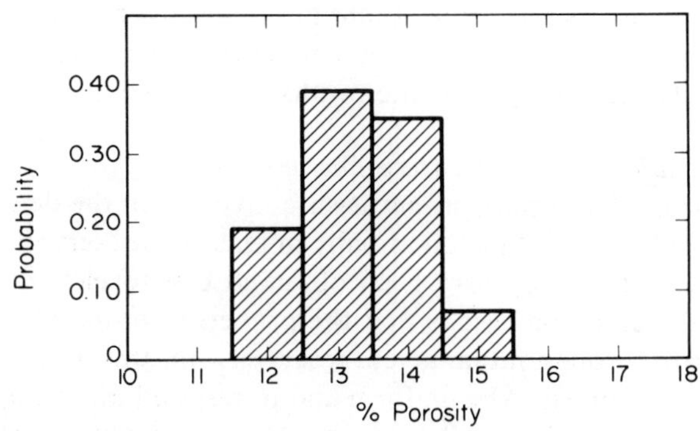

Figure 7.1. Calculated porosity (After Walstrom et al.[1])

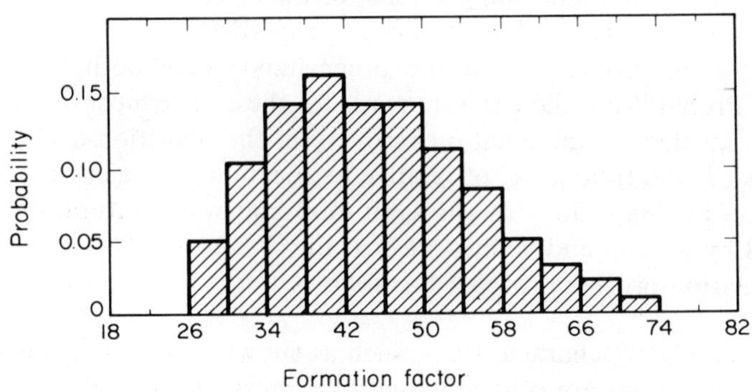

Figure 7.2. Formation factor. (After Walstrom et al.[1])

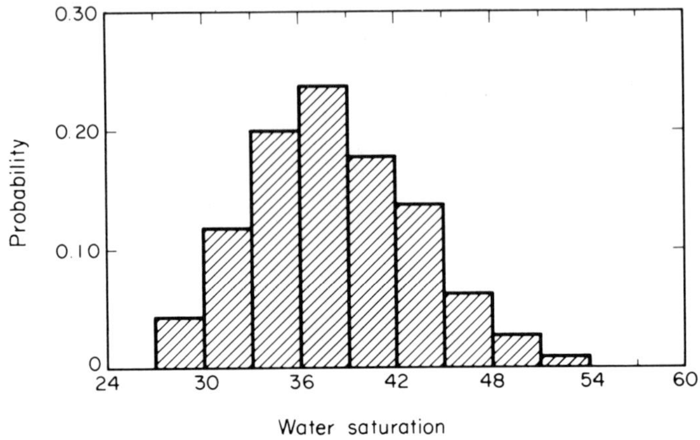

Figure 7.3. Calculated water saturation. (After Walstrom et al.[1])

through steps 1, 2, and 3. Here it may be noted that although uniform distributions were assumed for quantities entering the calculations, the resulting probability densities are not symmetrical. The functional relationships of the quantities may skew the results.

Example 2: *Calculating a Recovery Factor*[1]
Both reservoir fluid and rock properties are involved in the determination of the recoverable fraction of the oil initially in place in a reservoir. Although in some instances differences have been observed due to elevation or lack of equilibrium, the common case is that the properties of liquid petroleum are rather uniform throughout a given reservoir. In this example the fluid properties were assumed to be uniform and to respond uniformly throughout the reservoir, as shown in Fig. 7.4, as reductions in pressure accompanied the withdrawal of hydrocarbons. In any individual trial calculation within the simulation process, establishing a value of the reservoir pressure established simultaneously all the fluid hydrocarbon properties.
Reservoir rock properties, on the other hand, vary both vertically and laterally. Probabilistic choices were made for these. The most important rock property, insofar as fractional oil recovery in the depletion-drive process is concerned, is the ratio k_g/k_o of relative permeability to gas to that of oil. A correlation of this ratio with gas saturation and water saturation has been proposed by Knopp and is illustrated in Fig. 7.5.[2]
The simulation process proceeded as follows:

1. A set of *PVT* characteristics, such as shown in Fig. 7.4, was accepted. Empirical equations are generally used to express such correlations and data throughout a region of interest. By entering such empirical equations into the

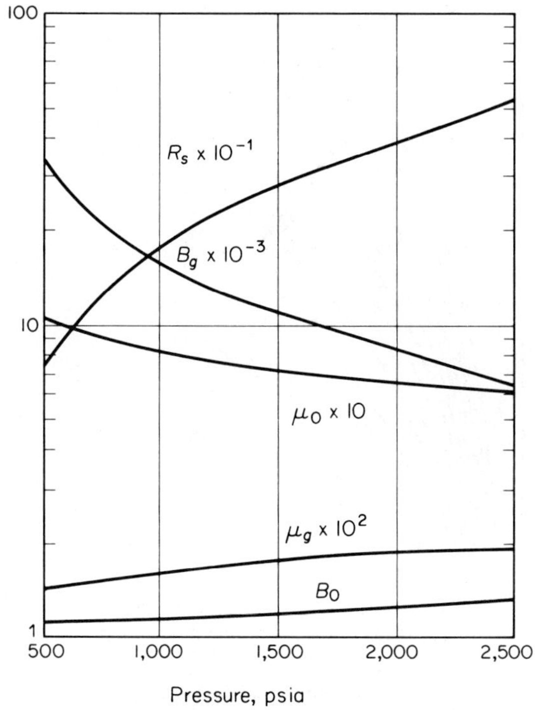

Figure 7.4. *PVT* data. (After Walstrom et al.[1])

computer program, the value of any quantity such as the oil formation volume factor B_o could be calculated for any assumed value of reservoir pressure. Use of empirical relations in such cases serves only to expedite the calculations. It is a fairly efficient means of getting necessary data into the computer.

2. A random selection from a predetermined distribution was made of a particular water saturation.

3. The water saturation determined the particular k_g/k_o curve, as illustrated in Fig. 7.5. Again, there occurs the use of an empirical equation to express a relationship between the ratio k_g/k_o and S_o, the oil saturation.

4. Since only a depletion-type recovery mechanism was assumed to be operating in the reservoir, the following equation from Muskat was solved to obtain the recovery[3]:

$$\frac{dS_o}{dP} = \frac{\dfrac{S_o B_g}{B_o}\dfrac{dR_s}{dP} - \dfrac{1 - S_o - S_w}{B_g}\dfrac{dB_g}{dP} + \dfrac{S_o}{B_o}\dfrac{\mu_o}{\mu_g}\dfrac{k_g}{k_o}\dfrac{dB_0}{dP}}{1 + \dfrac{\mu_o}{\mu_g}\dfrac{k_g}{k_o}}.$$

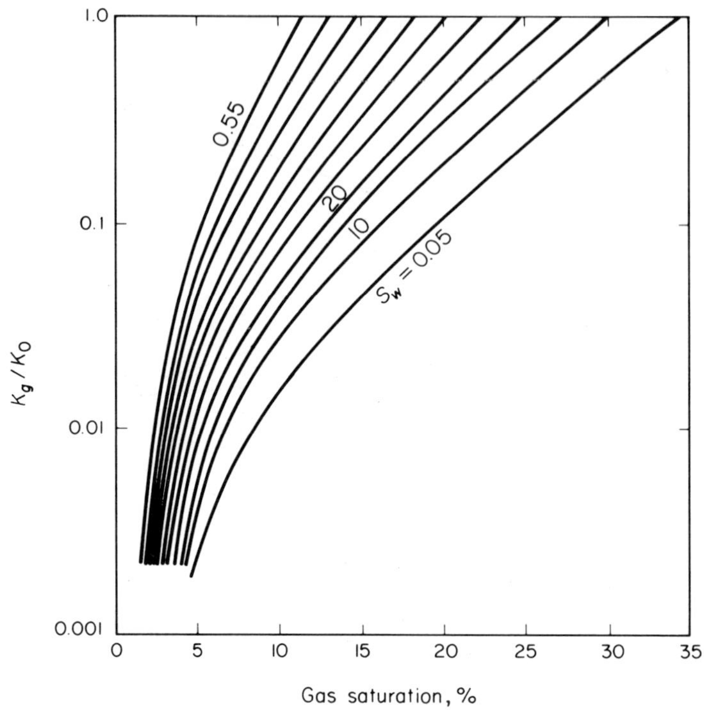

Figure 7.5. Relative permeability (After Knopp[2])

The method used in such cases is to assume a number of small finite pressure decreases and to calculate the change in S_o associated with each stepwise pressure reduction.

5. The calculations of steps 1 through 4 were repeated until a statistically significant answer was obtained. The results from 500 cases are shown in Fig. 7.6.

Example 3: *Petroleum Recovery from a Reservoir*

The amount of oil that will be recovered from a particular petroleum reservoir can be estimated from information that becomes available after the discovery well and several development wells have been drilled.[1] Information that becomes available at such time comes from electric and nuclear logs run inside the drilled wells and from analysis of fluid samples from the reservoir. It then becomes possible through interpretation of the logs to give estimates of such quantities as fractional porosity of the rock, fraction of the pore space filled with water, thickness of the oil zones, and the areal extent of the reservoir. In general, a single value is not obtained for a particular quantity, but rather a range of values is determined during the analysis. The relative

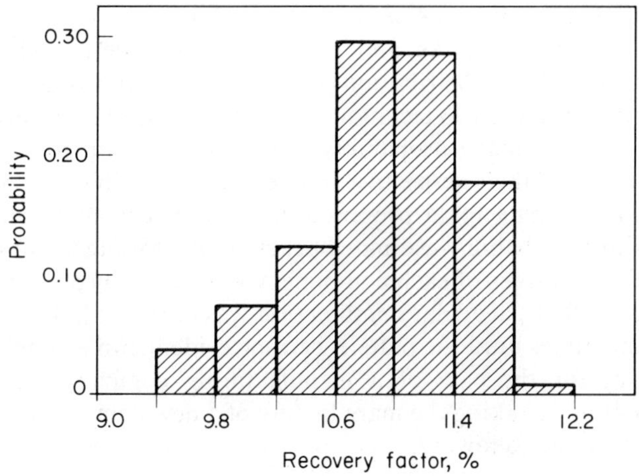

Figure 7.6. Calculated recovery factor, percentage of oil in place, at 500 psia. (After Walstrom et al.[1])

frequencies with which like values occur permit an estimate of a probability distribution of the various quantities.

Total ultimate recovery of oil from a reservoir depends upon the values of such physical quantities. For the most part, they are independent of each other, but this is not always the case. Combining different values of the quantities in random fashion by the Monte Carlo technique produces a probability distribution of the total ultimate recovery of oil. In turn, probability distributions of this nature can be used as the basis for economic decisions regarding whether or not to install facilities such as pipelines and processing plants. The use of the Monte Carlo technique in this area of decision making promises to increase very rapidly.

Recoverable oil in a reservoir was estimated by use of the following equation:

$$N_p = \frac{7{,}758 \cdot \varphi \cdot Ah(1 - S_w)E_R}{B_o}.$$

Symmetrical triangular distributions were assumed for all quantities except the recovery factor E_R. In the case of E_R, the triangular distribution was used, but the most likely value of the recovery factor was specified near the lower end of its range. Experience in similar reservoirs indicated that usually some recovery was by depletion drive rather than by complete high-pressure water influx. Reservoir area A and thickness h were estimated from structural interpretations based on seismic data and well logs. Reservoir rock porosity φ and connate water saturation S_w were estimated from log interpretation and core analysis. Oil formation volume factor B_o was estimated from general data and recombined samples.

Example 4: *Rate of Return from a Business Venture*
A method of analyzing risks and profits has been described by Hertz.[4] Estimates of probability distributions of component parts were combined by simulation, the Monte Carlo technique. The several experts within a company were called upon to make probability estimates of the range of values that might occur within their respective spheres of responsibility. For example, the marketing estimate for a proposed new product would include several demands, with a probability estimate of each. In the Monte Carlo simulation, randomly chosen values of each component were combined, and the profitability result of that particular combination was computed. The process was repeated many times until the pattern became evident. An example described by Hertz was for the case of a manufacturing company evaluating the profitability of undertaking the manufacture of a new item. Aspects that were considered were the following:

- A. Market analysis
  1. Market size
  2. Selling prices
  3. Market growth rate
  4. Share of the total market
- B. Investment cost analysis
  5. Investment required
  6. Salvage values
- C. Operating and fixed costs
  7. Operating costs
  8. Fixed costs
  9. Useful life of facilities

The results can be portrayed in a number of ways, such as undiscounted value, present value, or internal rate of return. The later was used by Hertz. From plots of the probability distribution of the rates of return that might be obtained from different prospective ventures, the management would have a basis for comparison and choice.

7.3 COMPARISON WITH OTHER METHODS

A variety of methods has been developed to express the anticipated outcome or value of an economic venture. Some of these methods attempt to cope with uncertainty, but most seem to be inadequate for providing a complete description.

1. *Expected values.* This is a deterministic method. It precludes making allowances for variations, which are necessary to compensate for the unfavorable cases. This method conceals much information. In Fig. 7.7 there are

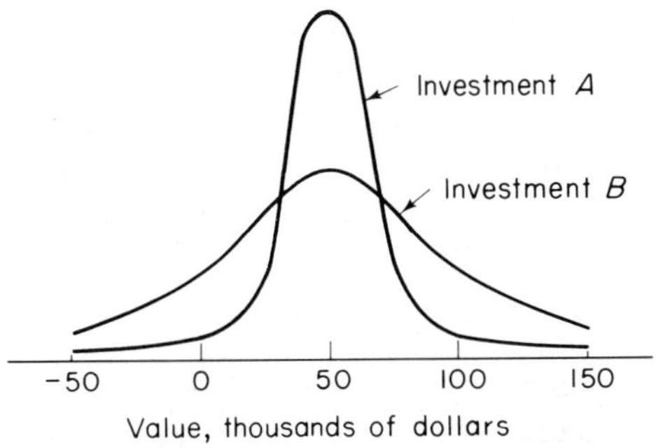

Figure 7.7. Comparison of probabilistic outcomes of net present worth of two investments.

shown two investments whose outcomes have the same expected value, \$50,000. In the case of investment *A* about 90% of the time the outcome will fall within a narrow range close to the expected value. The chance of losing money would be small, say, about 2%. On the other hand, there would appear to be about a 20% chance of actually losing money on investment *B*. This could be of concern to an investor with limited capital. It is of much less concern where capital resources are larger and the investor would average out over many investments. Nevertheless, investment *A* would generally be preferred because it would be free from the occurrence of a sequence of unfavorable outcomes. If the amounts risked or the variations in rewards are small compared to the operating capital, the use of expected values represents a single index for comparing alternative choices. But if these conditions are not met, use of the expected value or the expected rate of return is less satisfactory as a criterion.

2. *More accurate information.* Reduction of error in estimates might make it unnecessary to use probabilistic methods. Deterministic methods could then be used. However, the improvement of estimates takes both time and money. Opportunities may be lost while trying to refine the information. The problem of placing a monetary value on specific reductions in uncertainty is taken up in Chapter 5. In the end, some uncertainty will remain. The problem faced with simulation is to estimate the uncertainty that exists in a venture due to imperfect knowledge about its several component parts.

3. *Empirical adjustments.* Where certain estimates have consistently been too high, or too low, by about the same percentage, the use of an empirical adjustment would amount to improving the quality of the estimate. There is a tendency to adjust toward reducing out-of-pocket losses, which seem more tangible than the lost opportunities. Such adjustments tend to reflect bias. The functional responsibility of a department determines whether over-conservatism or strong optimism is the preferred virtue.

4. *Revising cutoff rates.* Rejecting investments whose expected values are

below a minimum reduces the risk of financial losses. Any such recommendation presumes there are a large number of opportunities immediately available. However, this may not be the best cutoff method. Investments *A* and *B* in Fig. 7.7 will both be accepted or both be rejected by such cutoffs, since they both have the same expected value. Investments typified by *B*, either singly or in groups, will always contain more risk of loss than equal number of investments such as *A*. Thus the use of this method may not accomplish the desired end.

5. *Three-level estimates.* This procedure consists of combining first all high values, next all medium values, then all low values of the several factors that enter into an estimate of value of a project. This is a step in the right direction but does not go far enough. For example, by using this method, values of −$50,000, $50,000, and $150,000 could be obtained for low, medium, and high in both investments *A* and *B* in Fig. 7.7 All the valuable information about the relative probabilities in the center of the range would be missing.

6. *Assigned probabilities.* This is the end toward which we have been striving. Ideally it would comprise information equivalent to a complete probability diagram. It is a matter of selecting the best way of accomplishing this. Probability diagrams may be crude or refined. Three-level estimates have been refined by assigning a probability of one-third to each of the three outcomes. Improvements in this have been made by weighting the middle estimate more heavily than the extremes.[5] Probability distributions of financial outcomes from drilling exploratory wells have been formed by assigning probabilities to specific outcomes.[6] Payoff tables for different levels of investment commitment have been constructed by assigning probabilities to various possible future economic growth rates.[7] However, where an outcome is the result of the interaction of several factors that can be broken out and estimated more accurately individually, which is usually the case, the best complete probability spectrum of the outcome can be formed by recombining them by simulation.

7.4 ENTERING VALUES INTO THE SIMULATION PROCESS

7.4-1 Choice of Ranges and Distributions

Variable quantities that enter into a simulation process are entered one by one, each chosen by a different random number. Three things are implied as necessary: definition of the range from which values of a variable will be selected, the specific probability distribution of the variable, and the random number used for the selection. The later are customarily generated within a computer as needed. Random decimal values between 0 and 1 are used, and a long sequence of these numbers forms a uniform distribution over the complete 0 to 1 range. This section is concerned with using a succession of such numbers for selecting specific values of the variable quantities.

The range of values of a quantity, porosity of a reservoir rock, for example, should reflect actual circumstances as near as possible. The range, from low to high, should satisfy one or more of the following objectives: (a) it should correspond to the range from which an actual choice will occur, (b) it should reflect the accuracy of present knowledge, or (c) it should reflect the best estimate of the future. When ranges are narrowed, when information becomes more complete and more precise, simulation should be repeated and up-dated.

Choice of a probability distribution should follow the same general objectives. It should correspond as closely as possible to the known information and conditions. The use of several distributions is discussed in the following sections.

7.4-2 Uniform Distribution

Use of the uniform distribution is indicated when the upper and lower limits of the range of a variable can be specified and when any of the values between these limits is as likely to occur as another value. The way in which the distribution is used is illustrated in Fig. 7.8. A solution is obtained by taking a one-to-one correspondence between the uniform (rectangular) distribution of the random number and the cumulative probability of the variable X.

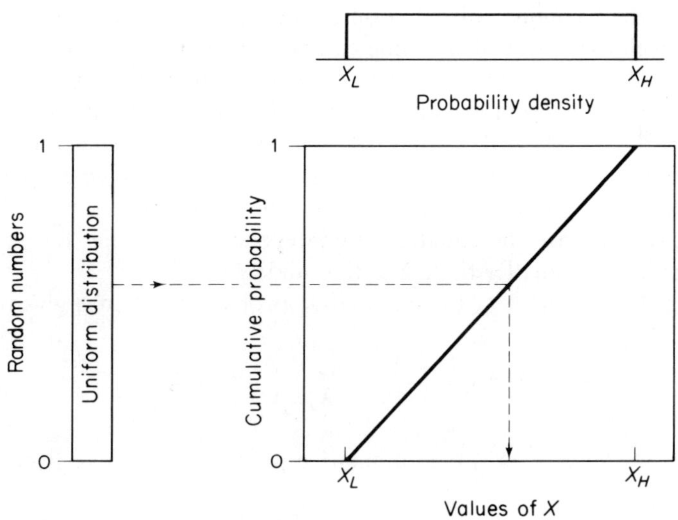

Figure 7.8. Selecting random values from a uniform distribution.

The cumulative probability of X is given by

$$F(X) = \frac{X - X_L}{X_H - X_L}. \tag{7-1}$$

Replacing $F(X)$ with R_N, the uniformly distributed random number, and solving for X gives

$$X = X_L + R_N(X_H - X_L). \tag{7-2}$$

This relation is rather obvious from the geometry of Fig. 7.8. For example, if X here were used for S_w, the connate water saturation, the high and low limits would be entered into Eq. (7-2). A random number would be generated within the computer and also used in Eq. (7-2), which then would yield the particular value of S_w to be used in the current case calculation. The simulation process consists, of course, of making many such case calculations.

A more explicit description of the process can be given as follows. Suppose that S_w is being used in an equation that gives the recoverable oil. Other quantities besides S_w would be involved in the calculation. However, each time a new calculation of the recoverable oil was made, the path of the calculations would pass through (1) the generation of a new random number to be used in Eq. (7-2), and (2) the use of Eq. (7-2) to calculate the current value of S_w.

7.4-3 Triangular Distribution

Use of the triangular distribution is indicated when the upper and lower limits as well as a most likely value can be specified. It is also apparent that the probability of an outcome occurring close to the limits of the range generally becomes progressively smaller, unless the distribution should happen to be highly skewed to one side. The manner of using the triangular distribution is illustrated in Fig. 7.9. Again, a one-to-one correspondence is taken between cumulative probability and the uniformly distributed random number. In this case the equation used depends on whether the value of X will be greater or smaller than X_M, the mode.

When $X_L \leq X \leq X_M$, the cumulative probability of X is given by

$$F(X) = \left(\frac{X - X_L}{X_M - X_L}\right)^2 \left(\frac{X_M - X_L}{X_H - X_L}\right). \tag{7-3}$$

When $X_M \leq X \leq X_H$, the cumulative probability of X is given by

$$F(X) = 1 - \left(\frac{X_H - X}{X_H - X_M}\right)^2 \left(\frac{X_H - X_M}{X_H - X_L}\right). \tag{7-4}$$

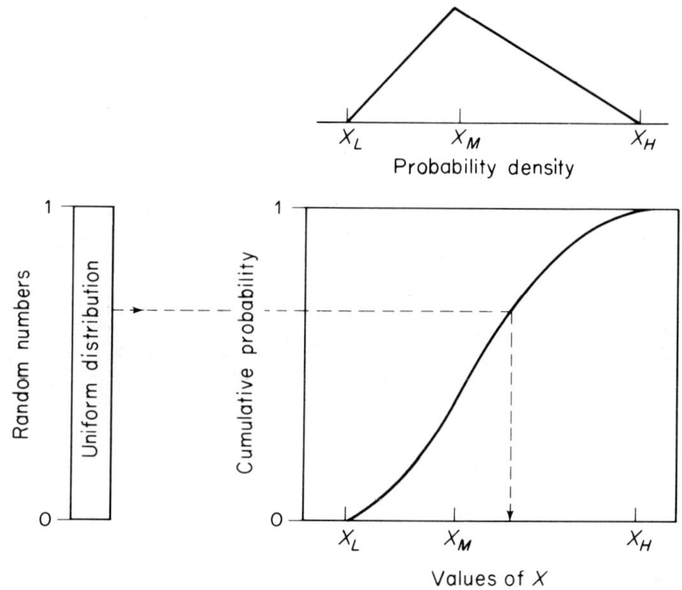

Figure 7.9. Selecting random values from a triangular distribution.

Replacing $F(X)$ with R_N, one has
If $R_N \leq (X_M - X_L)/(X_H - X_L)$:

$$X = X_L + \sqrt{(X_M - X_L)(X_H - X_L)R_N}. \tag{7-5}$$

If $R_N \geq (X_M - X_L)/(X_H - X_L)$:

$$X = X_H - \sqrt{(X_H - X_M)(X_H - X_L)(1 - R_N)}. \tag{7-6}$$

7.4-4 Discrete Distribution

Some probability distributions are not continuous; rather only particular values can occur. This is illustrated at the top of Fig. 7.10. In this case X can have values corresponding to X_1, X_2, X_3, or X_4, but it can have no other values beyond or between these specific numerical values. Besides the cases where the distribution actually is discrete, there are numerous cases where it is used simply because this distribution offers the most convenient way of describing the probabilistic nature of an event.

The manner of using the discrete distribution is indicated in Fig. 7.10 and in Table 7.2. R_N is, of course, a uniformly distributed random number with a range 0 to 1. Also, it is understood that the sum of the individual probabilities is unity.

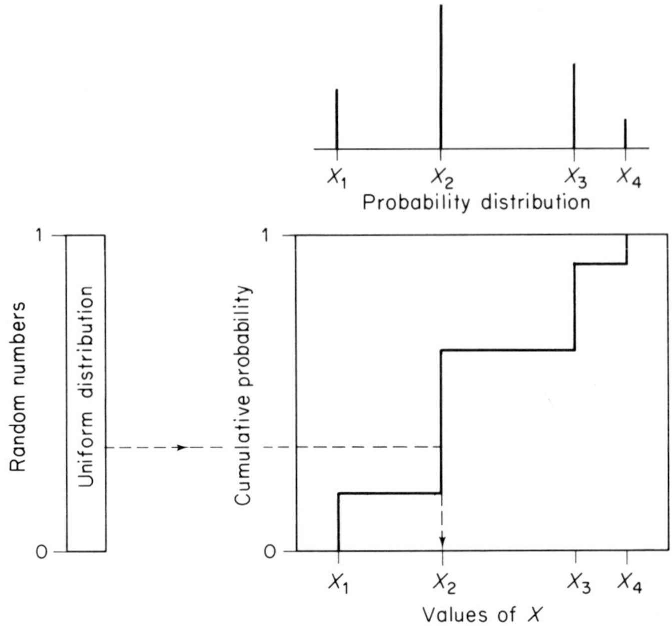

Figure 7.10. Random selections from a discrete distribution.

Table 7.2
Method of Choosing a Value in a Discrete Probability Distribution

Required Condition	*Choice of X*
$0 \leq R_N \leq p_1$	X_1
$p_1 < R_N \leq p_1 + p_2$	X_2
$p_1 + p_2 < R_N \leq p_1 + p_2 + p_3$	X_3
$p_1 + p_2 + p_3 < R_N \leq 1$	X_4

One method of performing this operation in the computer, using FORTRAN language, is as follows. Initially, the values of $X_1, X_2, \ldots, X_N$ and the corresponding values of $P_1, P_2, \ldots, P_N$ are read into machine memory (DIMENSION), where they can be relocated by subscript. The value of N, the total number of X-values, should also be available.

In any individual case calculation, in which a probabilistically chosen value of X is to be used, the particular value can be selected as follows. First, generate a decimal random number R_N. Then proceed to the following program segment:

```
      PCUM = 0.0
      DO    5    L = 1,N
      K = L
      PCUM = PCUM + P(L)
      IF (PCUM - RN) 5, 6, 6
 5    CONTINUE
 6    VALUE = X(K)
```

In successive passes through the DO loop, which ends as statement 5, index L has an initial value of 1 and can have a maximum value of N. The first time the IF statement is reached, PCUM has the value of p_1; the second time PCUM has the value of $p_1 + p_2$; and so on. As soon as PCUM reaches a value equal to or greater than RN, the IF statement sends control out of the DO loop to statement 6. The segment has supplied a value for K, such as the integer 3, which is the subscript for the randomly chosen value of X. Further calculations begin in statement 6, which will carry out the purposes of the complete program. The statement 6 given above is superfluous, since X_K is as readily available as would be a variable named VALUE for any calculations or printed output.

7.4-5 Histograms

The type of distribution shown in Fig. 7.11 is very often the best available description of a substance or one of its qualities. In the case illustrated, X is continuous throughout the range between X_1 and X_5. It would be entirely possible to have a zero probability of X occurring in parts of the total range, which would result in there being horizontal segments in the cumulative probability curve.

Simulating random sampling from this distribution by means of uniformly distributed decimal random numbers is done as follows. A one-to-one correspondence is taken between the random numbers and the cumulative probability. Interpolation on the cumulative probability curve is as follows. Let P_{CUM} be the cumulative probability at the upper end of the segment intersected by the dashed line. Let P_K be the probability associated with the intersected segment. Let X_L and X_K be the X-values corresponding to the upper and lower ends of the segment. Then, by straight-line interpolation,

$$X = X_K + \frac{R_N - [P_{\text{CUM}} - P(K)]}{P(K)}(X_L - X_K)$$

A FORTRAN segment that will perform the above calculation follows. Initially, the values of $X_1, X_2, \ldots, X_N$ and $P_1, P_2, \ldots, P_{N-1}$ are read into machine memory (DIMENSION), where they can be relocated by sub-

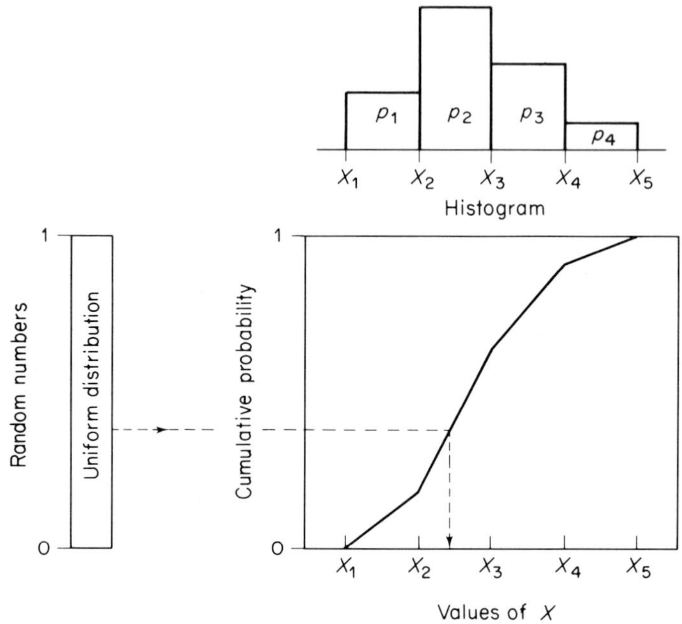

Figure 7.11. Random selections from a histogram.

script. The value of N, the total number of X-values, should also be available. Then, for any individual probabilistically chosen value of X, a decimal random number is generated and used in the following segment:

```
      PCUM = 0.0
      DO   5    J = 1,N
      K = J
      L = J + 1
      PCUM = PCUM + P(K)
      IF (PCUM - RN) 5, 6, 6
5     CONTINUE
6     V = X(K) + ((RN - PCUM + P(K)*(X(L) - X(K))/P(K)
```

In this segment it may be noted that P_{CUM} is initially set equal to zero. Then, in the repeating DO loop, which extends from "DO 5" to statement 5, the value of P_{CUM} is successively augmented by the value of the current P_K. As soon as the value of P_{CUM} is equal to or greater than R_N, the IF statement jumps control out of the DO loop and directs it to statement 6. The DO loop has supplied the proper subscripts K and L. In statement 6 the interpolated value of X is calculated, where it is called V. The reason for changing the name is that generally where memory space is allocated to the name X, all X-values must have a subscript. The simplest procedure is to change the name.

7.4-6 Dependent Distribution

An example of a dependent distribution is illustrated in Fig. 7.12. The case illustrated would apply to a partially developed oil field in which a sufficient number of wells had not yet been drilled to accurately determine either the horizontal extent or net pay thickness of the reservoir. The figure

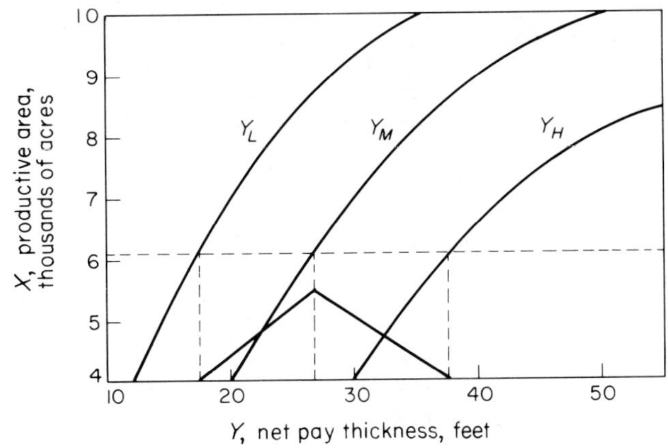

Figure 7.12. Use of a dependent distribution.

indicates a general increase in thickness with increasing areal extent, but no exact relation between the two can be estimated for the particular reservoir. A triangular distribution is indicated for the net thickness as a function of area. A rectangular distribution was used to describe the randomness of the areal extent of the reservoir.

In simulating this example, it was necessary to generate two random numbers. The first was used to select the choice of productive area; the second was used to select the choice of net pay thickness. In a computer all this must be done through the use of algebraic equations. Accordingly, it was necessary to develop empirical equations that fit the data.

The type of equation selected was

$$Y = B_0 + B_1X + B_2X^2$$

Data points used for fitting the constants are shown in Table 7.3.

Table 7.3
Data Points Used for Curve Fitting

Point	X	Y_L	Y_M	Y_H
1	4	12	20	30
2	7	20	30	42
(3)	8.5	—	—	55
3	10	35	50	—

Parameters for the Y_L relation were determined as follows:

$$\begin{aligned}
B_2 &= \frac{1}{X_3 - X_1}\left(\frac{Y_3 - Y_2}{X_3 - X_2} - \frac{Y_2 - Y_1}{X_2 - X_1}\right) \\
&= \frac{1}{10 - 4}\left(\frac{35 - 20}{10 - 7} - \frac{20 - 12}{7 - 4}\right) \\
&= 0.39, \\
B_1 &= \frac{Y_2 - Y_1}{X_2 - X_1} - B_2(X_2 + X_1) \\
&= \frac{20 - 12}{7 - 4} - 0.39(7 + 4) \\
&= -1.61, \\
B_0 &= Y_2 - B_1X_2 - B_2X_2^2 \\
&= 20 - (-1.61)(7) - (0.39)(7)^2 \\
&= 12.22.
\end{aligned}$$

The constants for the other curves were determined in a similar manner. This procedure produced the following set of relations:

$$\begin{aligned}
Y_L &= 12.22 - 1.61X + 0.39X^2, \\
Y_M &= 22.22 - 2.78X + 0.56X^2, \\
Y_H &= 43.04 - 7.41X + 1.04X^2.
\end{aligned}$$

At this point it became possible to assign the simulation task to the computer. Following through a typical individual case calculation illustrates how the process can be accomplished. The first random number accomplishes the choice of area and makes possible determination of the low, mode, and high values of the thickness. It is here taken that the first random number generated was 0.35. Then the following calculations occur:

$$\begin{aligned}
\text{area} = X &= X_{\min} - R_N(X_{\max} - X_{\min}) \\
&= 4 + (0.35)(10 - 4) \\
&= 6.1 \text{ thousands of acres,} \\
Y_L &= 12.22 - 1.61X + 0.39X^2 \\
&= 12.22 - (1.61)(6.1) + (0.39)(6.1)^2 \\
&= 16.91, \\
Y_M &= 22.22 - 2.78X + 0.56X^2 \\
&= 22.22 - (2.78)(6.1) + (0.56)(6.1)^2 \\
&= 26.10, \\
Y_H &= 43.04 - 7.41X + 1.04X^2 \\
&= 43.04 - (7.41)(6.1) + (1.04)(6.1)^2 \\
&= 36.54.
\end{aligned}$$

A second random number was then generated for the purpose of obtaining a random selection from the triangular distribution. It is here taken that the second random number had a value of 0.80. Then the following calculations occur:

$$R_N = 0.80,$$

$$\frac{Y_M - Y_L}{Y_H - Y_L} = \frac{26.10 - 16.91}{36.54 - 16.91} = 0.47,$$

$$R_N > 0.47.$$

It may be recalled that whether R_N is greater or smaller than the fractional position of the mode determines the choice of formula for the next step. In this case the following applies:

$$\begin{aligned}
h &= Y_H - \sqrt{(Y_H - Y_M)(Y_H - Y_L)}\sqrt{1 - R_N} \\
&= 36.54 - \sqrt{(36.54 - 26.10)(36.54 - 16.91)}\sqrt{1 - 0.80} \\
&= 30.14.
\end{aligned}$$

Inspection of Fig. 7.12 reveals that an upper limit of 55 should be placed on the net pay thickness. Proper statements can be placed in any program that would examine the calculated thickness, and if it were greater than 55, simply set it equal to 55.

The process that has been described has used two random numbers. It has provided a random choice of the reservoir area. It has also provided a random choice of the average reservoir thickness, and the second choice was made within limits that were consistent with and dependent upon the first choice.

7.5 A PROBABILITY DISTRIBUTION OF RATE OF RETURN

In the following example the answer is given in the form of a probability distribution of anticipated rate of return. A prospective venture, which is being considered for purchase, is a water-flooding project of an oil property that is essentially depleted of natural production (see Table 7.4). It can be

Table 7.4
Costs and Oil Recovery Data

Initial Costs:	
10 producers × 3,000 ft × $10.00	$ 300,000
30 inject wells × 3,000 ft × $6.50	585,000
6 supply wells × 1,000 ft × $12	72,000
Surface lines:	
6,000 ft 4 in. at $1.50	9,000
8,000 ft 2 in. at $1.00	8,000
Plant (5,000 bbl/day)($4.00):	20,000
Purchase lease:	50,000
	$1,044,000
Operating Expenses per Year:	
5,000 bbl/day at 1,200 psi	102 hp
Electricity (102 hp) ($70 per year):	7,140
Overhead (25 wells) ($40–$80):	1,000 to 2,000
Labor (600 to 1,200)(12 months):	7,200 to 14,400
Net Pay:	
10 to 20 ft, uniform probability distribution	
Porosity:	
12 to 22%, uniform probability distribution	
Recovery as a Percentage of Total Pore Space:	
20 to 40, uniformly distributed	
Area:	
500 acres	
Assume:	
$3.00 per barrel, crude price	
0.875 working interest	
5% state tax	

purchased for $50,000. Additional wells and a water-injection plant would have to be installed. Costs of the wells, pipelines, and plant are essentially deterministic. However, uncertainty exists regarding feet of net pay, average per cent porosity, fractional oil recovery, and rate at which recovery will take place (see Table 7.5), so these quantities are treated as probabilistic.

Table 7.5
Estimates of Rates of Oil Recovery

	*Production Schedules**			
Year	*1*	*2*	*3*	*4*
1	—	—	—	—
2	—	—	—	—
3	—	—	—	—
4	0.04	0.04	0.03	0.03
5	0.17	0.13	0.12	0.10
6	0.35	0.24	0.22	0.19
7	0.56	0.42	0.35	0.29
8	0.73	0.60	0.50	0.42
9	0.85	0.73	0.62	0.53
10	0.94	0.84	0.73	0.64
11	0.99	0.91	0.82	0.73
12	1.00	0.96	0.89	0.82
13		0.99	0.94	0.87
14		1.00	0.97	0.92
15			0.99	0.95
16			1.00	0.98
17				0.99
18				1.00

*The numbers are the fraction of the ultimate production that has been produced by the end of the year. The schedules are equally probable, 0.25 each.

A solution of the problem can be accomplished by following this sequence of steps:

1. Select the classes into which the answers will be grouped.
2. Calculate the initial investment, which is considered deterministic in this example.
3. Calculate the total recoverable oil, by making probabilistic choices according to random numbers.
4. Choose a production schedule by random numbers and calculate the yearly productions and convert into dollars.
5. Calculate the average yearly expenses, using random numbers, and subtract these from the yearly incomes found in step 4, to obtain the net cash flow in each year.
6. Calculate the rate of return for the particular individual trial and tally the answer into the appropriate class, as set up in step 1.
7. Return to step 3 and repeat through step 6 until a total of (for example) 200 tries have been made and tallied. After that go to step 8.
8. Calculate the fraction of the total cases tallied into each group, and print the answer.

Each of the foregoing steps will now be discussed individually.

1. *Selection of classes.* Valid answers in the Monte Carlo method can be obtained only by calculating a large number of individual trial solutions. It is generally impossible to interpret the answers unless they are organized into groups. There are two ways to do this. The first method is to set class boundaries. Then one finds out, for example, that 23% of the cases yield rates of return between 5 and 10%, 31% of the cases yield rates of returns between 10 and 15%, and so on. The second method is to accumulate all the answers in machine memory and later sort them. By this method one finds out, for example, that 10% of the cases yield rates of return less than 4.8%, 20% of the cases yield rates of return less than 7.2%, and so on. The first method is more direct and generally permits easier comparisons of one prospective venture with others.

Preliminary calculations show that, for this example, if the lowest values of all variables are used, the rate of return would be about 5%. If all highest values are used, the rate of return would be about 50%. The class boundaries of 5, 10, 15, 20, 25, 30, 35, 40, 45, and 50% were assumed to be sufficient for describing the outcome. Smaller divisions, such as going by steps of 2%, could have been used. However, several times as many trial calculations would need to be run to obtain a smooth answer.

The design is illustrated graphically in Fig. 7.13. In that example $X_1 = 0.05$, $X_2 = 0.10$, $X_3 = 0.15$, etc. The *K*-values are the tally values and initially must be set at zero. In the Monte Carlo process, each time a rate of

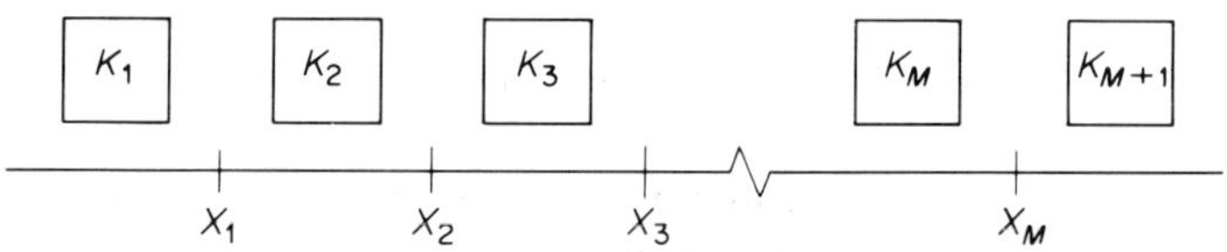

Figure 7.13. Class boundaries (*X* values) and class tallies (*K* values).

return calculates to be less than 0.05, K_1 will be increased by 1. Each time the rate of return calculates to be between 0.05 and 0.10, K_2 will be increased by 1. In the example there are 10 class boundaries. To have a flexible program, there should be one more tally than there are class boundaries. The first tally receives all cases below the first class boundary, and the last tally receives all cases greater than the highest class boundary. The actual tallying is discussed in step 6.

2. *Initial investment.* In this example the calculation of the initial investment is shown in Table 7.3. The amount of $1,044,000 is assumed to be spent, or committed, as of the beginning of the first year. This simplifying assumption is assumed to be sufficiently accurate for the present purposes.

3. *Calculation of total recoverable oil.* Several probabilistic choices are to be combined in this calculation. The basic formula is

$$N_p = 7{,}758 \cdot (\text{area}) \cdot h \cdot \varphi \cdot R_F,$$

where

N_p = total recovery, barrels of oil,
7,758 = barrels per acre-foot,
area = area, 500 acres,
h = thickness of the porous zone, feet,
φ = fractional porosity,
R_F = recovery factor, expressed as a fraction of total pore space.

Probabilistic choices must be made individually for values of h, φ, and R_F. In this example the rectangular distribution has been used for each of these quantities. The general model for this is

$$V = \text{low} + R_N(\text{high} - \text{low}),$$

where

V = value used in the trial case,
low = low estimate of the property,
high = high estimate of the property,
R_N = decimal random number, distributed uniformly between 0 and 1.

4. *Calculating gross income per year.* Four production schedules were assumed to be equally probable. A random number was used to select the schedule used for a particular trial. Multiplying the total recoverable oil by the cumulative fraction gave the total oil produced by the year's end. Subtracting the previous total gave the oil produced during the year. This was converted to dollars at \$2.49 gross revenue per barrel.

5. *Yearly operating expenses.* The yearly operating expenses were determined from within a given range by random-number selection. The choice made was used throughout a particular trial case. Another way of doing this, which might be better in some cases, would be to make a probabilistic choice for each year. This second approach would produce less variation in the final results, since the several yearly choices would tend to average out to the mean. Subtracting the yearly expenses from the gross income determined in step 4 gave the net cash flow for each year.

6. *Discounted-cash-flow rate of return.* The DCF rate of return was calculated by finding the interest rate that would reduce the future net income to a value equal to the required initial investment. Total present value of the future net cash flow is given by

$$\text{P.V.} = \sum \frac{S_L}{(1 + r)^L},$$

$$\text{net present value} = \text{P.V.} - \text{initial investment},$$

$$\text{slope} = \frac{d(\text{P.V.})}{dr} = \sum \frac{(-L)(S_L)}{(1 + r)^{L+1}}.$$

Figure 7.14 shows how an initial trial value of $r = 0$ can be used to find the value of r that reduces the net present value to zero. This is the *Newton–Raphson method* of iterating to find a numerical answer. The answer was assumed to be sufficiently close when there was less than 0.1% difference between successive answers.

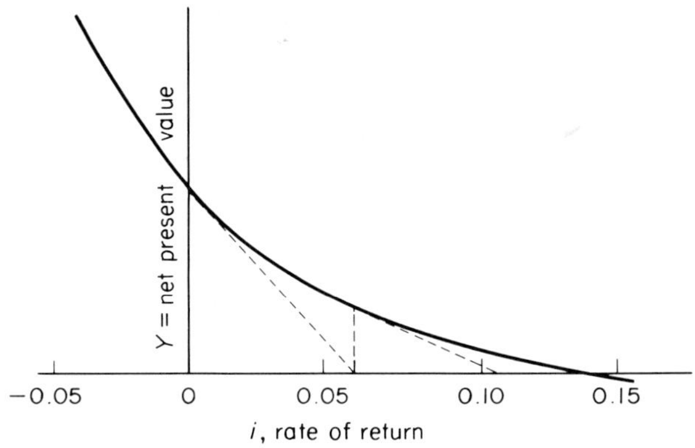

Figure 7.14. Numerical trial-by-error solution for rate of return. The dotted lines show the first and second approximations, starting from $i = 0$, by using Newton's method.

After calculating a rate of return, it was tested against the class boundary values, successively. In this process, when a class boundary value exceeded the calculated rate, the trial case was simply counted as falling into the class below that boundary; or tallied above the highest boundary.

Random numbers were brought into the program by calling a subroutine. The process statements are:

```
CALL RANDU (N1, N2, YFL)
SUBROUTINE RANDU (IX, IY, YFL)
```

There is a one-to-one correspondence between the positions within the two sets of parentheses. The value supplied as N1 in the program becomes IX in the subroutine, the value of IY is returned as N2 in the program, and YFL is the generated decimal random number.

Where listed, or generated, random numbers are used to select a group of factors which are then combined, the original random quality of the list will be reduced. This is to be expected. The question now raised is whether or not groups of successive numbers in a table would give results of smaller variance. This would be in comparison to the variance achieved by using independent sequences of random numbers. For example, numbers going down one column could be used to choose porosity; numbers going down another column could be used to choose pay thickness, and so on. The equivalent of the latter technique was used in this example.

Five different original numbers were supplied for starting five different series of random numbers. It was determined that the given numbers are separated by at least 50 numbers. That is, by chance, one could put in numbers that are adjacent or very close to each other in the generator's sequence of random numbers. The five numbers supplied were referred to as N1, N3, N5, N7, and N9. These are all used in the same relative position inside the parentheses when calling subroutine RANDU. This technique produces five independent sequences of YFL, the decimal random number.

The computer program and input data are given in Table 7.6, and the computer output data are given in Table 7.7. A list of symbols used within the program is provided as Table 7.8. Figure 7.15 is a histogram of the results obtained after running 200 trial case calculations in the computer.

Table 7.6
Computer Input Data and Program

```
                INPUT DATA

NUMBER OF CLASS BOUNDARIES AND NUMBER OF TRIAL RUNS

   10   200

VALUES OF THE CLASS BOUNDARIES, RATES OF RETURN

 0.05 0.10 0.15 0.20 0.25 0.30 0.35 0.40 0.45 0.50

SEED VALUES FOR FIVE DIFFERENT RANDOM NUMBER SERIES

  583 1265 3117 2439 8297

INITIAL     ESTIMATED LOW AND HIGH     GROSS REVENUE
 COST          OPERATING EXPENSES          PER BBL.

1044000.00    8200.00   16400.00         2.49

AREA, RANGES OF THICKNESS, POROSITY AND RECOVERY

 500.00   10.00   20.00    0.12    0.22    0.20    0.40

NO. OF TIME SCHEDULES, GREATEST NO. OF YEARS

   4    18

TOTAL YEARS IN INDIVIDUAL TIME SCHEDULES

   12    14    16    18

PROBABILITY OF EACH TIME SCHEDULE

       0.25        0.25        0.25        0.25
```

FRACTIONAL CUMULATIVE PRODUCTION BY YEARS

0.00	0.00	0.00	0.00
0.00	0.00	0.00	0.00
0.00	0.00	0.00	0.00
0.04	0.04	0.03	0.03
0.17	0.13	0.12	0.10
0.35	0.24	0.22	0.19
0.56	0.42	0.35	0.29
0.73	0.60	0.50	0.42
0.85	0.73	0.62	0.53
0.94	0.84	0.73	0.64
0.99	0.91	0.82	0.73
1.00	0.96	0.89	0.82
0.00	0.99	0.94	0.87
0.00	1.00	0.97	0.92
0.00	0.00	0.99	0.95
0.00	0.00	1.00	0.98
0.00	0.00	0.00	0.99
0.00	0.00	0.00	1.00

```
$JOB             ID,KP=26,TIME=120
C       DISCOUNTED CASH FLOW RATE OF RETURN, PROBABILITY DISTRIBUTION
C       MONTE CARLO METHOD
        DIMENSION X(15),K(16),PK(16),NYF(6),PSCH(6),C(40,6),S(40)
        READ(5,104)MX,N
        READ(5,105)(X(J),J=1,MX)
        READ(5,104)N1,N3,N5,N7,N9
        READ(5,107)COST,OPL,OPH,GRPB
        READ(5,108)AREA,HL,HH,PFL,PFH,RFL,RFH
        READ(5,104)M,NYRS
        READ(5,104)(NYF(J),J=1,M)
        READ(5,107)(PSCH(J),J=1,M)
        DO 2 L=1,NYRS
      2 READ(5,107)(C(L,J),J=1,M)
    104 FORMAT(16I5<
    105 FORMAT(16F5.2<
    107 FORMAT(8F10.2<
    108 FORMAT(10F7.2<
        MK=MX+1
        DO 3 J=1,MK
      3 K(J)=0
C       BEGINNING RANDOM TRIALS, DETERMINING TOTAL OIL RECOVERED
        NTAL=0
      4 NTAL=NTAL+1
        CALL RANDU (N1,N2,YFL)
        N1=N2
        H=HL+YFL*(HH-HL)
        CALL RANDU (N3,N4,YFL)
        N3=N4
        PF=PFL+YFL*(PFH-PFL)
        CALL RANDU (N5,N6,YFL)
        N5=N6
        RF=RFL+YFL*(RFH-RFL)
        FNP=7758.0*AREA*H*PF*RF
C       DETERMINING J, SUBSCRIPT OF PRODUCTION SCHEDULE
        CALL RANDU (N7,N8,YFL)
        N7=N8
        PCUM=0.0
        DO 11 LJ=1,M
        J=LJ
        PCUM=PCUM+PSCH(J)
        IF(PCUM-YFL)11,12,12
     11 CONTINUE
C       PROBABILISTIC CHOICE OF OPERATING EXPENSES
     12 CALL RANDU (N9,N10,YFL)
        N9=N10
        OPXP=OPL+YFL*(OPH-OPL)
C       CALCULATING NET CASH FLOW FOR EACH YEAR
```

```
      S(1)=C(1,J)*FNP*GRPB-OPXP
      NF=NYF(J)
      DO 14 L=2,NF
      LD=L-1
   14 S(L)=(C(L,J)-C(LD,J))*FNP*GRPB-OPXP
C     CALCULATING RATE OF RETURN, SOLUTION FROM I#0 BY NEWTONS METHOD
      XI=0.0
      LTAL=0
   20 LTAL=LTAL+1
      IF(LTAL-30)21,21,23
   21 ZI=XI
      PV=0.0
      DYDX=0.0
      DO 22 L=1,NF
      LU=L+1
      VL=L
      PV=PV+S(L)/(1.0+XI)**L
   22 DYDX=DYDX-VL*S(L)/(1.0+XI)**LU
      Y=PV-COST
      XI=XI-Y/DYDX
      IF(ABS(XI-ZI)-0.001)24,20,20
   23 WRITE(6,217) XI
  217 FORMAT(30H EXCEEDED 30 ITERATIONS AT I #,F15.8<
      GO TO 50
C     TALLYING RATE ANSWER INTO A CLASS
   24 DO 25 J=1,MX
      JK=J
      IF(XI-X(J))26,25,25
   25 CONTINUE
      JK=JK+1
   26 K(JK)=K(JK)+1
C     CALCULATING FRACTION OF RESULTS IN EACH CLASS
      IF(NTAL-N)4,30,30
   30 VN=N
      DO 31 J=1,MK
      VK=K(J)
   31 PK(J)=VK/VN
C     PRINTING THE RESULTS
      WRITE(6,200)
  200 FORMAT(10X,36H DISCOUNTED CASH FLOW RATE OF RETURN,//<
      WRITE(6,201)
  201 FORMAT(15X,22H BY MONTE CARLO METHOD,//<
      WRITE(6,202)
  202 FORMAT(20X,17H CLASS BOUNDARIES,/<
      WRITE(6,203)(X(J),J=1,MX)
  203 FORMAT(6X,15F8.2<
      WRITE(6,204)NTAL
  204 FORMAT(/,20X,30H CLASS PROBABILITIES, TOTAL OF,I6,8H  TRIALS,/<
      WRITE(6,205)(PK(J),J=1,MK)
  205 FORMAT(3X,16F8.4<
   50 STOP
      END
```

```
$EXEC
 DISCOUNTED CASH FLOW RATE OF RETURN


        BY MONTE CARLO METHOD


              CLASS BOUNDARIES

   0.05     0.10     0.15     0.20     0.25     0.30     0.35     0.40     0.45     0.50

              CLASS PROBABILITIES, TOTAL OF    200   TRIALS

0.0000  0.0000  0.0250  0.1250  0.2700  0.2800  0.2050  0.0600  0.0300  0.0050  0.00
```

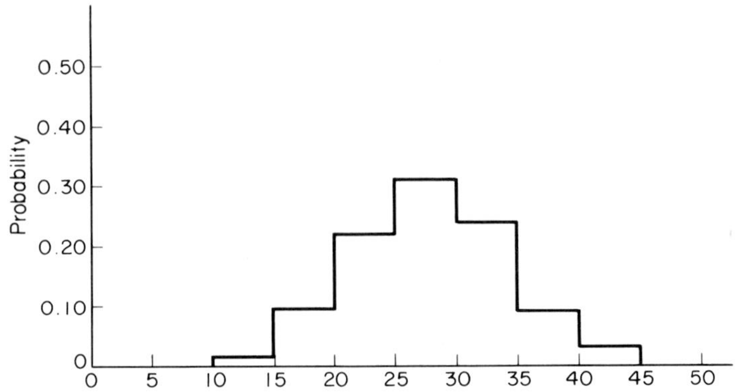

Figure 7.15. Discounted cash flow rate of return, percent.

Table 7.7
Computer Output Data for Discounted-Cash-Flow Rate of Return

Rate-of-Return Class Boundaries	*Fraction of Trials in Class*
–0.05	0.000
0.05–0.10	0.000
0.10–0.15	0.025
0.15–0.20	0.125
0.20–0.25	0.270
0.25–0.30	0.280
0.30–0.35	0.205
0.35–0.40	0.060
0.40–0.45	0.030
0.45–0.50	0.005
0.50–	0.000
	1.000

Table 7.8
List of Symbols Used on Computer Program

X(J) = one of the class boundaries, such as 0.05, 0.10, 0.15, etc.,
MX = number of class boundaries,
K(J) = number of outcome cases that fall into the *J*th class,
VK = *K*(*J*) value (real or decimal-type number),
PK(J) = relative number of outcome cases that fall into the *J*th class,
MK = total number of *K*(*J*) classes into which the results are tallied,
N = total number of trial cases run,
VN = total trials run (a real or decimal-type number),
NTAL = tally for keeping count of the number of trials that have been run,
COST = initial investment, dollars,

Table 7.8 (Cont.)

OPXP	= operating expenses per year,
OPL	= low estimate of yearly operating expenses,
OPH	= high estimate of yearly operating expenses,
Y	= net present value, dollars,
DYDX	= dY/d(XI), slope of the present-value curve,
XI	= rate of return, decimal,
ZI	= former value of the rate of return, in the trial-by-error solution,
LTAL	= tally for counting the number of trial solutions for the rate of return; program stops if iteration goes beyond 30 trials,
M	= number of different production schedules,
PSCH(J)	= probability of encountering the *J*th production schedule,
PCUM	= sum formed by adding probabilities of successive production schedules,
NYF(J)	= total years in the *J*th production schedule,
C(L, J)	= cumulative fraction of the ultimate production produced by the end of the *L*th year in the *J*th production schedule,
AREA	= productive area, acres,
H	= average net pay thickness, feet,
HL	= low estimate of net pay thickness,
HH	= high estimate of net pay thickness,
PF	= average porosity, decimal,
PFL	= low estimate of average porosity,
PFH	= high estimate of average porosity,
RF	= average recovery factor, expressed as a fraction of the total pore space,
RFL	= low estimate of average recovery factor,
RFH	= high estimate of average recovery factor,
FNP	= cumulative oil recovered, barrels,
GRPB	= gross revenue per barrel, dollars/barrel,
S(L)	= net cash flow for the *L*th year, dollars,
PV	= present value of the net operating income,
NYRS	= total years in the production schedule of maximum length,
L	= years (index, integer number),
VL	= years (real or decimal-type number),
N1, N3, N5, N7, N9	= numbers supplied to the random-number generator,
N2, N4, N6, N8, N10	= random integers returned from the random-number generator,
YFL	= decimal random number returned from the random-number generator.

REFERENCES

1. Walstrom, J. E., T. D. Mueller, and R. C. McFarlane, "Evaluating Uncertainty in Engineering Calculations." *Journal of Petroleum Technology*, Vol. 19, No. 12 (Dec. 1967), p. 1595.

2. Knopp, C. R., "Gas–Oil Relative Permeability Ratio Correlations from Laboratory Data." *Journal of Petroleum Technology*, Sept. 1965, p. 111.

3. Muskat, M., *Physical Principles of Oil Production*. New York: McGraw-Hill Book Company, 1949, p. 408.

4. HERTZ, D. B., "Risk Analysis in Capital Investment." *Harvard Business Review*, Vol. 42, No. 1 (Jan.–Feb. 1964).
5. TOWNES, H. L., "Using Economics in Exploration Decisions." Presented at Oklahoma City Section of SPE of AIME, Oklahoma City, Dec. 5, 1967.
6. GRAYSON, C. J., Jr., *Decisions Under Uncertainty*. Boston: Division of Research, Harvard Business School, 1960.
7. BENION, E. G., "Capital Budgeting and Game Theory." *Harvard Business Review*, Nov.–Dec. 1956.

PROBLEMS

7-1. An appreciation of the Monte Carlo process can be obtained by making a few trials manually using a table of random numbers. The following problem can also be solved by rolling a die, using the low value of the range if a 1 is obtained, the high value if a 6 is obtained, and proportional values for numbers in between. By this means one can approximate a uniform distribution. The monthly operating profit is given by the relation

$$P = NS - C,$$

where N is the number of items sold, S is the selling price, and C is the operating costs. N varies over the range from 200 to 300, S varies from \$10 to \$15, and C varies from \$1,500 to \$2,000. If a die is used, the first roll provides a value of N, the second roll provides a value of S, and the third roll provides a value of C. For example, if the numbers 2, 5, and 3 were rolled, one has

$$\begin{aligned} P &= (200 + 20)(\$15 - \$1) - [\$1{,}500 + (2)(\$100)] \\ &= \$1{,}380. \end{aligned}$$

About 30 to 50 trials are suggested. The results should be grouped into classes and plotted as a frequency distribution. Relative frequency distributions so obtained are in a sense independent of the total number of trial cases that were run. However, smooth results depend upon running a large number of trials. Also, where few trials are run, the total number of classes must be small. Where many trials are run, the class intervals can be made smaller.

7-2. It was desired to approximate the normal distribution by using the triangular distribution. The upper and lower limits of the triangular distribution were set at $+3\sigma$ and -3σ of the normal distribution, with the mode set at zero, of course. Determine the per cent error in the cumulative probability as calculated by the triangular distribution at the following values of standard deviation σ: -3, -2.5, -2, -1.5, -1, -0.5, 0, 0.5, 1, 1.5, 2, 2.5, and 3. At which points is there no error? (Areas under the normal probability curve are to be found from tabulated values.)

7-3. A well is to be drilled on 80-acre spacing. Net pay thickness can be represented by a uniform distribution between the limits of 24 and 32 ft. Porosity can be represented as a triangular distribution where low, mode, and high are 0.08, 0.15, and 0.20. Water saturation can be represented by a uniform distribution between the limits of 0.20 and 0.40. The recovery factor can be represented by a uniform distribution between the limits of 15 and 30%. Design a series of equations (program) for determining the probability distribution of the amount of oil to be recovered by the well. The following relation applies:

$$\text{recovery(bbl)} = (7{,}758)(80 \text{ acres})h\phi(1 - S_W)R_F.$$

7-4. In a certain reservoir the water saturation is partially dependent upon the porosity. The porosity varies from 0.08 to 0.22. If the porosity is 8%, the water saturation varies between 40 to 60%. If the porosity is 22%, the water saturation varies between 15 and 30%. Intermediate positions are linearly related. Both the porosity and water saturation are considered to be uniformly distributed. Design a series of calculations (program segment) that will supply interdependent values of porosity and water saturation according to two random numbers. The random numbers will be decimal numbers uniformly distributed between 0 and 1.

7-5. Construct by simulation a probability diagram of the amount of oil that may be recovered from a 160-acre tract. Net pay thickness is estimated to be between 25 and 35 ft. The recovery will be between 150 and 300 bbl/acre-ft, with the most likely value at 200 bbl/acre-ft.

8

EVALUATION OF EXPECTED DISCOVERIES IN MATURE REGIONS

8.1 INTRODUCTION

8.1-1 Economic Significance

Mature regions are those which have been extensively tested and production obtained from them. A mature region must be defined not only as to geographical area but also in terms of a depth interval. Many times activities in an area have been rekindled by the discovery of deeper petroleum deposits, resulting in increased interest and values. Less spectacular but nonetheless valid reasons for regional reevaluation also occur. These include less expensive testing procedures, such as leasing and exploratory well costs, and more effective ways of locating test wells. Improved recovery methods, such as the development of reservoir rock-fracturing processes, also justify reevaluation of a region.

Figure 8.1 shows the common experience encountered in an oil play. A *play* consists of the search and successive discovery of a group of petroleum reservoirs located in similar or related geologic formations. Discovery effort may not be linear with time; rather, it tends to be punctuated with wildcat wells drilled at irregular time intervals. At some point sufficient information is obtained so that the geologic features that control accumulation of oil and gas can be delineated. Additional discoveries are then made in rapid order. Eventually most of the reservoirs are discovered and the pace of discovery slows. At some point the economic limit is reached, where the

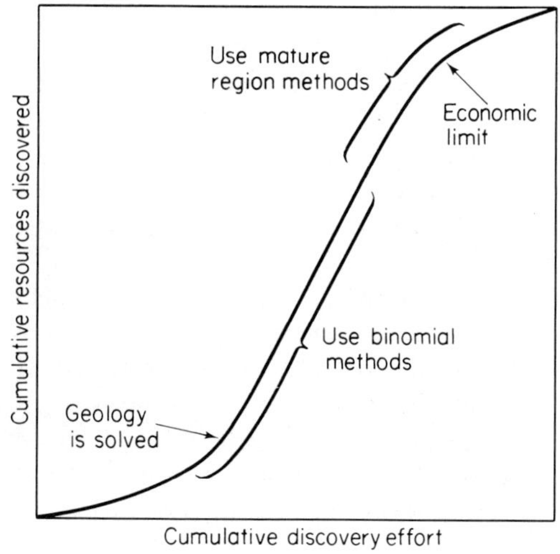

Figure 8.1. Cumulative resources discovered versus cumulative discovery efforts. (After R. W. Rausch[7])

effort and cost of discovery exceeds the value of the additional resources that may be discovered.

For any given set of economic conditions, it is important to know whether or not additional exploratory efforts are justified. Economic models are used when making such estimates. These models most often take the form of a set of mathematical expressions that relate to the real world. The objective is to quantify such relationships. We should like to know how many barrels of oil or cubic feet of natural gas, on the average, we can expect to discover per exploratory well. If this can be estimated, and assuming that the total cost per exploratory well is known, it becomes possible to calculate the cost of discovering a unit of oil or natural gas.

It is apparent that economic models are a very powerful tool. They can be used to point out areas in which exploratory efforts are still worthwhile. Also they can point out when further efforts should cease in a certain region. The effects of the costs of seismic surveys, drilling, and steel pipe can be calculated. Of national importance, the relation between future available reserves and the price of oil or gas can be predicted. Thus it is to be expected that the use of such models will greatly increase in the future.

8.1-2 Methods of Estimating Expected Discoveries

The chief thrust of this present chapter is to present some prediction methods based on trend analysis. These are included in category 4 in the following classification. Before discussing these, some of the other methods for estimating expected discoveries will be listed and interrelationships noted.

1. *Volume of sediments.* During the early stages of development of a geologic basin or region, estimates of the total cubic miles of potentially favorable sediments have been multiplied by a unit recovery factor per cubic mile. Such recovery factors are based on analogy to more mature petroleum provinces. The method has also been used to estimate reserves of natural gas.[3] The depth of the basement rocks, or, conversely, the height of the sedimentary column, is prime information to have when using this method.

2. *Constant binomial probability.* Included in this general class are the common binomial calculations in which all exploratory wells are assumed to have a success probability of 0.2, for example, and all successful wells discover the same amount of reserves. The method is improved by assuming a log-normal or triangular or other distribution which better describes the possible range of reserves that may be discovered in a successful well. Chapters 4 and 5 contain applications of these methods which allow for statistical variations in results.

3. *Assumed number of fields.* An estimate of the total number of fields present may be based on the extent of a region. The number of fields can also be estimated from the number of geological and/or geophysical prospects, where experience would indicate that some fraction of these leads will eventually prove to be dry. Furthermore, the prospects can be divided into probable size classes. If one is willing to assume that a region contains M_1 fields of size class 1 and M_2 fields of size class 2, and so on, then it is possible to calculate the probable results to be obtained from a group of exploratory wells. These methods are described in Chapter 4 under the multinomial and hypergeometric distributions. One of the Bayes applications in Chapter 9 describes ways of improving the first estimates of the numbers of fields originally present by interpretation of exploratory results.

4. *Trend predictions.* These methods are analogous to the use of decline curves to predict future production in an already developed reservoir. They depend upon choosing some suitable factor, or factors, with which past performance can be correlated.[5,6] The assumption is then made that performance in the future will follow the same trends as in the past. Methods used in this chapter do not incorporate prior estimates of the number of fields or the amount of reserves originally present. Rather, these parameters are determined when applying the methods.

8.1-3 Bases of Estimation for Trend Predictions

Trend analyses methods depend upon formulating adequate models. The objective, of course, is to find a relationship that has been followed in the past and then to apply it to predict future performance. One general relationship seems evident:

$$\text{effectiveness} \propto (\text{discovery efforts}) \cdot (\text{target area}). \tag{8-1}$$

The measure of effectiveness is directly in terms of the barrels of oil or cubic feet of gas discovered. Normally this would be the producible reserves, whose recovery can be anticipated by current practices and economic conditions. That is, the data entered into the calculations would normally be those listed as proved reserves. The expected discoveries, as predicted by the model, would then be in these terms.*

Discovery efforts encompass many activities. Financial arrangements, organizational and staff commitments, geological and geophysical research, engineering capabilities, and other resources are required. Any quantity used as a measure of discovery effort implicitly depends upon all of these. It should, however, reflect the extent of the actual probing of the earth's crust for petroleum deposits.[2] Some factors that have been considered but not generally used include total expenditures for exploration, total wells drilled, total footage drilled, and total exploratory footage with dry holes allocated between oil and gas. Time has been used as the independent variable but is not considered entirely satisfactory because exploratory effort is not always proportional to time.[4,2] One yardstick that has been successfully used is simply the number of exploratory wells which have been drilled.[1] It was applied where reservoirs, which were within the same geologic formation were widely scattered throughout a geographic area. Another yardstick that has been successfully used is the total exploratory footage drilled.[2] The later can be used where a study is being made of an entire province or national area, where many different formations are found at various different depths, and where there are multiple zones at different depths within one field.

Target area is used in Eq. (8-1) both literally and symbolically. In some studies, where the chance of finding a field depended largely upon the geographical extent of the field, the actual areas of the fields have been used as the measure.[1] In some cases it has been possible to develop a relationship between area and recoverable reserves. In many cases, data on the areal extent of the reservoirs are not readily available, and the recoverable reserves themselves are used as a measure of the extent of the target. In any case, the effectiveness depends upon the remaining target area, which is the difference between the original and what has already been discovered. This is, of course, a decreasing quantity, and the relations that are developed later reflect a diminishing effectiveness of discovery efforts.

*Total acre-feet of productive formations has been used where past production history was not available.[8]

8.2 EXPECTED DISCOVERIES ESTIMATED BY AREAS OF PRODUCING FIELDS

8.2-1 Ideal Regional Qualities of Study Areas

Economic studies based primarily on areas of individual fields are greatly enhanced when costs of wells and geologic conditions controlling accumulations are the same throughout the study region. For making an initial study, it will be assumed that these conditions are approximately satisfied in a region of interest. Generally this means that only a limited number of producing formations at approximately the same depth are included in the study. Wells that are drilled to approximately the same depth and completed in similar formations tend to cost about the same. Where the geologic conditions that caused the accumulation of minerals are of similar nature throughout a region, the relative difficulty of locating the reservoirs should not change. That is, if the same technology is applied, the effectiveness of future discovery efforts should be a continuation of trends exhibited by past efforts.

The area method presumably was developed for studying the Denver–Julesburg Basin.[1] The study area was comprised of 5,700,000 acres located in Colorado and Nebraska. Formations of interest were the D and J zones of the Dakota Sandstone series. On the east flank these formations slope gently westward at about 25 ft/mile. Oil accumulation is in sand lenses. The oil fields were classed as 8.4% caused by structural, 23.1% combined structural–stratigraphic, and 68.5% caused by purely stratigraphic characteristics of the sediments. In 1958 drilling costs were \$2 per foot. A dry exploratory well could be drilled, logged, and, if necessary, straddle-packer-drill-stem-tested for about \$20,000. If the well were completed, the total cost would rise to about \$60,000. The bulk of the difference went to pay for oil string casing, tubing, cementing, perforating, and surface producing equipment.

The economic attractiveness depends upon the total number of wells that must be drilled. In the Denver–Julesburg Basin, the spacing had been established at 40 acres per well. For that region a good correlation was found between the size of a field and the number of dry wells drilled around the periphery, as follows:

$$\text{(number of dry extension wells)} = 0.76A^{0.345},$$

where A is the area of the field in acres. The study of that particular region was enhanced by finding a good correlation between the size of a field and the ultimate recovery, as follows:

$$[\text{ultimate recovery (bbl oil)}] = 530A^{1.275}.$$

Rather interestingly, the latter equation indicates that effective recovery per acre increases in the larger fields. This is explained by the fact that the sand lenses tend to be thicker with increasing size. Also, they tend to contain proportionately less shale, which permits higher oil recoveries.

8.2-2 Discovery Effectiveness Relations

In order to develop an economic relationship for predicting expected discoveries, Eq. (8-1) is herein written in terms of individual oil fields.

$$\begin{pmatrix}\text{fields}\\ \text{discovered}\end{pmatrix} \propto \begin{pmatrix}\text{exploratory}\\ \text{wells drilled}\end{pmatrix} \cdot \begin{pmatrix}\text{target}\\ \text{area}\end{pmatrix}. \tag{8-2}$$

This can be written on a per-exploratory-well basis as follows:

$$\begin{pmatrix}\text{fields discovered}\\ \text{per exploratory well}\end{pmatrix} \propto \begin{pmatrix}\text{number of}\\ \text{undiscovered fields}\end{pmatrix} \cdot \begin{pmatrix}\text{area}\\ \text{per field}\end{pmatrix}. \tag{8-3}$$

If the fields are divided into class sizes, so each class contains only fields of approximately the same size, the relation can be written in differential form, for the *i*th class, as follows:

$$\frac{dK_i}{dW} = C_1(K_{oi} - K_i)A_i. \tag{8-4}$$

Integration yields the following useful relationship:

$$K_i = K_{oi}(1 - e^{-C_1 A_i W}), \tag{8-5}$$

where

K_i = number of fields, in the *i*th class, which have been discovered in the region,

K_{oi} = number of fields, in the *i*th class, originally present in the region,*

A_i = average areal extent of fields within the *i*th class, acres,

W = total number of exploratory wells that have been drilled in the region,

C_1 = proportionality constant.

In obtaining Eq. (8-5), the constant of integration is evaluated by setting $K_i = 0$ when $W = 0$, when no wells were drilled.

Something about the nature and value of the constant C_1 can be learned

*The Symbol K_{oi} and the symbol M_i, which was used in discussing the multinomial and hypergeometric distributions, both refer to number of fields, within a certain size range, originally present in a region.

by considering the case of purely random drilling within the trend area. Such happening might result from chance, from promotional deals, from expiring leases. For such a case, the probability of discovering a field should be equal to the ratio of the combined areas of the undiscovered fields divided by the total area. If the area of the already discovered fields is small compared to the total area of the region, or basin, one may write

$$\begin{pmatrix}\text{probability of}\\ \text{a discovery}\end{pmatrix} = \frac{\text{area of undiscovered fields}}{\text{total area of the region}}, \tag{8-6}$$

$$\frac{dK_i}{dW} = \frac{(K_{oi} - K_i)A_i}{B}. \tag{8-7}$$

Integration of this equation yields

$$K_i = K_{oi}(1 - e^{-A_iW/B}), \tag{8-8}$$

from which it is apparent that, for the case of random drilling,

$$C_1 = \frac{1}{B},$$

where B = area of the entire region, or basin, acres.

Some limiting values are brought to light by considering pattern drilling. For any part of a region where wells have been uniformly spaced near the center of every 640-acre tract, all the fields that are 640 acres or more in size must have been discovered. That is,

$$K_{oi} = K_i$$

where $A_i > B/W$.

Most drilling is done on geological or geophysical leads. The efficiency should be better than for random drilling. To account for this, an effectiveness factor J will be introduced, which gives the relation

$$C_1 = \frac{J}{B}, \tag{8-9}$$

where

J = relative effectiveness of discovering an oil field compared to random drilling.

Equation (8-5) can then be written as

$$K_i = K_{oi}(1 - e^{-JA_iW/B}). \tag{8-10}$$

Determining the proper value of J is one of the problems involved in the use of this method. For random drilling, $J = 1.0$. Statistical studies of the American Association of Petroleum Geologists indicate that between 1944 and 1956 the success ratio of wells located on geological and/or geophysical advice was 2.75 greater than for random drilling. Values of $J = 2.0$ and $J = 3.0$ have been used. In a particular situation, an estimate is possible of whether the effectiveness J should be greater or less than the national average. If sufficient data are available, then values of J can be determined that best fit the data.

8.2-3 Calculation of Anticipated Discoveries

Two procedures for estimating expected discoveries by the area method will be discussed. They vary only in detail. First the essential steps used by Arps and Roberts can be enumerated as follows[1]:

1. Group the data by field sizes. Successive class boundaries were increased by a factor of 2. When there is a good correlation between area and ultimate recovery, it makes little difference which is used. In the example cited, the ultimate recovery was preferred.
2. Plot A_i versus R_{Ui}, where R_{Ui} is the ultimate recovery of the average-sized field in the ith class.
3. Plot K_i versus R_{Ui}.
4. Calculate K_{oi} for each class of fields. Equation (8-10) is used for this purpose. In this calculation, use a smoothed value of A_i from step 2 and a smoothed value of K_i from step 3, located at the R_{Ui} class mark of each group.
5. Calculate a new K_i value for each class, for an assumed new and greater value of W.
6. Calculate the total expected ultimate recovery discovered by all exploratory wells up through the new value of W. The following applies:

$$R_U = \sum K_i R_{Ui}. \tag{8-11}$$

7. The value of interest is ΔR_{Ui}, which is the difference between the new R_U value from step 6 and the original R_U value corresponding to the former value of W. Thus one can calculate $\Delta R_U/\Delta W$, the expected discoveries per exploratory well. Dividing this figure by the cost per exploratory well gives the discovery cost in dollars per barrel.

Some of the values obtained by this method are shown in Table 8.1. Following is a calculation made for class 7 after 3,705 exploratory wells had been drilled and 36 fields of this size had been discovered.

Table 8.1
Calculation of Ultimate Number of Fields in Each Size Class to be Found in the Denver–Julesburg Basin

Class	*Number of Fields After 3,705 Wildcats*	*Number of Fields (Smoothed Data)*	*Estimated Average Ultimate Recovery (M bbl)*	*Average Production Area (acres)*	*Estimated Ultimate Number of Fields*
1	3	3.0	1.7	2.5	967.7
2	8	7.4	3.0	3.9	1,453.8
3	14	13.0	5.3	6.1	1,651.8
4	23	25.0	12.2	11.7	1,656.7
5	26	35.0	23.4	19.4	1,401.1
6	52	43.0	47.6	34.0	994.4
7	36	46.0	87.5	54.7	669.4
8	45	45.0	181.3	97.3	378.7
9	38	40.5	390.7	178.2	195.8
10	34	34.0	791.0	390.8	102.6
11	28	27.5	1,420.7	490.5	58.3
12	17	17.0	2,944.7	868.5	25.1
13	10	10.0	4,962.0	1,306.0	12.2
14	3	3.2	10,987.0	2,442.0	3.3
15	1	0.5	50,750.0	8,108.0	0.5

Source: After Arps and Roberts.[1]

$$\frac{JA_iW}{B} = \frac{(2.0)(54.7\text{ acres})(3{,}705)}{5{,}700{,}000\text{ acres}}$$

$$= 0.0711,$$

$$K_{oi} = \frac{K_i}{1 - e^{-JA_iW/B}}$$

$$= \frac{36}{1 - e^{-0.0711}}$$

$$= 669.4 \text{ total fields of this size originally present in the region.}$$

Next an estimate is made of the number of fields in this same size range which will be discovered by drilling an additional 1,000 exploratory wells.

$$\frac{JA_iW}{B} = \frac{(2.0)(54.7\text{ acres})(4{,}705)}{5{,}700{,}000\text{ acres}}$$

$$= 0.0903,$$

$$K_i = K_{oi}(1 - e^{-JA_iW/B})$$

$$= 669.4(1 - e^{-0.0903})$$

$$= 57.8 \text{ fields of size class 7.}$$

The increase in recovery from fields in this class size, discovered by the next 1,000 exploratory wells, would be

(57.8 − 36.0 fields)(87.5 M bbl) = 1,908 M bbl from fields of size class 7.

In order to complete the solution, similar calculations would have to be made for all 15 classes listed in Table 8.1. That is, 15 different values of A_i would be used in the calculations. This would permit adding together the new reserves from fields in each class size to determine the total new reserves found by drilling the additional exploratory wells.

Diminishing returns are shown by continuing the drilling of more exploratory wells. For example, the next increment of 1,000 exploratory wells, which would result in $W = 5{,}705$, would yield an additional 1,015 M barrels of reserves from fields in the size class 7. This is about half of that obtained from the immediately preceding group of 1,000 exploratory wells. Of the discoveries made, fewer and fewer are in the larger field classes as exploration is continued within a region. The average size of a discovery, as well as the number of discoveries, decreases. These relations are illustrated in Fig. 8.2,

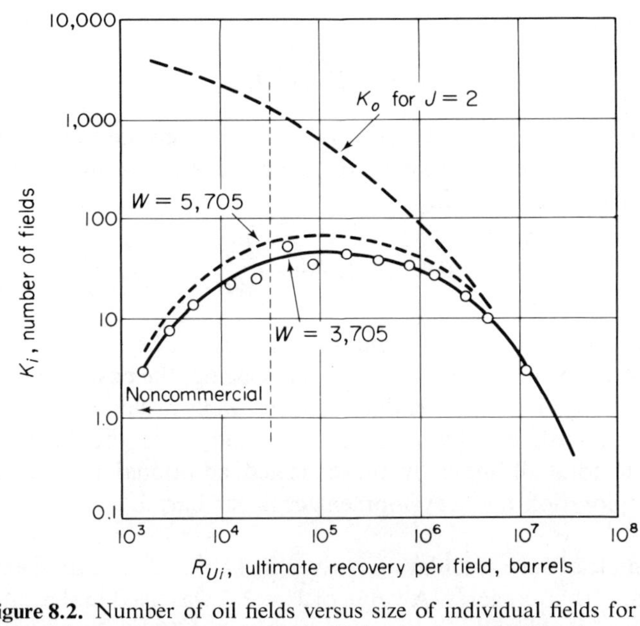

Figure 8.2. Number of oil fields versus size of individual fields for east flank of the Denver-Julesberg Basin. (After Arps and Roberts[1])

where it may be observed that the limiting number for larger field has already been reached, while many additional smaller ones remain to be discovered.

The total drilling costs include the exploratory wells, some of which are completed as producers. They also include development wells which are completed as producers, whose number depends upon the acreage sizes of the fields and the well spacing. Also there are the development wells which turn out to be dry and are not completed. In addition, there are the leasing costs, geological and geophysical costs, and overhead costs. All contribute to the exploration and development costs.

A second calculation procedure, which is intentionally equivalent to the preceding, has been given by Rausch.[7] The areal size of fields is used throughout as the basis of calculation. The steps can be enumerated as follows.

1. Group the data into classes according to the areal size of the fields. Let A_i be the class mark (average size, in acres) of the fields in the ith class.

2. Calculate K_{oi} for each class of fields. Equation (8-10) is used for this purpose.

3. Calculate a new K_i value for each class, for an assumed new and greater value of W. Let ΔW be the total number of new exploratory wells drilled. Let ΔK_i be the number of new fields expected to be discovered in each size class.

4. The total new productive acreage expected to be discovered can be calculated as follows:

$$\Delta A = \sum \Delta K_i A_i. \tag{8-12}$$

5. Where the recovery per acre is approximately constant within any particular areal size class of fields, the expected increase in reserves can be calculated from the following:

$$\Delta R_U = \sum \Delta K_i A_i \frac{R_{Ui}}{A_i}. \tag{8-13}$$

6. Total exploration and development costs can be estimated, based on ΔW and ΔA. Of prime interest are the well spacing, the cost of dry and completed wells, and the number of dry extension wells drilled around each field. Leasing, geological, geophysical, and overhead costs should be included. Dividing the total of these by the expected additional reserves gives the expected exploration and development cost per barrel.

An example of such calculation is given in Table 8.2. The total exploration region covers 100,000 acres. A value of $J = 2.5$ was used in the calculations. At the time of making the calculations, a total of 50 exploratory wells had been drilled in the region. The calculation is carried through step 4, for the purpose of determining the total number of producing acres that can be projected for expected discovery by the next 50 exploratory wells.

Table 8.2
Calculation of Expected Discovery Acreage for Drilling 50 Additional Exploratory Wells

	(Find K_{oi} for each A_i.)		
A_i	3,000	1,000	500
W	50	50	50
JA_iW/B	3.75	1.25	0.625
K_i	2	3	5
K_{oi}	2.0482	4.2046	10.7573
	(Find new K_i for $W = 100$.)		
JA_iW/B	7.50	2.50	1.25
K_i	2.0471	3.8595	7.6753
	(Find incremental acres.)		
ΔK_i	0.0471	0.8595	2.6753
ΔK_iA_i	141	860	1,338
	Total $= 141 + 860 + 1{,}338 = 2{,}339$ acres		

Source: After Rausch.[7]

8.3 EXPECTED DISCOVERIES ESTIMATED BY TOTAL EXPLORATORY FOOTAGE

8.3-1 Advantages of a Footage Model

Exploratory effort can be equated to the number of exploratory wells drilled. Also it can be reckoned by the total exploratory footage drilled.[2] Where the exploratory wells are all of approximately the same depth, as for testing the same formation, either basis would be exactly equivalent to the other. However, where one well is drilled to test a formation at 6,000 ft, another to test a formation at 10,000 ft, and in another well both of these are tested, it becomes difficult to convert from one to the other. Quite evidently the footage model is more flexible. It has an advantage where the geology remains obscure during the exploration period. It has kinship to the concept of estimating possible reserves on the basis of a cubic mile of sediments, since both depth and area are involved in cubic measure.

Ultimately, to be useful any relationships developed must be converted into exploration costs in terms of dollars per barrel of oil discovered. Where exploration costs differ among several regions, the economic limits also vary, and exploratory drilling will first cease in the higher-cost regions. Thus studies by individual geographical regions are indicated. Individual studies by depth intervals are also indicated, since drilling costs often increase rapidly with depth. This permits developing the detailed economic information necessary to justify exploratory drilling to a given depth.

8.3-2 Initial Evaluation of Discoveries

The first relationship needed concerns the effectiveness of finding new discoveries. These will first be considered as they are evaluated in their discovery year. This problem can be approached by writing Eq. (8-1) in terms of resources discovered per foot of exploratory drilling:

$$\begin{pmatrix}\text{reserves discovered}\\ \text{per exploratory}\\ \text{foot drilled}\end{pmatrix} \propto \begin{pmatrix}\text{reserves}\\ \text{remaining to}\\ \text{be discovered}\end{pmatrix}. \tag{8-14}$$

If the ultimate reserves R_U are achieved after an ultimate growth ratio r_U, the ultimate reserves as discovered are expressed by the ratio R_U/r_U. This permits the relation

$$\frac{dR_D}{dF_D} = a\left(\frac{R_U}{r_U} - R_D\right), \tag{8-15}$$

where

R_D = cumulative resources discovered, as they were evaluated in their discovery years, barrels,

F_D = total exploratory footage drilled at the time of discovery, feet,

a = proportionality constant,

R_U = total resources ultimately to be discovered, including their subsequent growth after discovery, barrels,

r_U = ultimate growth ratio of the discovered resources.

Integration of Eq. (8-15) is evaluated by letting $R_D = 0$ when $F_D = 0$, to give

$$R_D = \frac{R_U}{r_U}(1 - e^{-aF_D}). \tag{8-16}$$

A convenient expression for the effectiveness of exploration can be obtained by differentiating the above equation, to obtain

$$\frac{dR_D}{dF_D} = \frac{aR_U}{r_U}e^{-aF_D}. \tag{8-17}$$

Since dR_D/dF_D is the effectiveness of discoveries per exploratory foot, and since the coefficient on the right must be the initial effectiveness when $F_D = 0$, the equation can be written

$$E = E_0 e^{-aF_D}. \tag{8-18}$$

The following relation is noted also:

$$E_0 = \frac{aR_U}{r_U}. \tag{8-19}$$

The decline in discovery effectiveness versus exploratory footage drilled in the United States is shown in Fig. 8.3.

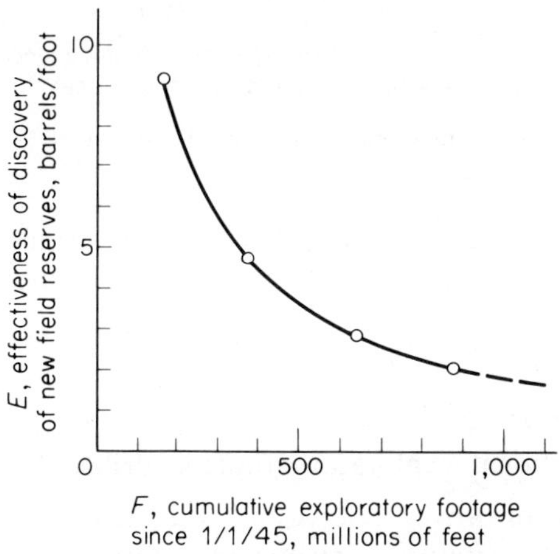

Figure 8.3. Effectiveness of exploration, crude oil, United States excluding Alaska. (After Arps et al.[2])

8.3-3 Growth of Proved Reserves After Discovery

The proved reserves of a new field tend to grow. An initial evaluation is made in the discovery year. The reserves reported in later years include new reservoirs found by extending the area laterally by drilling outpost extension wells. The reserves may be increased by finding deeper reservoirs and also by exploiting shallower reservoirs. This is shown in Fig. 8.4. The variable quantity of interest is the additional exploratory footage drilled since the initial discovery. A relationship can be developed as follows:

$$\begin{pmatrix}\text{increase in growth}\\ \text{ratio per additional}\\ \text{exploratory foot drilled}\end{pmatrix} \propto \begin{pmatrix}\text{remaining possible}\\ \text{increase in}\\ \text{growth ratio}\end{pmatrix}, \tag{8-20}$$

$$\frac{dr}{d(F - F_D)} = b(r_U - r). \tag{8-21}$$

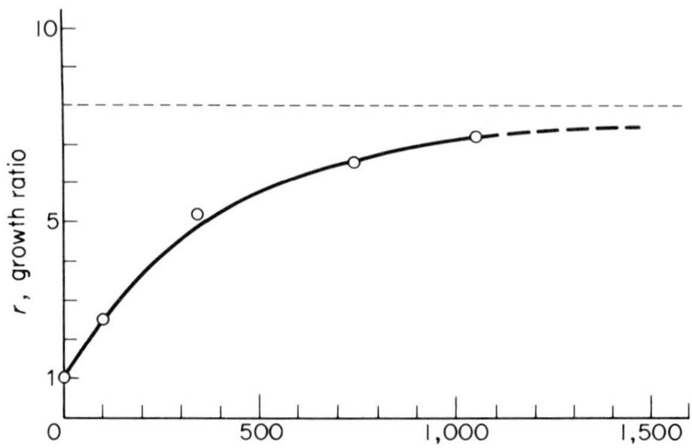

Figure 8.4. Growth of new field reserves, crude oil, United States excluding Alaska. (After Arps et al.[2])

The constant of integration can be evaluated, since $r = 1$, when the variable quantity $(F - F_D)$ is equal to zero. The integrated relation becomes

$$r = r_U - (r_U - 1)e^{-b(F-F_D)}, \tag{8-22}$$

where

$F - F_D$ = additional exploratory footage drilled beyond discovery,

r = ratio of proved reserves as evaluated at footage F divided by proved reserves as evaluated at discovery footage F_D,

b = proportionality constant.

8.3-4 Evaluation of Total Proved Resources

Consider a region where the total exploratory footage now drilled has a value of F. Total resources amounting to R have been developed. This has resulted both from initial discoveries and from their subsequent growth. During the time when F_D, the discovery footage, varied from zero to F, the following equation describes the accumulating process:

$$dR = Er\,dF_D, \tag{8-23}$$

where R = total resources as evaluated at total exploratory footage F.

Discoveries are accumulated between $F_D = 0$ and $F_D = F$. The growth footage on any discovery is $F - F_D$. The following equation expresses these ideas:

$$\int_0^R dR = \int_0^F (E_0 e^{-aF_D})[r_U - (r_U - 1)e^{-b(F-F_D)}]\,dF_D.$$

Integration gives the relation

$$R = R_U[1 + C_2 e^{-bF} - (1 + C_2)e^{-aF}], \tag{8-24}$$

where

$$C_2 = \frac{a}{b-a}\frac{r_U - 1}{r_U}. \tag{8-25}$$

8.3-5 Concept of Rolled-Back Reserves

Insight into the total discovery process can be gained by a backward calculation. At total exploratory footage F, the resources discovered at footage F_D have grown considerably. By taking this latter growth into account, one can calculate backward and determine how much oil, or other resource, must have been located and essentially discovered at exploratory footage F_D. For this calculation, Eq. (8-23) is again integrated. This time, the discoveries are permitted to accumulate only to footage F_D. The discoveries grow, however, from their initial discovery footage up to total footage F. The following equation expresses this:

$$\int_0^{R_{RB}} dR = \int_0^{F_D} (E_0 e^{-aF_D})[r_U - (r_U - 1)e^{-b(F-F_D)}]\, dF_D.$$

Integration yields

$$R_{RB} = R_U[1 + C_2 e^{-bF} - e^{-aF_D}(1 + C_2 e^{-b(F-F_D)})], \tag{8-26}$$

where

R_{RB} = rolled-back reserves, cumulative discoveries at discovery footage F_D adjusted for growth to total discovery footage F.

8.3-6 Application of the Exploration Model

In order to use any of the various relations for predicting future performance, it is necessary first to obtain a match with available historical data. These historical data consist of a series of yearly estimates of the proved reserves, a series of successive values of R. Also available is a record of the corresponding exploratory footage drilled during each yearly period. These, however, are not values of F, but incremental values of F. Thus a starting value F_0 is one of the parameters that must be determined in fitting the historical data. The parameters that must be determined are F_0, a, b, r_U, and R_U. The values chosen should minimize the standard deviation between the historical data and the mathematical model. A method that has been recommended consists of successive numerical approximations.[2,8] The optimum value of each parameter was obtained by scanning the entire range of that parameter, using suitable increments and holding other parameters constant. This scanning narrows the range in which each parameter should be located. The procedure is repeated using a smaller range and smaller increments for each parameter in successive passes. With the aid of a computer to perform the calculations, the method converges rapidly.

Figure 8.5 shows an analysis of exploration in the United States (excluding Alaska). Data used was for the period from 1945 to 1969. A Rolled-Back Reserves curve is shown from the final data point. There is, of course, an

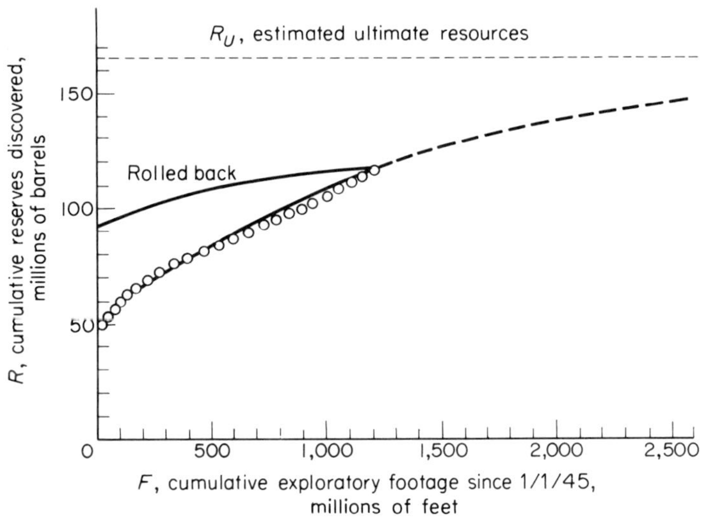

Figure 8.5. Exploration model, crude oil, United States excluding Alaska. (After Arps et al.[2])

infinite number of these, and a Rolled-Back Reserves curve can be calculated from any point on the Cumulative Resources curve. Figure 8.6 illustrates a case where the effectiveness of discovery has changed. This condition can occur when one or more new geologic basins is developed within the geographical area. In such a case the old and the new regions can be analyzed separately and then added together to obtain the final results. Improved technology in locating wells and exploiting reservoirs, particularly if the improvements occur suddenly, could result in a discontinuity. For such cases a new set of parameters would need to be calculated for the data following the discontinuity. Such analyses must, of course, be recognized as trend analyses whose validity in forecasting depends upon a continuation of conditions controlling the original data.

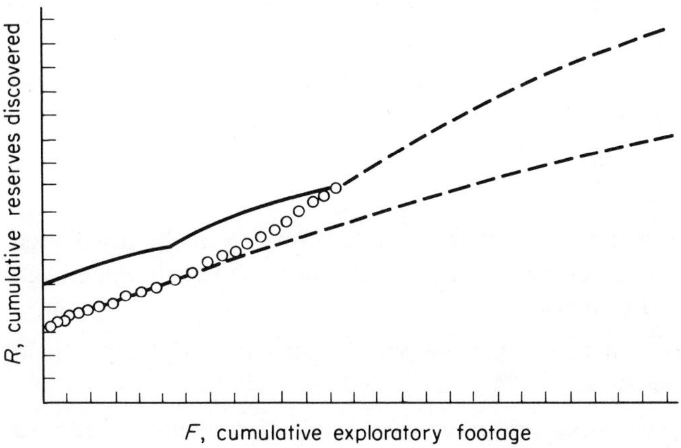

Figure 8.6. Exploratory model, illustrating a discontinuity in discovery rates. (After Arps et al.[2])

NOMENCLATURE

A = area of an oil field, acres,

A_i = average area of fields in the ith class, acres,

a = proportionality constant for new discoveries,

B = total area of a region, acres,

b = proportionality constant for growth of reserves,

C_1 = proportionality constant for area model,

C_2 = constant in footage model,

E = new discoveries effectiveness for footage model,

E_0 = original new discoveries effectiveness for footage model,

F = total exploratory footage drilled, feet,

F_D = total exploratory footage drilled at the time of a discovery, feet,

F_0 = total prior exploratory footage drilled, choosen to fit present data, feet,

i = subscript denoting the ith class,

J = relative effectiveness of discovering an oil field compared to random drilling,

K_i = total number of fields discovered in a region, which are the ith class size,

K_{oi} = total number of fields originally present in a region, which are in the ith class size,

R = cumulative producible resources discovered, evaluated at cumulative exploratory footage F, barrels,

R_D = cumulative producible resources discovered, evaluated in their discovery year at footage F_D, barrels,

R_U = total cumulative resources ultimately to be discovered at infinite exploratory footage, including their subsequent growth after discovery, barrels,

R_{Ui} = ultimate cumulative production from an average field in the ith class size,

R_{RB} = rolled-back reserves, cumulative discoveries at exploratory footage F_D, adjusted for growth to total exploratory footage F, barrels,

r = growth ratio of cumulative producible resources evaluated at footage F, divided by their estimated value at discovery footage F_D,

r_U = ultimate growth ratio of the discovered resources,

W = total number of exploratory wells drilled within a region.

REFERENCES

1. ARPS, J. J., and T. G. ROBERTS, "Economics of Drilling for Cretaceous Oil on East Flank of Denver–Julesburg Basin." *Bulletin of the AAPG*, Vol. 42, No. 11 (Nov. 1958), pp. 2549–2566.

2. ARPS, J. J., M. MORTADA, and A. E. SMITH, "Relationship Between Proved Reserves and Exploratory Effort." *Journal of Petroleum Technology*, June 1971, pp. 671–675.

3. POTENTIAL Gas Committee: "Potential Supply of Natural Gas in the United States as of Dec. 31, 1968." Golden, Colo.: Potential Gas Agency, Mineral Resources Institute, Colorado School of Mines Foundation, Inc., Oct. 1969.

4. HUBBERT, J. K., "Energy Resources" (Publication No. 1000-D). Washington, D.C.: National Academy of Science, National Research Council, 1962.

5. MOORE, C. L., "Method for Evaluating U.S. Crude Oil Resources and Projecting Domestic Crude Oil Availability." Washington, D.C.: Office of Oil and Gas, U.S. Department of Interior, May 1962.

6. RYAN, J. M., "Limitations of Statistical Methods for Predicting Petroleum and Natural Gas Reserves and Availability." *Journal of Petroleum Technology*, Mar. 1966, pp. 281–287.

7. RAUSCH, R. W., "Risk Analysis in Drilling Ventures." Presented at Oklahoma City Section of SPE of AIME, Oct. 28, 1969.

8. RIOS, N. G., "An Exploratory Model for the Prediction of Future Recoverable Reserves in the Seminole Region of Oklahoma." M.Sc. Thesis, University of Oklahoma, 1972.

PROBLEMS

8-1. To date 40 wildcats have been drilled in a 1,500-mile^2 target area. The following fields have been found:

2 fields of 4,000 acres,

3 fields of 1,000 acres,

6 fields of 500 acres.

Drilling is on seismic leads, with an estimated efficiency factor of 3.0. Estimate how many acres will be found productive by drilling an additional 40 wildcat wells.

8-2. It may be noted that dK_i/dW is the probability of discovering a field of the ith class size, per exploratory well. Referring to Problem 8-1, at the time when the first 40 wildcats have been drilled, what are the probabilities that a well will discover a field of 4,000 acres, a field of 1,000 acres, a field of 500 acres? What is the probability that the well will be dry? Repeat these calculations for the time just after completing the second 40 wildcat wells.

9

BAYES STRATEGIES AND ESTIMATES OF VALUE

9.1 INTRODUCTION

In this chapter the elementary Bayes probability concepts are reviewed. These are the methods that point out the most probable cause of an event, where there are several possible causes. Bayes strategies are concerned with the likelihood of making a wrong choice and the economic loss associated with making the wrong choice. Procedures are formulated that minimize the expected costs of error in routine operations. Decision rules are formulated for a special problem in petroleum exploration activities. An example is given on judging the quality of the information that becomes available during the development period of a petroleum producing region.

9.2 BAYES PROBABILITY METHODS

Bayes methods are characterized by introducing initial probabilities that certain events, or states of nature, are true, and then revising these estimates as more information becomes available. Consider a case where three boxes are received from New York, Chicago, and Los Angeles, respectively. If the receiving section somehow removed all the labels, the initial or a priori probability that any particular one came from New York would be one-third. But as one opens the boxes and begins to sample the contents, the probabilities become more and more certain regarding which box is which. The revised estimates are quite commonly referred to as the a posteriori probabilities. In a certain prospective oil-producing region, it might be estimated initially that 20% of the exploratory wells would be productive. But as drilling proceeds, the additional information gained would permit revision of the initial estimates. The economic consequences of the various possible actions is also a factor considered in Bayesian analysis, and the Bayes estimate is the one characterized by the least expected cost of error.

The concept of revising prior estimates so that they reflect new information is attributed to Thomas Bayes, an English clergyman. His basic formula is portrayed geometrically in the Venn diagram shown in Fig. 9.1. In this

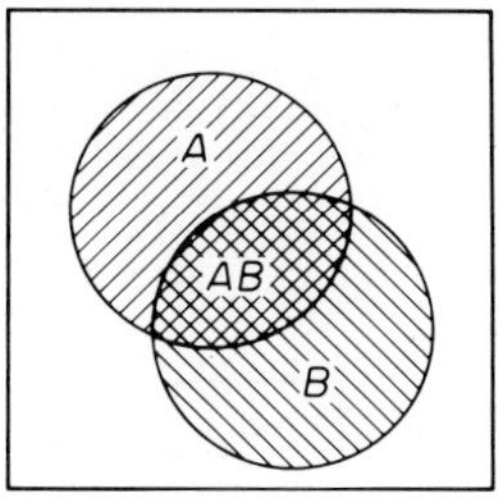

Figure 9.1. Conditional probability relations.

figure the total area enclosed in the square is unity, the total sample space. The areas labeled A and B represent the respective probabilities of events A and B, and their intersection AB indicates that there is some probability that the outcome of a trial will include the occurrence of both events A and B. The following relations apply:

$$p(AB) = p(B \mid A)p(A). \tag{9-1}$$

This reads: the probability of both events A and B occurring equals the probability of B given that A has occurred times the probability that A will occur. Geometrically, the fractional area AB equals the product of two fractions, first the fraction of B within A, and second the fraction of A within the total sample space.

This can as well be written as

$$p(AB) = p(A \mid B)p(B),$$

and the equality

$$p(B \mid A)p(A) = p(A \mid B)p(B)$$

is obtained, which can be written as

$$p(B \mid A) = \frac{p(A \mid B)p(B)}{p(A)}. \tag{9-2}$$

A more productive display of the concept is shown in Fig. 9.2, where three mutually exclusive events B_1, B_2, and B_3 are illustrated. These might be, for example, the outputs of three machines which produce identical bolts. The probability of event B_1 would be interpreted geometrically as the area assigned to B_1 divided by the total areas of B_1, B_2, and B_3. Event A is an event that can happen in any of the B spaces. For example, any one of the machines might produce a defective part. Given that one is concerned only with machine B_1, the conditional probability of a defective part is the intersection area AB_1 divided by the total area B_1. Physically, it is the fraction of the output of machine B_1 that is defective, such as 0.5% defective of all the items produced by machine B_1.

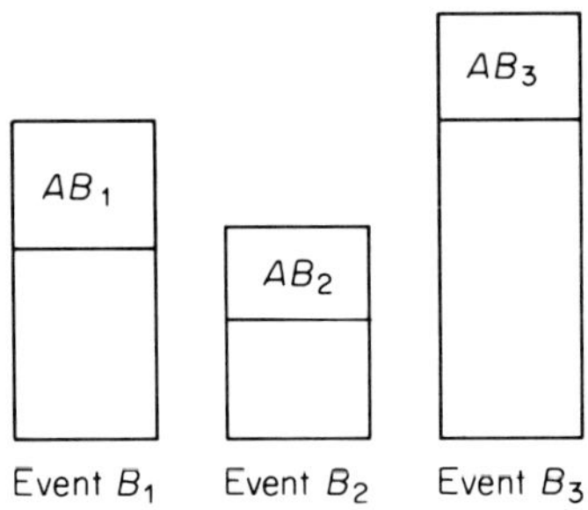

Figure 9.2. Conditional probability within several mutually exclusive events.

Development of a relationship that includes all the independent events, or rather their probabilities, along with the conditional probabilities within each independent event, can be developed as follows:

$$\text{Total sample space} = 1 = p(B_1) + p(B_2) + \cdots + p(B_N),$$

$$p(A) = p(A \mid B_1)p(B_1) + p(A \mid B_2)p(B_2) + \cdots + p(A \mid B_N)p(B_N),$$

$$p(B_1 \mid A) = \frac{p(A \mid B_1)p(B_1)}{p(A)}. \tag{9-3}$$

In more general terms, this relation may be written

$$p(B_i \mid A) = \frac{p(A \mid B_i)p(B_i)}{\sum_{i=1}^{N} p(A \mid B_i)p(B_i)}. \tag{9-4}$$

The last relation is known as *Bayes Theorem*, and its use is illustrated by the following two problems.

Example 1: There are two boxes, B_1 and B_2, from which the labels have been removed. However, it is known that B_1 contains 2 white and 3 black balls; and that B_2 contains 1 white and 4 black balls. A sample of 2 balls is drawn from one of the boxes and it turns out that one is white and one is black. What is the probability that the box from which the sample was drawn is B_1?
Solution: There is originally no basis other than a 50% chance that the box is, in fact, B_1.

$$p(B_1) = 0.5,$$
$$p(B_2) = 0.5.$$

Event A is drawing 1 white and 1 black, in either order:

$$p(A \mid B_1) = \tfrac{2}{5}\tfrac{3}{4} = 0.3,$$
$$p(A \mid B_2) = \tfrac{1}{5}\tfrac{4}{4} = 0.2,$$
$$p(B_1 \mid A) = \frac{p(A \mid B_1)p(B_1)}{p(A \mid B_1)p(B_1) + p(A \mid B_2)p(B_2)}$$
$$= \frac{(06)(0.5)}{(0.6)(0.5) + (0.4(0.5)}$$
$$= 0.6.$$

Therefore, after drawing the specified sample from one of the boxes, it now appears that there is a 60% probability of that box being box B_1.

Example 2: Three machines, B_1, B_2, and B_3, are making identical bolts in a factory. The output of B_1 averages 0.3% defectives, B_2 averages 0.8%, and B_3 averages 1.0% defectives. A total of 1,000 bolts from the three machines are mixed in the same bin, 500 from B_1, 350 from B_2, and 150 from machine B_3. What are the probabilities that a defective bolt came from each particular machine?
Solution: The same equation is used to obtain the desired answers. It is often convenient to set such solutions in the form of a table. From Table 9.1 it may

Table 9.1
Solution of Bayes Theorem

Machine	$p(B_i)$	*A Priori Conditional* $p(A \mid B_i)$	*Unconditional* $p(A \mid B_i)p(B_i)$	*A Posteriori Conditional* $p(B_i \mid A)$
B_1	0.50	0.003	0.0015	0.26
B_2	0.35	0.008	0.0028	0.48
B_3	0.15	0.010	0.0015	0.26
	1.00		0.0058	1.00

be seen that there is a 26% chance that the defect came from B_1, a 48% chance that it came from B_2 and a 26% chance that it came from machine B_3.

9.3 THE MAXIMUM-LIKELIHOOD METHOD

There sometimes occurs the problem of identifying the state of nature from information obtained from a sample. Consider a geologic region in which the formations are faulted. There are three limestone formations, which are all partially converted to dolomite. A well has been drilled through one of these formations, but it is not known which one. Thus, given the presence of the limestone, which one is it? Mathematically, if Y stands for a limestone formation, the quantity of interest is $p(Y_i \mid Y)$. The maximum-likelihood method consists of estimating the value of Y to be that one of its possible values which maximizes the likelihood function.

The three limestone formations have been studied by analyzing logs of wells already drilled. The amount of dolomitization varies in different locations. However, in formation A most of the dolomite present occurs in the top portion, in formation B most of the dolomite present occurs in the middle portion, and in formation C most of the dolomite present occurs in the bottom portions of the formation. The average results are shown in Table 9.2.

Table 9.2
Location of Dolomite in Three Limestone Formations

	Fraction of Dolomite Present		
	In Top	*In Center*	*In Bottom*
Formation A	0.75	0.15	0.10
Formation B	0.35	0.50	0.15
Formation C	0.15	0.25	0.60

Geological formations vary laterally of course. However, the information displayed in the table gives some basis for suggesting that the formation penetrated by the drill is either formation A, or B, or C, depending upon whether the greatest fraction of the dolomite in the formation penetrated is in the top, the middle, or the bottom of the formation.

9.4 BAYES STRATEGIES

9.4-1 Losses Due to Errors

A Bayes strategy consists of selecting from among alternative choices the one that has the least expected cost of error. Let Y represent the unknown state of nature, where Y can assume one of two possible values, Y_1 or Y_2. In many instances a choice must be made, even though it proves to be wrong; otherwise, all operations would cease. The particular choice that might be made is of no consequence unless some loss is associated with the wrong choice, either an out-of-pocket loss or a lost opportunity for gain. Such a case is illustrated in Table 9.3. If the state of nature is actually Y_1; and the

Table 9.3
Costs of Error Where There Are Two Possible Choices (in dollars)

Unknown State of Nature	*Choice (or Estimate) of Decision Maker*	
	Y_1	Y_2
Y_1	0	300
Y_2	500	0

decision maker chooses Y_1, there is no error and no loss due to error. But if he should choose Y_2, there would be a loss due to error. If the probabilities of error can be calculated, the expected loss of error can be calculated in each case, and by choosing the smallest expected loss, the cost of error will be less in the long run. That is, the strategy pays off over a series of repeated trials.

9.4-2 Minimizing Losses in Maintenance Operations

Losses can be minimized in cases where mechanical failures occur from time to time, and successive inspections must be made until the particular item that failed has been located. Choosing to inspect an item that has not failed represents an error, and the cost of the inspection is a cost of error. The costs arise from direct labor and materials used for the inspection and the loss due to the additional time that the facility is out of service, the downtime. The problem here is to determine the expected costs of errors, and then rank them from lowest to highest, so that the procedural order in which the several items should be inspected becomes evident. This strategy of inspecting the items in the order of increasing expected costs of error will result in the lowest costs of error in the long run.

The case that will be considered concerns the failures of control mechanisms on four separate streams of hydrocarbons as they flow through processing facilities in a natural gasoline plant. Each stream is controlled by a separate thermostat, a pressure regulator, and a flow regulator. Failure of any one of these items results in shutting down the stream, shutting down the particular operating unit. Then it is necessary to determine by inspection which of the items serving that particular stream actually failed.

Records kept in the plant provide the basis for the information shown in Table 9.4. This shows that 20% of all shutdowns were due to failures of thermostats, and 50% were due to failures of flow regulators. Furthermore, 30% of all thermostats replaced were used in stream *A*, and 40% of the thermostats were used in stream *C*. Table 9.5 shows the corresponding

Table 9.4
Conditional A Priori Probabilities

Item	*Relative Failures by Items*	*Relative Replacements of Each Item in the Individual Operating Units*			
		Stream A	*Stream B*	*Stream C*	*Stream D*
Thermostat	0.20	0.30	0.15	0.40	0.15
Pressure regulator	0.30	0.20	0.30	0.30	0.20
Flow regulator	0.50	0.40	0.20	0.10	0.30

Table 9.5
Unconditional Probabilities

Item	*Stream A*	*Stream B*	*Stream C*	*Stream D*
Thermostat	0.06	0.03	0.08	0.03
Pressure regulator	0.06	0.09	0.09	0.06
Flow regulator	0.20	0.10	0.05	0.15
	0.32	0.22	0.22	0.24

unconditional probabilities, which are obtained by multiplying the relative failures of items times the relative replacements of that item in each stream. Table 9.6 shows the a posteriori probabilities, where it may be seen that if a

Table 9.6
A Posteriori Probabilities of Failure of an Item

Item	*Stream A*	*Stream B*	*Stream C*	*Stream D*
Thermostat	$\frac{6}{32}$	$\frac{3}{22}$	$\frac{8}{22}$	$\frac{3}{24}$
Pressure regulator	$\frac{6}{32}$	$\frac{9}{22}$	$\frac{9}{22}$	$\frac{6}{24}$
Flow regulator	$\frac{20}{32}$	$\frac{10}{22}$	$\frac{5}{22}$	$\frac{15}{24}$

failure occurs in stream *A*, the probability is $\frac{20}{32}$ that the flow regulator failed. But if a failure occurred in stream *C*, the probability is only $\frac{5}{22}$ that the flow regulator failed.

It now becomes necessary to again consider the definition of error in the case at hand. Only by inspecting an item that has not failed can an error be made. The probability that an item has not failed is (1—the probability that it did fail). This, then, is the probability of error. These values are shown in Table 9.7, and they correspond to the a posteriori probabilities wherein

Table 9.7
A Posteriori Probabilities of Error

Item	*Stream A*	*Stream B*	*Stream C*	*Stream D*
Thermostat	$\frac{26}{32}$	$\frac{19}{22}$	$\frac{14}{22}$	$\frac{21}{24}$
Pressure regulator	$\frac{26}{32}$	$\frac{13}{22}$	$\frac{13}{22}$	$\frac{18}{24}$
Flow regulator	$\frac{12}{32}$	$\frac{12}{22}$	$\frac{17}{22}$	$\frac{9}{24}$

the failure is known to be in a particular stream. The costs of error are shown in Table 9.8, which represent all the costs associated with inspecting each item. In Table 9.9 are shown the expected costs of error, which are the product of the probability of error times the cost of error for each item.

Table 9.8
Costs of Error (in dollars)

Item	*Stream A*	*Stream B*	*Stream C*	*Stream D*
Thermostat	200	150	250	175
Pressure regulator	250	300	150	200
Flow regulator	350	350	400	400

Table 9.9
Expected Costs of Error (in dollars)

Item	*Stream A*	*Stream B*	*Stream C*	*Stream D*
Thermostat	163	129	159	153
Pressure regulator	203	177	74	150
Flow regulator	131	191	308	110

The end result of this calculation is shown in Table 9.10. This shows the recommended order for inspecting the various items, according to each stream. The order is, of course, that of the expected costs of error. For example, if a failure occurs in stream *A*, the flow regulator should be inspected first, next the thermostat, and, if neither of these are faulty, the pressure regulator must be inspected and repaired. However, if a failure occurs in stream *B*, the thermostat should be inspected first, then the pressure regulator, and finally the flow regulator.

An analysis of the results that could be expected from following the rule derived for stream *A* is shown in Table 9.11. The items are arranged in the

Table 9.10
Recommended Order for Inspecting Items in Each Stream

Item	*Stream A*	*Stream B*	*Stream C*	*Stream D*
Thermostat	2	1	2	3
Pressure regulator	3	2	1	2
Flow regulator	1	3	3	1

Table 9.11
Analysis of Costs Resulting from Following the Recommended Inspection Order for Stream *A*

Order	*Item Inspected*	*Number of Failures*	*Costs of Unnecessary Inspection* ($)	*Total Cost of Error* ($)
1	Flow regulator	20	0	0
2	Thermostat	6	350	2,100
3	Pressure regulator	6	350 + 200	3,300
		32		5,400

$$\text{Average cost of error} = \frac{\$5{,}400}{32} = \$169$$

order of successive inspections. When any item fails, the cost of error is the cost of inspecting the previous items in the list. The recommended order for each stream results in the lowest cost for that stream. Where there are many items that could have failed, the costs of inspections to determine which item needs to be repaired or replaced becomes formidable. This seems to happen occasionally with one's automobile. Such expenses justify the costs of instruments that can precisely locate the failure location.

9.4-3 Justifying Experimental Methods and Design

One of the responsibilities of management is innovation. Unless new products or improved processes or better ways of doing things are developed, economic competition tends to severely limit profits. Risk is inherent in the failure to innovate, to diversify, and to otherwise maximize opportunity. Changes in demands, in technology, and in preferences all require flexibility. Risks must be taken before new ideas can be converted into new facilities, new plants, and new products or services. Risks are taken for profit, for personal satisfaction in the results, for career advancement, among other reasons. Money is risked, and personal safety, time, and reputation are subject to risk. Some means is needed to measure and compare proposed innovations against present conventional procedures.[1]

Comparison of new proposals to conventional or proved methods can be developed from the Bayes concept of minimizing losses. This is illustrated in the following relation:

$$\begin{pmatrix}\text{risk/gain}\\ \text{ratio}\end{pmatrix} = \frac{\begin{pmatrix}\text{possible loss from}\\ \text{trying new method}\end{pmatrix} \cdot \begin{pmatrix}\text{probability}\\ \text{of failure}\end{pmatrix}}{\begin{pmatrix}\text{potential savings by}\\ \text{trying new method}\end{pmatrix} \cdot \begin{pmatrix}\text{probability}\\ \text{of success}\end{pmatrix}}. \tag{9-5}$$

If this ratio turns out to be unity, then whether or not the new method is attempted is a matter of indifference. Greater than unity is, of course, unfavorable. Since there is, of necessity, some uncertainty in the estimates of quantities as well as probabilities, it is not until the ratio would be less than perhaps 0.5 that the odds would definitely appear to favor experimentation and development of the new method. The equation can be expanded into the following form:

$$r = \frac{(C_{\text{NEW}} + C_{\text{OPN}} + C_{\text{MOD}} - C_{\text{SAL}} - C_{\text{OLD}})(1 - p)}{(C_{\text{OLD}} + C_{\text{OPO}} - C_{\text{NEW}})p}. \tag{9-6}$$

The capital equivalents of the differences in operating revenues can be estimated from the following two relations. Present value factors could be included in them if desired:

$$C_{\text{OPN}} = N_{\text{MOD}}[(R_{\text{NEW}} + Y_{\text{MT}})(1 - t_x) - t_x Y_{\text{DP}}], \tag{9-7}$$

$$C_{\text{OPO}} = N_{\text{OLD}}[(R_{\text{OLD}} - Y_{\text{MT}})(1 - t_x) + t_x Y_{\text{DP}}], \tag{9-8}$$

where

r = risk/gain ratio, fraction,
C_{NEW} = initial investment for new type equipment, dollars,
C_{OLD} = initial investment for conventional-type equipment, dollars,
C_{MOD} = cost of installations and modifications required if the new method is not satisfactory, dollars,
C_{SAL} = salvage from the new equipment, if it is not satisfactory, dollars,
C_{OPN} = all net revenue losses if the new method fails, dollars,
C_{OPO} = all net revenue losses by continuing the old method, dollars,
N_{MOD} = number of years required to modify or replace the new equipment if it fails,
N_{OLD} = payout time of the old-type equipment, years,

R_{NEW} = net operating profit lost, if the new method fails, dollars/year,
R_{OLD} = net operating profit lost by continuing the old-type method, dollars/year,
Y_{MT} = difference in the indirect costs of maintenance, property taxes, and insurance, calculated for new minus old types, dollars/year,
Y_{DP} = difference in depreciation, calculated for new minus old types, dollars/year,
p = probability of the new method succeeding,
t_x = income tax rate, fraction.

A first example concerns the installation of an experimental facility for continuous treatment of plant wastes, which would cost $45,000. If the new method should fail, a drain worth $22,500 could be used in the conventional treating basin that would then need to be installed. The conventional basin would cost $300,000. Any differences in operating revenues were ignored. The new-type facility was given only a 50% chance of working. Substituting this information into Eq. (10-6) produces the following calculation:

$$r = \frac{(45{,}000 + 0 + 300{,}000 - 22{,}500 - 300{,}000)(1 - 0.50)}{(300{,}000 + 0 - 45{,}000)(0.50)}$$

$$= 0.0883, \text{ or } 8.83\%$$

This is a favorable risk ratio which indicates that the experimental facilities should be installed.

A second example concerns a reactor vessel that must be installed by an oil refinery. Value of the product is $20 MM per year after allowing for raw material, labor, and other operating costs. A somewhat experimental single high-pressure vessel will cost $1.5 MM, and if it could not be used at the higher pressure there would be an 8% loss in capacity and an additional $720,000 would be required for a supplementary vessel, which would require 1.25 years for installation. A conventional-type vessel would cost $2.0 MM, would have a 4-year payout, but would lack 1% of having the desired capacity. Maintenance is estimated at 4% and depreciation at 5% yearly. Income tax rate is 52%. The new design is given a rating of 90% probability of success.

Difference in maintenance costs per year:

$$Y_{MT} = (0.04)(1.5\text{ MM} - 2.0\text{ MM}) = -\$28{,}000/\text{yr}.$$

Difference in depreciation per year:

$$Y_{DP} = (0.05)(1.5\text{ MM} - 2.0\text{ MM}) = -\$35{,}000/\text{yr}.$$

Operating income lost if the new method fails:

$$R_{\text{NEW}} = (0.08)(\$20\text{ MM}) = \$1.6\text{ MM/yr.}$$

Operating income lost by conventional equipment:

$$R_{\text{OLD}} = (0.01)(\$20\text{ MM}) = \$0.2\text{ MM/yr.}$$

Losses in total net operating revenues:

$$\begin{aligned} C_{\text{OPN}} &= 1.25[(1{,}600{,}000 - 28{,}000)(1 - 0.52) - (0.52)(-35{,}000)] \\ &= \$965{,}950, \\ C_{\text{OPO}} &= 4[(200{,}000 - (-28{,}000))(1 - 0.52) + (0.52)(-35{,}000)] \\ &= \$364{,}960, \\ r &= \frac{(1{,}500{,}000 + 965{,}950 + 720{,}000 - 0 - 2{,}200{,}000)(1 - 0.90)}{(2{,}200{,}000 + 364{,}960 - 1{,}500{,}000)(0.90)} \\ &= 0.103, \text{ or } 10.3\%. \end{aligned}$$

This is a favorable risk ratio which indicates 10: 1 odds in favor of the new method.

9.5 CHOOSING DECISION RULES IN PETROLEUM EXPLORATION

9.5-1 Economic Background and Cost of Errors

Petroleum exploration efforts eventually arrive at the point where exploratory wells must be drilled. Prior investigations, principally geologic in nature, can point out favorable new regions and horizons in which to drill test wells. But these are expensive, and after some number of them are drilled, the decision must be made whether to continue or whether to stop so that efforts can be moved elsewhere. Two conditions which would indicate that it would be best to stop are (1) if it appears that no successful wells will be found, and (2) if it appears that the proportion of successful wells is too low to make the total effort profitable. Suitable decision rules would therefore, be helpful.[2]

The case to be considered concerns the exploration program of an oil company in a new geological region. If the proportion of successful wells should prove to be sufficiently high, the program would be successful, but the probability P of a successful well is unknown. It has been decided to drill 10 wells, which would give a sample size of $n = 10$. Then the outcome of the sample will be used to estimate the true value of P for the geologic region.

Economic data that apply are summarized as follows. The present value of the operating profit per successful well is \$1,250,000. The cost per well is \$250,000. Where N represents a large number of wells drilled and P_B is the breakeven proportion of wells that are successful, the following relation applies:

$$0 = P_B N(\text{present value of operating profits}) - NC,$$

whence

$$P_B = \frac{C}{\text{present value of op. profits}} = \frac{250{,}000}{1{,}250{,}000} = 0.2.$$

Here C represents the cost of each well. The following apply also:

$$\begin{aligned}\text{net profit from } N \text{ wells} &= (\text{net value of } N \text{ wells}) - (0)\\ &= \left(PN\frac{C}{P_B} - NC\right) - \left(P_B N\frac{C}{P_B} - NC\right),\end{aligned}$$

whence

$$\begin{aligned}\text{net profit per well} &= (P - P_B)\frac{C}{P_B}\\ &= (P - 0.2)(\$1{,}250{,}000).\end{aligned}$$

Where P is less than 0.2, the equation gives the loss from drilling, an out-of-pocket loss. Where P is greater than 0.2 and the company has failed to pursue the exploration program, the equation gives the opportunity loss. Values calculated with this equation are given in Table 9.12. These values are conditional upon the particular value of P being valid.

Table 9.12
Conditional Losses for Various Values of *P*

P	*Loss of Drilling* (\$)	*Loss of Not Drilling* (\$)
0.0	250,000	0
0.1	125,000	0
0.2	0	0
0.3	0	125,000
0.4	0	250,000
0.5	0	375,000

9.5-2 Decision Rules

The problem of primary concern is to choose the best decision rule for estimation of the unknown value of P, the unknown state of nature. There may be errors from the sampling, because the sample is not large enough. That is, the sample will rarely agree exactly with the state of nature. One must be concerned about the probabilities of such discrepancies. Therefore, the chances that a particular decision rule will lead to a wrong decision must be investigated.

Any one of the following decision rules might be used. The 10,0 rule means that unless there are more than zero good wells in the sample of size 10, the program will be abandoned. The 10,1 rule means that unless there is more than one good well in the sample, the program will be abandoned. The 10,2 rule means that unless there are more than two good wells in the sample, the program will be abandoned.

A decision process can be formulated as follows. In the oil industry, erroneously abandoning an area would generally be considered more serious than would erroneously continuing testing. Accordingly, the null hypothesis assumed for testing purposes will be that the area is favorable. In Table 9.13, x refers to the number of successful wells.

Table 9.13
Analysis of Decision Rules

H_0: $P \geq 0.2$, *the Null Hypothesis*

Decision rules (let x = number of successful wells):
10,0 rule: Reject H_0 if $x = 0$.
10,1 rule: Reject H_0 if $x = 0$ and if $x = 1$.
10,2 rule: Reject H_0 if $x = 0$, if $x = 1$, and if $x = 2$.

Type I Error—Erroneous Rejection

Example, where $P = 0.3$ (There can be no erroneous acceptance.)
10,0 rule: There will be erroneous rejection if $x = 0$. Probability of this error $= p(x = 0)$.
10,1 rule: There will be erroneous rejection if $x = 0$ and if $x = 1$. Probability of this error $= p(x = 0) + p(x = 1)$.
10,2 rule: There will be erroneous rejection if $x = 0$, if $x = 1$, and if $x = 2$. Probability of this error $= p(x = 0) + p(x = 1) + p(x = 2)$.

Type II Error—Erroneous Acceptance

Example, where $P = 0.1$ (There can be no erroneous rejection.)
10,0 rule: H_0 is accepted if $x > 0$. Probability of this error $= 1 - p(x = 0)$.
10,1 rule: H_0 is accepted if $x > 1$. Probability of this error $= 1 - [p(x = 0) + p(x = 1)]$.
10,2 rule: H_0 is accepted if $x > 2$. Probability of this error $= 1 - [p(x = 0) + p(x = 1) + p(x = 2)]$.

9.5-3 Probabilities of Wrong Decisions

Binomial probabilities can be used to estimate the probabilities of errors, since a well is considered either as good or as bad in this analysis, if it is assumed that the tests are independent and that the probability P of obtaining a good well remains constant throughout the testing. Table 9.14 gives some

Table 9.14
Some Binomial Probabilities

n	P	x	*Probability of x*	*Cumulative Probability**	1 − *Cumulative Probability*†
10	0	0	1.0000	1.0000	0.0000
	0.1	0	0.3487	0.3487	0.6513
		1	0.3874	0.7361	0.2639
		2	0.1937	0.9298	0.0702
		3	0.0574	0.9872	0.0128
	0.2	0	0.1074	0.1074	0.8926
		1	0.2684	0.3758	0.6242
		2	0.3020	0.6778	0.3222
		3	0.2013	0.8791	0.1209
	0.3	0	0.0282	0.0282	0.9718
		1	0.1211	0.1493	0.8507
		2	0.2335	0.3828	0.6172
		3	0.2668	0.6496	0.3504
	0.4	0	0.0060	0.0060	0.9940
		1	0.0403	0.0463	0.9537
		2	0.1209	0.1672	0.8328
		3	0.2150	0.3822	0.6178
	0.5	0	0.0010	0.0010	0.9990
		1	0.0098	0.0108	0.9892
		2	0.0439	0.0547	0.9453
		3	0.1172	0.1719	0.8281

*Probabilities for type I errors were taken from this column.
†Probabilities for type II errors were taken from this column.

of the binomial probabilities as might be obtained in a table or calculated directly. In Table 9.15 the appropriate values extracted from Table 9.14 have been arranged with the several decision rules and with the appropriate type of error. The values in Table 9.15 are plotted in Fig. 9.3.

The developments made to this point represent the usual extent of a classical statistical analysis. It is desirable to have the probabilities of both type I and type II errors as low as possible. But type I is considered the more serious error (failing to adequately test the region). So attention is focused on the type I error characteristics. Usually, it is taken that the probabilities of such errors should be held below 0.05, although the decision maker should choose a significance level appropriate to the amount of harm that would be suffered as a consequence of the error. Even the 10,0 rule does not satisfy this condition when $P = 0.2$, but this is the best that can be done if the sample

Table 9.15
Probabilities of Wrong Decisions Based on Binomial Probabilities of Specific Test Outcomes*

		Probability of Error		
Type of Error	*P*	10,0 *Rule*	10,1 *Rule*	10,2 *Rule*
Type II	0.0	0.0000	0.0000	0.0000
(erroneous	0.1	0.6513	0.2639	0.0702
acceptance)	0.2	0.8926	0.6242	0.3222
Type I	0.2	0.1074	0.3758	0.6778
(erroneous	0.3	0.0282	0.1493	0.3828
rejection)	0.4	0.0060	0.0463	0.1672
	0.5	0.0010	0.0108	0.0547

*Economic breakeven point occurs when $P = 0.2$.

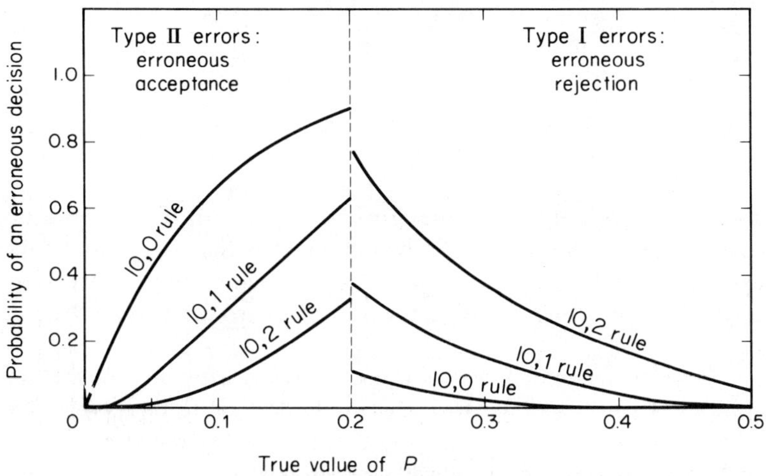

Figure 9.3. Error characteristics.

size is no larger than 10, and the recommendation would no doubt be to adopt the 10,0 rule.

9.5-4 Bayesian Analysis

Bayesian analysis is primarily concerned with the expected financial losses that might result from certain procedures. The Bayes estimate is the one that entails the smallest expected losses. Furthermore, the method uses preassigned estimates of the probabilities of various possible outcomes, which should be made by those who are knowledgeable in the area. In the present example this refers to preassigned probabilities of encountering various possible values of *P*, whose true value will be known only after extensive drilling. Finally, the method provides for the revision of the pre-assigned, or a priori, estimates as more data becomes available.

The Bayesian approach is further developed in Table 9.16. Economic information from Table 9.12 is combined with the probability of making a

Table 9.16
Conditional Losses Expected by Various Decision Rules

		10,0 *Rule*		10,1 *Rule*		10,2 *Rule*	
P	*Conditional Loss* | *P* ($)	*Probability of Wrong Decision*	*Expected Loss* | *P* ($)	*Probability of Wrong Decision*	*Expected Loss* | *P* ($)	*Probability of Wrong Decision*	*Expected Loss* | *P* ($)
0.0	250,000	0.0000	0	0.0000	0	0.0000	0
0.1	125,000	0.6513	81,410	0.2639	32,990	0.0702	8,780
0.2	0	0.1074	0	0.3758	0	0.6778	0
0.3	125,000	0.0282	3,530	0.1493	18,660	0.3828	47,850
0.4	250,000	0.0060	1,500	0.0463	11,580	0.1672	41,800
0.5	375,000	0.0010	380	0.0108	4,050	0.0547	20,510

wrong decision. Both of these values are conditional, for they depend upon a particular value of *P* being valid. Their product is the conditional expected loss associated with following a particular decision rule. The results are plotted in Fig. 9.4, which shows the expected losses for various true values of *P*.

The Bayesian technique makes use of a priori estimates of the relative probabilities of various values of *P* ultimately proving to be correct. That is, since *P* is unknown, it is taken as a random variable; and the probabilities of encountering several representative values of *P* are estimated on the best basis available, which in the present example might be experience gained in similar geologic regions. Table 9.17 shows that initially it has been assumed that there is a 40% probability that $P = 0$ will ultimately prove to be the case. It has also been assumed that there is a 20% probability that $P = 0.3$ will ultimately prove correct. The product of these probability fractions and the expected losses for various possible values of *P* give the unconditional expected losses according to the various decision rules. The expected unconditional losses can then be summed up to give the total expected loss for each decision rule. Table 9.17 shows that the 10,0 decision rule gives the lowest value of $5,120 versus $8,110 and $24,420 for other rules, so the 10,0 rule would be chosen by the Bayes method. Incidentally, in this example the 10,0 rule was also chosen by classical methods, but the two methods do not always lead to the same choice.

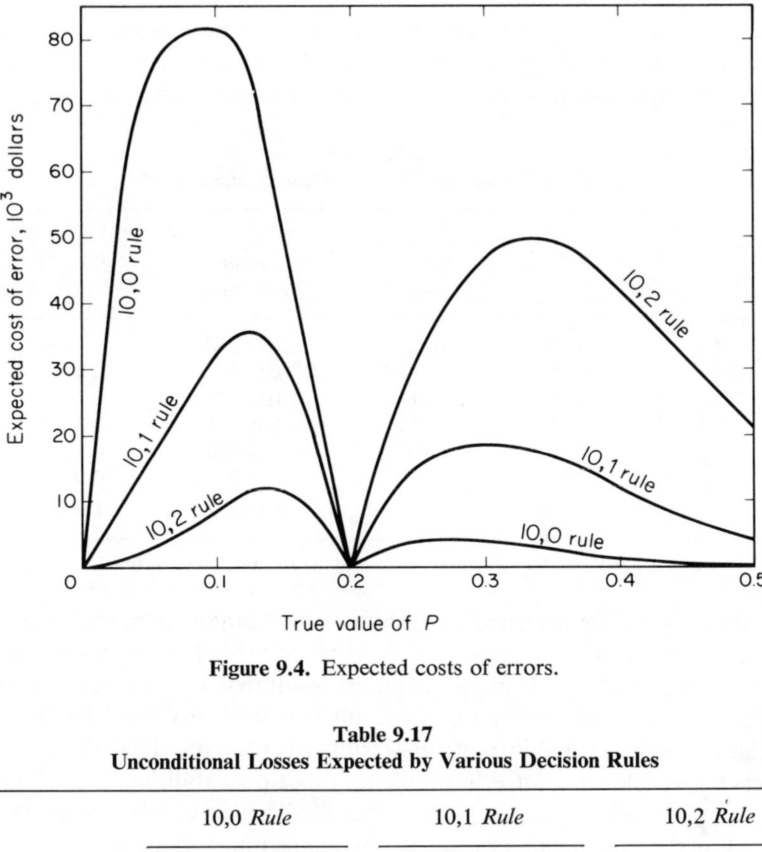

Figure 9.4. Expected costs of errors.

Table 9.17
Unconditional Losses Expected by Various Decision Rules

		10,0 *Rule*		10,1 *Rule*		10,2 *Rule*	
P	*p(P)*	*Expected Loss | P* ($)	*Unconditional Expected Loss* ($)	*Expected Loss | P* ($)	*Unconditional Expected Loss* ($)	*Expected Loss | P* ($)	*Unconditional Expected Loss* ($)
0.0	0.40	0	0	0	0	0	0
0.1	0.05	81,410	4,070	32,990	1,650	8,780	440
0.2	0.05	0	0	0	0	0	0
0.3	0.20	3,530	710	18,660	3,730	47,850	9,570
0.4	0.20	1,500	300	11,580	2,320	41,800	8,360
0.5	0.10	380	40	4,050	410	20,510	2,050
			5,120		8,110		20,420

9.5-5 Revising the Initially Assumed Probabilities

It remains to illustrate the method of revising the a priori probabilities as later additional information is obtained. Let it be assumed that a sample group of 10 wells was drilled and that one of these proved successful. Obviously, the preassigned probability of 0.40 that $P = 0$ is ruled out by this result. Revisions are in order. Table 9.18 illustrates a method of utilizing

Table 9.18
Revising Estimates of Various Degrees of Success*

P	*Prior* $p(P)$	$p(x = 1 \mid P, n)$ *Binomial Probabilities*	*Joint Probabilities*	*Revised* $p(P)$
0.0	0.40	0.0000	0.00000	—
0.1	0.05	0.3874	0.01937	0.2971
0.2	0.05	0.2684	0.01342	0.2059
0.3	0.20	0.1211	0.02422	0.3715
0.4	0.20	0.0403	0.00806	0.1239
0.5	0.10	0.0097	0.00097	0.0016
			0.06519	1.0000

*The assumption is that 1 good well was discovered in a sampling group of 10 wells.

both the original preassigned probabilities and binomial probabilities that best fit the new result, to obtain a revised set of probabilities to assign to the various values of P. For example, given the result that $x = 1$ successful well in 10, the binomial probability of this result is 0.3874 provided that $P = 0.1$, but the binomial probability of this result is 0.1211 provided that $P = 0.3$. For each possible value of P, multiplying those probabilities times the originally assigned values gives a joint probability, which can be normalized by proportional adjustment so that the sum equals unity. This process gives the new set of probability values to assign to the various possible values of P. This would, of course, permit a revision of Table 9.17 and a new consideration of the various decision rules. The same steps as those described would be used in this process.

The method that has been discussed has shortcomings. It does not consider the cost of the sample itself, which perhaps cannot be ignored in some instances. The method offers a means of choosing the best decision rule, but it does not provide an a priori answer to the question of whether or not a company should attempt to operate, or prefer not to operate, in a particular region. It does not handle the case where there are variations in outcomes, for example, where there are probability distributions of sizes of discoveries. The method gives equal weight to losses and gains, although it refers to the latter as lost opportunity. By referring to utility theory, one senses that this can be true only in the region near the origin. More directly here one would say that there is in all this the implicit assumption that the financial resources of the company must be quite large in comparison to the costs of the proposed exploratory venture.

9.6 ESTIMATING THE DISTRIBUTION OF SIZES OF PETROLEUM DISCOVERIES

9.6-1 A Bayesian Approach

In the preceding section it was assumed that all discoveries were equally good. For such a case, the important quanity to determine is the probability of hitting one of the good wells. However, in most cases, the discoveries are not of equal size or value. Some fall short of paying for the full costs of exploration and development. Some discoveries make a rather good profit, and only relatively few are highly successful.*

The problem considered in this section is that of estimating the nature of the size distribution of petroleum reservoirs and their value. It applies where there are continuing exploratory efforts in a geologic region. It is economically important to form an estimate of the type of distribution that is being probed by the exploratory wells.[3]

The method discussed here consists of drawing up several typical representative distributions. These should be broad enough in scope, between the best and the worst, to provide a satisfactory economic analysis. They should reflect general patterns of similar regions. They should include dry holes, as well as several representive class sizes of discoveries, and of course the probabilities associated with each of these classes. Then the test results obtained from drilling a sample group of exploratory wells are analyzed. Comparisons are made between the probabilities of obtaining the actual sample from each of the prior assumed distributions. By inference, this leads to the relative probabilities of which of the distributions best resembles the one actually present.

The method permits a more detailed economic analysis than is possible by use of the binomial distribution, where wells are simply called good or bad. The method discussed here uses either the multinomial or the multiple hypergeometric distribution to calculate the probabilities, where several types of results are recognized.

9.6-2 Representative Distributions and Estimates of Value

Table 9.19 shows three assumed distributions of the sizes of target reservoirs. Experience indicates that one of these might prove to closely

*See also the sections in Chapter 4 on multinomial, hypergeometric, and lognormal distributions and their applications.

Table 9.19
Expected Value per Well for Different Representative Distributions of Discovery Sizes

Distribution	*Kind of Well*	*Probability*	*Value per Well* ($)	*Conditional Expected Values* ($)
1*a*	*A*	0.60	−150,000	−90,000
	B	0.22	100,000	22,000
	C	0.12	500,000	60,000
	D	0.06	2,000,000	120,000
		1.00		112,000
2*a*	*A*	0.76	−150,000	−114,000
	B	0.14	100,000	14,000
	C	0.06	500,000	30,000
	D	0.04	2,000,000	80,000
		1.00		10,000
3*a*	*A*	0.90	−150,000	−135,000
	B	0.06	100,000	6,000
	C	0.02	500,000	10,000
	D	0.02	2,000,000	40,000
		1.00		−79,000

approximate the distribution in a particular petroleum exploration area. The exact nature of the distribution of reservoir sizes will not be known with complete certainty until after the entire area is completely drilled. However, it is important from an economic standpoint to discern as early as possible the general nature and value of the actual distribution.

In Table 9.19 the possible discovery sizes are grouped into four classes. Type *A* is a dry hole that gives a loss of $150,000 per well. Types *B*, *C*, and *D* are increasingly successful. Three distributions are shown in the table. Distributions 1, 2, and 3 contain decreasing amounts, or fractions, of good wells. The expected value, for an average well in each distribution, is calculated according to the probabilities within that distribution. These are called conditional expected values, conditional upon the particular distribution being the one that actually applies.

Table 9.20 illustrates how an original estimate of the unconditional expected value can be calculated. This requires an estimate of the probabilities of each distribution being the one that best represents the unknown state of nature. In the absence of better information, it was assumed that there is a one-third probability of each distribution ultimately proving to be the one that is present. Therefore, one-third of the expected values within each distribution are added together to give the unconditional expected value per well.

Table 9.20
A Priori Assigned Probabilities $p(D_i)$ and Unconditional Expected Value

Distribution, D_i	*Assigned Probability,* $p(D_i)$	*Conditional Expected Value* (\$)	*Unconditional Expected Value* (\$)
1*a*	0.333	112,000	37,300
2*a*	0.333	10,000	3,300
3*a*	0.333	−79,000	−26,300
	1.000		14,300

9.6-3 Samples and Their Probabilities

Table 9.21 shows 4 possible outcomes from drilling a group of 10 exploratory wells. Sample 1 is the case where all 10 wells would prove to be dry, of

Table 9.21
Anticipated Possible Outcomes from a Sample Group of 10 Wells

Sample	*Type*				*Total Sample*
	A	*B*	*C*	*D*	
1	10	0	0	0	10
2	9	1	0	0	10
3	8	1	1	0	10
4	5	2	2	1	10

type *A*. In sample 2, 9 are dry and 1 is of type *B*, with a value in the \$100,000 bracket. Samples 3 and 4 also contain various numbers of wells of types *A*, *B*, *C*, and *D*. We shall next face the problem of calculating the probabilities of obtaining these exact samples. This must be done for each of the representative distributions.

Table 9.22 is a generalized presentation of the representative distributions chosen for analysis purposes. Distribution 1*a* gives the fractions of types *A*, *B*, *C*, and *D* that are contained in the total prospective drilling locations. In distributions 1*b*, 1*c*, and 1*d*, the fractional proportions are the same, but the data are given in terms of the actual numbers of drilling locations. Where the data are in fractional form, as in distribution 1*a*, the multinomial distribution can be used to calculate the probabilities of obtaining specific samples. Where the data are in the form of actual numbers of prospective drilling locations, as are distributions 1*b*, 1*c*, and 1*d*, the preferred method would be

Table 9.22
Anticipated Possible Distributions

Distribution	Type A	Type B	Type C	Type D	*Total Prospective Well Locations*
1*a*	0.60	0.22	0.12	0.06	∞
1*b*	120	44	24	12	200
1*c*	60	22	12	6	100
1*d*	30	11	6	3	50
2*a*	0.76	0.14	0.06	0.04	∞
2*b*	152	28	12	8	200
2*c*	76	14	6	4	100
2*d*	38	7	3	2	50
3*a*	0.90	0.06	0.02	0.02	∞
3*b*	180	12	4	4	200
3*c*	90	6	2	2	100
3*d*	45	3	1	1	50

to use the multiple hypergeometric distribution for calculating probabilities of obtaining specific samples.

Table 9.23 shows the probabilities of obtaining the exact samples from each of the representative distributions. These are conditional probabilities, based in each case upon the particular distribution being the one present.

Table 9.23
Conditional Probabilities of Obtaining the Sample, Given the Distribution, $p(S_i \mid D_i)$ Values

Distribution	Sample 1	Sample 2	Sample 3	Sample 4
1*a*	0.0061	0.0222	0.0399	0.0246
1*b*	0.0052	0.0205	0.0395	0.0266
1*c*	0.0044	0.0188	0.0390	0.0289
1*d*	0.0029	0.0153	0.0376	0.0343
2*a*	0.0643	0.1184	0.0842	0.0054
2*b*	0.0597	0.1169	0.0877	0.0056
2*c*	0.0551	0.1152	0.0915	0.0058
2*d*	0.0460	0.1111	0.1000	0.0062
3*a*	0.3487	0.2325	0.0465	0.0001
3*b*	0.3398	0.2384	0.0499	0.0001
3*c*	0.3305	0.2448	0.0537	0.0001
3*d*	0.3106	0.2588	0.0630	0.0000

In the cases of distributions 1*a*, 2*a*, and 3*a*, the probabilities were calculated according to Eq. (4-19) for the multinomial distribution. In all other cases, the probabilities were calculated according to Eq. (4-20) for the multiple hypergeometric distribution. In subsequent calculations, the only probabilities used are those calculated for distributions 1*a*, 2*a*, and 3*a*. This, in effect, is equivalent to assuming that there are a large number of prospective drilling locations.

9.6-4 Revised Probabilities and Estimates at Value

Table 9.24 shows the revised probabilities that are assigned for each distribution eventually to be proved to be the one that prevails in the region. Calculations are made for each of the four samples that might be obtained. The revised probability estimates that are assigned to each distribution

Table 9.24
Revised or A Posteriori Probabilities and Unconditional Expected Values per Well, Calculated for Each Anticipated Possible Sample

Distribution	*A Priori* $p(D_i)$	$p(S_i \mid D_i)$	*Joint Probability*	*A Posteriori* $p(D_i \mid S_i)$	*Conditional Expected Value* ($)	*Unconditional Expected Value* ($)
			Calculations according to sample 1			
1	0.333	0.0061	0.00201	0.0144	112,000	1,600
2	0.333	0.0643	0.02143	0.1535	10,000	1,500
3	0.333	0.3487	0.11622	0.8321	−79,000	−65,700
			0.13966	1.0000		−62,600
			Calculations according to sample 2			
1	0.333	0.0222	0.00739	0.0594	112,000	6,600
2	0.333	0.1184	0.03947	0.3175	10,000	3,200
3	0.333	0.2325	0.07748	0.6231	−79,000	−49,200
			0.12434	1.0000		−39,400
			Calculations according to sample 3			
1	0.333	0.0399	0.01330	0.2340	112,000	26,200
2	0.333	0.0842	0.02805	0.4934	10,000	4,900
3	0.333	0.0465	0.01550	0.2726	−79,000	−21,500
			0.05685	1.0000		9,600
			Calculations according to sample 4			
1	0.333	0.0246	0.00819	0.8161	112,000	91,400
2	0.333	0.0054	0.00180	0.1796	10,000	1,840
3	0.333	0.0001	0.00004	0.0043	−79,000	− 300
			0.01003	1.0000		92,900

differ, of course, according to which sample is obtained. In the calculations, the products of the a priori probabilities $p(D_i)$ and the conditional probabilities $p(S_i \mid D_i)$ give a joint probability. This joint probability is normalized (proportionally adjusted so that its sum equals unity). These are the calculations explicitly given by Eq. (9-4), the Bayes Theorem. This results in the a posteriori probabilities $p(D_i \mid S_i)$.

The revised probabilities assigned to each distribution permit a recalculation of the unconditional expected value per exploratory well location. The expected value within each distribution has not changed, but the probability of realizing each particular distribution has changed, according to light shed by the particular sample. The entire region then looks accordingly either better or worse than the original estimates.

In interpreting such results, it should be said that a sample tends to indicate the general quality of distribution present. From observing the a posteriori probabilities in Table 9.24, it is evident that the obtaining of sample 1 or 2 would almost preclude the possibility that the actual distribution resembles distribution 1. However, obtaining sample 4 would tend to indicate that the actual distribution probably resembles distribution 1, at least much more than it resembles distribution 3. Regardless of whether or not the assumed distributions exactly coincide with that which actually prevails, the method will provide numerical answers. Therefore, it is best to consider the results as qualitative guides rather than as absolute values. Repeated applications of this method to an area, as more and more wells are drilled, tends finally to give the correct answer for that area, because the unknowns become exhausted. A weakness of this method is that it has no formal provisions for taking into account the size of the sample.

9.7 EVALUATION OF REGIONAL MAPS AND DRILLING DECISIONS

Opinions and decisions are based upon information, and the quality of the latter determines the quality of ensuing decision and actions. Confidence limits and risks that are inherent within information must inevitably be carried over into decisions, where both apply to the same problem and environment. Judging the quality of information is then a function of considerable importance.*

Exploration and development of oil production within a region is a process that generally extends over a period of time. At the start, the information may be quite meager. But as drilling proceeds, information and interpretation of the region become more complete. Geologic processes and petroleum accumulations are not entirely random and haphazard. Rock horizons are formed by chains of events through space and time wherein the

*The problem studied here was posed by J. P. Dowds.[4]

deposition of each grain of sand is influenced by preceding events and in turn influences the deposition and events that follow. Knowledge gained in one well can be extended into adjacent areas. The passage of time becomes an integral part of the decision process. As knowledge becomes more complete, the accuracy of the projections increases.

The problem studied in this section concerns information available at four successive points in time, T_1, T_2, T_3, and T_4. At time T_1, information available would permit drawing a crude map of the region, outlining the discovery prospects according to the information then available. At time T_2, more information would be available. The information and corresponding map of time T_1 must then be revised in part, and the relative amount of revision is an index of the quality of the information as of time T_1. At time T_3, additional information would again be available, and the quality of the information as of time T_2 could then be judged. The process can be repeated. At successive time intervals, which are probably best set by the number of additional wells drilled or the amount of acreage tested rather than strictly by the calendar, the relative improvement in information should become less. Relative improvement in the quality of information during the past, extended to the present, should be a good indication of the quality of present information.

The development history of an oil field is portrayed in Figs. 9.5 and 9.6 (see Table 9.25 for map symbols). These are isopach maps drawn at different

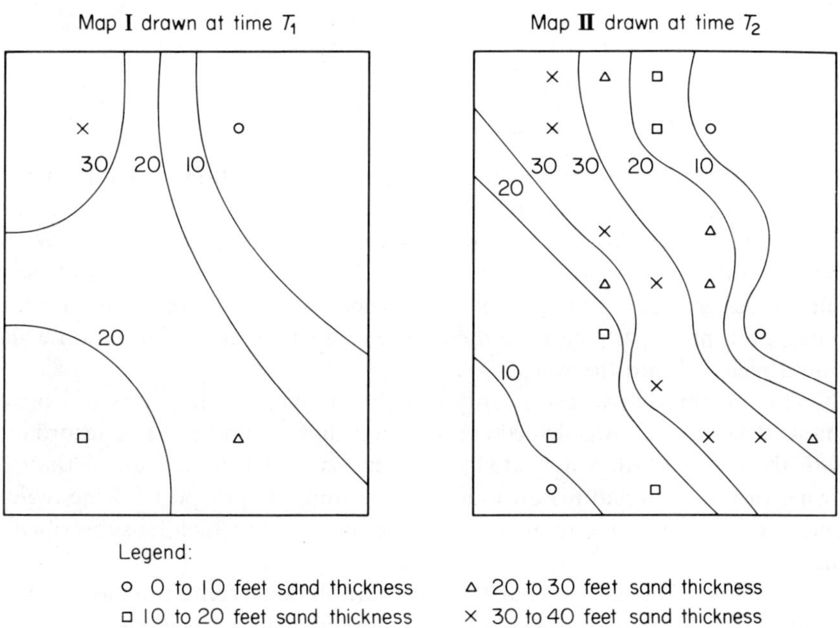

Figure 9.5. Isopach maps of an area, drawn at different times.

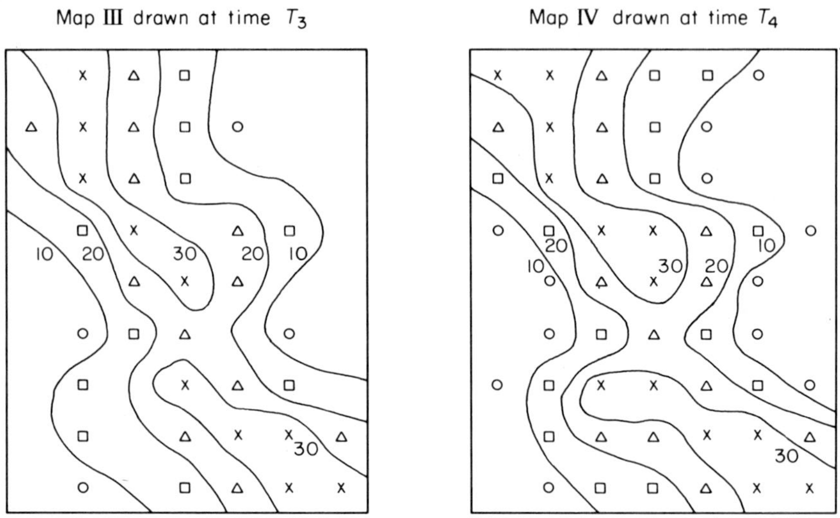

Figure 9.6. Isopach maps of an area, drawn at different times.

Table 9.25
Symbols Used in Maps and Tables

Symbol	*Sand Thickness (ft)*
○	0–10
□	10–20
△	20–30
X	30 or more

times, and they correspond to the amounts of information available at each point in time. The information came from the wells drilled, and the number of wells from which information was available is shown on each map. Where the sand thickness is less than 10 ft, the wells were an economic loss. At sand thicknesses greater than this, wells were successively better from an economic standpoint, principally because there were more acre-feet of formation which would yield oil into the wells.

The information whose quality is to be analyzed is the series of isopach maps. Where a well was drilled at a location that would not be in accordance with the interpretation indicated on the maps, it will be presumed that the owner of that well had his own ideas concerning the prospects of the region. One should analyze the quality of the information to which he subscribed at the time.

One way of indicating that the information concerning a region appears to be perfect is shown in Table 9.26. In order to attach significance to this

Table 9.26
Ideal Case of No Revisions in Information

Early Classification	*Later Classification*			
	○	□	△	X
○	1	0	0	0
□	0	1	0	0
△	0	0	1	0
X	0	0	0	1

apparent indication, we shall specify that in this particular case an additional 30% of the acreage was tested. By way of explanation, let the third row represent 1,000 acres with a net pay thickness between 20 and 30 ft. The 1 in position (3, 3) indicates that, after the additional testing, the same 1,000 acres remained in the same sand-thickness classification. Suppose, however, after the additional testing, that of the original 1,000 acres, 200 acres were shown to be in the 30- to 40-ft pay thickness, and 300 acres were shifted down into the 10- to 20-ft class. Then row 3 would become (0, 0.3, 0.5, 0.2). The 0.5 on the main diagonal (3, 3) position indicates that the original information on that category was only 50% correct.

Table 9.27 shows the changes that were made when the sand thickness contours were redrawn. Areas are given directly in acres in the left half of the table and in fractional form on the right half. The top part of the table shows changes between maps I and II. On the left margin are the areas in each sand-thickness class as shown on map I. The extent to which these were changed when drawing map II is shown in the body of the table. On the bottom margin, the sums of the columns gives the total acreages as drawn on map II. These, of course, become the original acreages that are shown on the left margin of the following stage of the table.

The cost-of-error function that was used in computing Table 9.28 is as follows. The absolute values of the errors were used, so that both increases and decreases in acre-feet were counted as errors rather than permitting any masking by averaging positive and negative errors. Consider row 2, designated by the □ symbol, at the top of Table 9.27. Originally there had been, in map I, a total of 790 acres in this sand-thickness area. In map II, there remained 380 of these same acres still classified by the same □ symbol. There was no error connected with classifying these 380 acres. But 220 acres from this area were reclassified in map II as having the thickness designated by the ○ symbol, which is 10 ft less in thickness for an error of $(10 \times 220) = 2{,}200$ acre-ft. Also, there were 170 acres changed to the △ thickness classification, which is 10 ft greater in thickness, for an error of $(10)(170) = 1{,}700$ acre-ft. Finally, 20 acres were changed to the X classification, which

Table 9.27
Changes in Areas Between Sand-Thickness Contours as Information Became More Complete

Thick-ness	*Original Area (acres)*	*Acres*				*Fractional*			
		○	□	△	X	○	□	△	X
				Changes from map I to map II					
○	550	440	100	10		0.800	0.182	0.018	
□	790	220	380	170	20	0.279	0.481	0.215	0.025
△	910		240	280	390		0.264	0.308	0.428
X	270		10	70	190		0.037	0.259	0.704
	2,520	660	730	530	600				
				Changes from map II to map III					
○	660	630	30			0.955	0.045		
□	730	210	480	40		0.288	0.658	0.054	
△	530		130	390	10		0.245	0.736	0.019
X	600			210	390			0.350	0.650
	2,520	840	640	640	400				
				Changes from map III to map IV					
○	840	790	50			0.940	0.060		
□	640	100	470	70		0.156	0.734	0.110	
△	640		50	500	90		0.078	0.781	0.141
X	400			20	380			0.050	0.950
	2,520	890	570	590	470				

is taken as 20 ft greater in thickness for an error of $(20 \times 20) = 400$ acre-ft. Thus, in the original 790 acres, there was an apparent total error of $2{,}200 + 1{,}700 + 400 = 4{,}300$ acre-ft. This last number is shown in the second row in Table 9.28.

Reduction to dollars of the error given in acre-feet was estimated for purposes of an illustration as follows. A recovery of 100 barrels of oil per acre-foot of reservoir rock was assumed to be representative of the particular rock formation. Next, a value of $1 per barrel of oil reserves was assumed to be a fair approximation for the particular case. This gives an error of $100 per acre-foot change in classification. This was converted to error per well location within the original area, using a spacing of 40 acres per well. Such a basis is held to be desirable, at least for some purposes, since the investment increment is more or less on a per well basis.

The selection of a suitable areal basis upon which to distribute the monetary value of the total error is not as easy as it might appear of first. It is desirable from the strategic viewpoint to calculate the relative amount of error within each classification group, since choosing a location within one

Table 9.28
Errors in Acre-Feet of Sand and in Dollars per Well Location Associated with the Shifting of Sand-Thickness Contours as New Information Became Available

Map	*Sand Thickness*	*Area (acres)*	*Well Locations, at 40 Acres*	*Total Error (acre-ft)*	*Average Error per Well Location* ($)
I	○	550	13.75	1,200	8,750
	□	790	19.75	4,300	21,800
	△	910	22.75	6,300	27,800
	X	270	6.75	900	13,300
					20,200
II	○	660	16.50	300	1,820
	□	730	18.25	2,500	13,700
	△	530	13.25	1,400	10,500
	X	600	15.00	2,100	14,000
					10,000
III	○	840	21.00	500	2,380
	□	640	16.00	1,700	10,600
	△	640	16.00	1,400	8,750
	X	400	10.00	200	2,000
					6,030

group would be preferred to a location within another group. One basis that suggests itself is to distribute the error within any classification group according to the number of new wells that were drilled within that group. But in the case of the problem at hand, wells drilled within one classification can result in changed contours in adjacent classifications without any wells being drilled in the adjacent areas. Therefore, distributing the error on a per-new-well basis drilled within a classification falls short of offering a suitable solution. The method of allocating the error within a classification group according to the total acreage or possible well locations, which was used in Table 9.28, may be satisfactory in some instances. In the present example it is probably satisfactory in the cases where the sand is thicker, which generally tends to be areas enclosed in expanses of thinner or less permeable rock. For the enclosed areas, the method is feasible. But for the surrounding areal expanse, it would appear in general that only the adjacent peripheral locations should be included within any analysis.

Improvement in the estimates over time is shown graphically in Fig. 9.7. The curves for the several individual classifications show peculiar behavior in some cases. Not much confidence should be placed in the curve associated with the ○ symbols, which represent generally the surrounding barren areas, for reasons noted previously. However, the composite curve shows a smooth

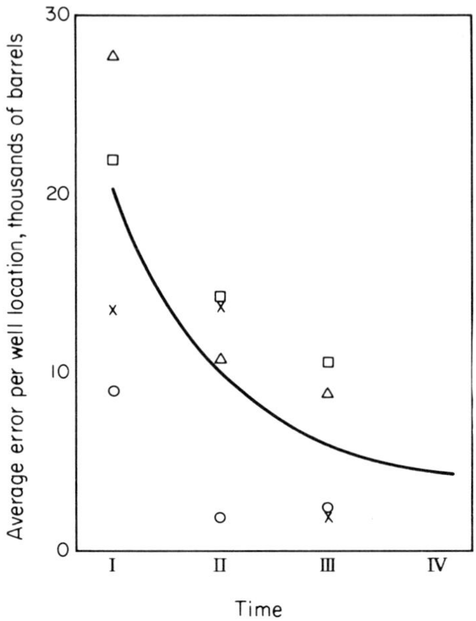

Figure 9.7. Average error of estimate of reserves per well location versus time.

trend, which indicates a gradual reduction in uncertainty with time. It is well to recall here that the information available at time T_4 was used to estimate the errors as of time T_3. Now, however, it may be seen on Fig. 9.7 that it possible to project the trend of the errors as of T_1, T_2, and T_3, so that in most cases a good estimate can be made regarding the probable error, as of time T_4, which is within the information available at the present.

9.8 COST-OF-ERROR FUNCTIONS

In the examples in this chapter, the costs of error have been assumed to be proportional to the discrepancy between the estimate and the later proved value of the event. In the last section, this practice permitted the error to be expressed in acre-feet, and the cost of error was reckoned as some constant times the number of acre-feet. Where W is a cost-weighting function, this can be expressed mathematically as

$$W = k\,|X - X_{\text{est}}|.$$

The use of the absolute value of the difference is recommended by Hoel, who also recommends the possible use of the squared function,[5]

$$W = k(X - X_{\text{est}})^2.$$

The latter would be used where the consequences of large errors are very serious, and the former would be used where they are less serious. The implication here is, of course, that the economic consequences should enter into the calculations exactly as they would affect the future of the economic unit.

REFERENCES

1. STUHLBARG, DAVID, "Calculating the Calculated Risk." *Chemical Engineering*, July 15, 1968, pp. 152–154.
2. GRAYSON, C. J., Jr., "Bayesian Analysis—A New Approach to Statistical Decision Making." *Journal of Petroleum Technology*, Vol. 14, No. 6 (June 1962), pp. 603–607.
3. ROTHROCK, R. R., "An Application of Statistical Decision Theory to Petroleum Exploration." M.E. Thesis, University of Oklahoma, 1969.
4. DOWDS, J. P., "Statistical Geometry of Petroleum Reservoirs in Exploration and Exploitation." Presented at SPE Annual Meeting, Sept. 1968.
5. HOEL, P. G., *Introduction to Mathematical Statistics*. New York: John Wiley Sons, Inc., 1962, 3rd ed., p. 367.

PROBLEMS

9-1. There are three unmarked urns. It is known that originally urn *A* contained 3 white and 3 black balls, *B* contained 2 white and 4 black, and *C* contained 2 white and 5 black. Three balls, 2 white and 1 black, have been drawn from one of the urns, without replacement. What are the probabilities of urn *A*, urn *B*, and urn *C* being the one from which the balls were withdrawn?

9-2. There are three estimates of the prospects of a small geologic basin. Estimate *A* is that it contains 6 prospective structures, 4 of which should prove to be oil fields. Estimate *B* is that it contains 6 prospective structures, 3 of which should prove successful. Estimate *C* is that it contains 5 structures, 2 of which should prove successful. To date, 3 structures have been tested and only 1 has proved productive. What are the probabilities of each one of the three original estimates being the correst estimate, the one that most closely resembles the actual conditions underground?

9-3. Three machines are turning out identical brackets. During a certain day machine 1 turned out 400, machine 2 turned out 350, and machine 3 turned out 250 brackets. The records kept on these machines indicate that they produce 0.6%, 1.0%, and 1.2% faulty brackets, respectively. What are the probabilities that a particular faulty piece was produced by each one of the machines?

9-4. A control system contains three electronic controls, units *A*, *B*, and *C*. Failure of any one of them stops the entire production system. In the event of a failure, inspections and replacements are done by the regular crew. However, downtime while awaiting repairs costs $1,800 per hour. The following table includes the numbers of each of the items that have been replaced during the past 3 years. Based on the available data, calculate the expected cost of error, and prepare a schedule giving the order in which the units should be inspected in case of a failure.

Unit	*Number Replaced*	*Inspection Time (min)*	*Additional Installation Time (min)*	*Cost of Unit ($)*
A	15	30	20	2,800
B	45	45	15	1,800
C	90	50	10	2,400

9-5. A new unit for treating injection water must be installed. Conventional equipment would cost $45,000. A newer type is available at a cost of $25,000. The use of the newer type in the particular area would have to be classed as experimental. However, it is given a 60% chance of success. If it fails, the salvage value is considered to be $15,000, since it could be used in other areas. Operating costs may be neglected. Calculate the risk/gain ratio. Is installation of the newer equipment favorable?

9-6. A company is operating 30 wells on the north flank of the Star Trend area. These wells were drilled to the *D* Sand and are rapidly approaching their economic limit. In other sections in the Trend, the *B* Sand has proved to be sufficiently well developed to make recompletion of the wells in that shallower sand attractive. In such cases the present value of the operating profits is estimated to average $100,000. However, the development of the sand is irregular and about 40% of the attempted workovers in the other sections have been unsuccessful. The workovers cost $40,000. Management has decided to attempt recompletions in five wells, after which the workover program will be reevaluated. Following is the consensus estimate of chances for various degrees of success.

Fraction of Workovers, P	*Probability of the Particular Value of P Being Correct, $p(P)$*
0.0	0.10
0.2	0.10
0.4	0.40
0.6	0.20
0.8	0.10
1.0	0.10

(a) Formulate decision rules for interpretation of the outcome of the five scheduled workovers. For the several possible values of P, calculate the probabilities of type I and type II errors associated with each decision rule.

(b) Calculate the conditional and the unconditional expected losses for the various possible values of P, for each decision rule. Include both out-of-pocket losses and lost opportunities.

(c) For the event that four of the five scheduled workovers are successful, calculate the revised estimates of $p(P)$, the probabilities of various degrees of success.

9-7. There are 20 wells in a recompletion program. After a workover is completed, an estimate of the present value of the well in that particular area can be made from the initial capacity of the well to produce oil. For purposes of estimating the value of the workover program, the prospective recompletions have been grouped into four different classes, according to their value. Also, for analysis purposes, three different distributions, of the proportions of wells that will fall in each class, have been assumed.

Class	*Initial Potential (bbl/day)*	*Present Value* ($)	*Distributions, Fraction of Wells in Each Class* I	II	III
A	60	190,000	0.20	0.10	0.05
B	40	100,000	0.20	0.10	0.05
C	20	20,000	0.40	0.50	0.50
D	0	−50,000	0.20	0.30	0.40

(a) Calculate the unconditional expected value per well, assuming that distributions I, II, and III have probabilities of 30, 40, and 30%, respectively, of most closely representing the actual underground conditions.
Ans.: $27,750 per well, expected value.

(b) Revise the estimate of the average value per well for the event that 5 workovers have been completed, with 2 of the wells falling in class *B*, 2 in class *C*, and 1 proving to be dry.
Ans.: $44,800 per well, expected value.

9-8. Drilling is under way on a gas field at the 10,000-ft depth. The areal extent of the field is 2,000 acres. A recovery of 1.4 MM ft^3 is expected per acre-foot of net pay. After allowing for royalties, taxes, and production costs, an operating profit of $200 per MM ft^3 will be realized. There will be a 4-year delay before pipeline connections are completed, followed by a 20-year life. An average rate of return of 15% (before federal income taxes) is desired, which, combined with the time pattern of the sales, yields a present-value factor of 0.18. Total costs will include 3 dry holes for delineating the field limits at $200,000 each. There will be 8 producing wells at $400,000 each. Production facilities and other costs amount to an additional $500,000. The total net pay is difficult to estimate. For evaluation purposes, three representative distributions have been assumed:

Class	*Net Pay* (ft)	*Number of Wells* *Dist. I*	*Dist. II*	*Dist. III*
A	120	2	1	0
B	90	2	2	2
C	60	2	2	2
D	30	2	3	4

Initially it was assumed that the probabilities of distributions I, II, and III being the most representative of underground conditions are 0.3, 0.4, and 0.3, respectively.

(a) Assume that equal portions of the areal extent of the reservoir are associated with each well. Calculate the unconditional expected present value of the entire field.
Ans.: \$2.20 MM.

(b) For the event where 4 wells have been drilled into the reservoir, with 1 in class *B*, 2 in *C*, and 1 in *D*, revise the estimate of the unconditional expected present value of the entire field. Again assume that equal portions of the areal extent of the entire reservoir are associated with each of the wells already drilled.
Ans.: \$1.95 MM.

10

CONTROL OF OPERATIONS: CPM AND PERT METHODS

10.1 OBJECTIVES

The larger and more significant projects are often one of a kind. In some cases they are intermittent in their occurrence. Examples are the construction of a building or plant and the periodic maintenance of a machine or facility. For each such project there is a beginning and an end. In many cases the time in which the project can be completed is of controlling interest, and one seeks to find the shortest possible time. In other cases the objective is to find a suitable time–cost tradeoff, a balance between higher operating costs associated with rapid completion methods and the longer time associated with normal methods. A third type of solution is that of leveling of resources, for example, utilizing a constant number of men on a project rather than reassigning personnel or other necessary resources. A fourth type is concerned with finding the best scheduling procedure when insufficient resources and facilities limit the number of jobs that can be undertaken simultaneously. The problem that will be considered first is that of finding the shortest possible time in which a project can be completed. Both the CPM and PERT methods were devised for this purpose.

The *critical path method* (CPM) was first used about 1959 by DuPont to control the major overhaul of a neoprene plant. The estimated savings were very large, reported to approach $1 million. A similar scheme, the *program evaluation and review technique* (PERT), was used by the U.S. Navy to manage development of the Polaris submarine missile system. It has been estimated that use of the method shortened the development period by 2 years. Rapid development and adoptions of the methods occurred in the period 1960–1962 as a result of the reported successes and savings in time and money.

Distinction between CPM and PERT methods is the contrast between deterministic and probabilistic models. The CPM method is deterministic, in that the estimated completion times required for individual components are assumed to be definitely known or to be subject to precise control. The PERT method provides for the use and combining of estimated completion times of component parts wherein the time estimates admittedly contain a probable amount of error or variance. Both methods are oriented toward finding the path or sequence of component jobs (activities) whose consecutive completions ultimately control the final completion date of the entire project.

Advantages that can be gained from using critical path planning stem from the fact that it offers a logical method for setting down and organizing a plan of action. It facilitates additional breakdown of activities into smaller component activities and thereby aids in additional study in planning a project. The additional study is the source of any benefits gained rather than the outline of a method for listing individual activities. However, the method tends to permit fuller utilization of individual capacities for planning and organization. It points out those activities which are on the critical path, those whose delay would result in delay of completion of the entire project. Thus critical path planning can help to point out where best efforts will be the most rewarding. However, without effective follow-through in the organization and control of the actual physical performance of the work, very little could be gained.

Another advantage can be realized when a computer is utilized to make time network computations. As individual activities are completed, they can be assigned zero duration time, and a new calculation yields an up-to-date report of the status of the project. Also, any new complications that develop can be entered into the calculations. A running account of the progress and current status can be periodically calculated and printed out.

10.2 DESCRIPTION OF CRITICAL PATH METHOD

The basic concept of CPM consists of representing, as a network of arrows, the individual activities that go to make up a total project. This is equivalent to a road map which must be followed. However, all the activities or jobs must be performed, and the network map shows the sequences in which they fit together. In any extensive project there are different activities that can be done concurrently. Thus parallel paths appear in the arrow diagrams. One of such parallel or concurrent paths, which represents a series of one or more consecutive activities, will generally require a longer time for its completion than is required for the others. For the entire project, that series of consecutive activities which requires the longest time, and therefore determines the completion date, is referred to as the *critical path.*

There are two ways of drawing such arrow diagrams.[1] One is called the *activity-on-node diagram.* An example is shown in Fig. 10.1. This method of representation is fast and readily shows the relationship between major stages of a project, thus can be recommended for initial planning stages. The arrows merely show direction, thereby indicating sequences. Each node

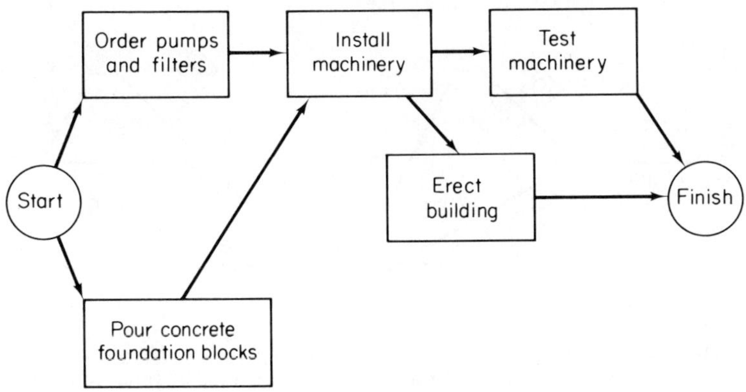

Figure 10.1 Activity on node diagram. Initial planning for a water-injection plant.

contains a description of the activity that is to be performed. The time required for completion of the activity could also be noted inside each node. An identifying number could be given to each node. Such an identifying number might be entirely satisfactory for a visual diagram but is less satisfactory for computer processing, since there is no connection other than the drawn arrow either from or to other activities.

The second type of diagram is referred to as the *activity-on-arrow diagram.* An unsatisfactory example of this is shown in Fig. 10.2. Here the arrows

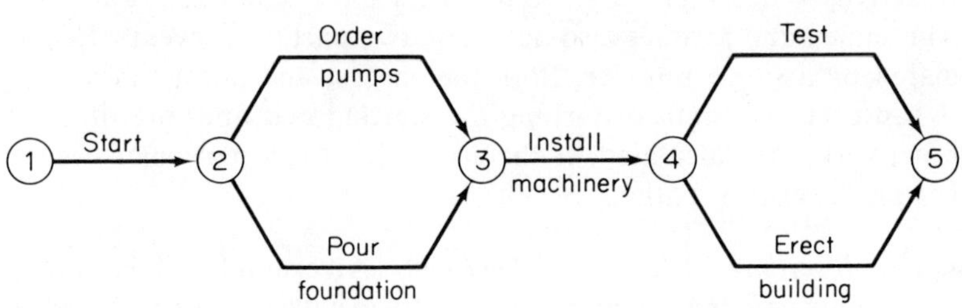

Figure 10.2. Activity on arrow diagram. Example is unsatisfactory because two node numbers designate more than one activity.

start and end at numbered circles. The circles are referred to as *events*. Figure 10.2 is unsatisfactory because more than one arrow (activity) would be designated by the same two numbers. This has been remedied in Fig. 10.3 through the addition of dummy arrows (dummy activities), so that only one activity is designated by a particular pair of numbers. The direction of the arrow designates at which number the activity begins and at which event number the activity ends. It is, of course, not necessary that the smaller number be at the beginning, although such arrangement is more convenient

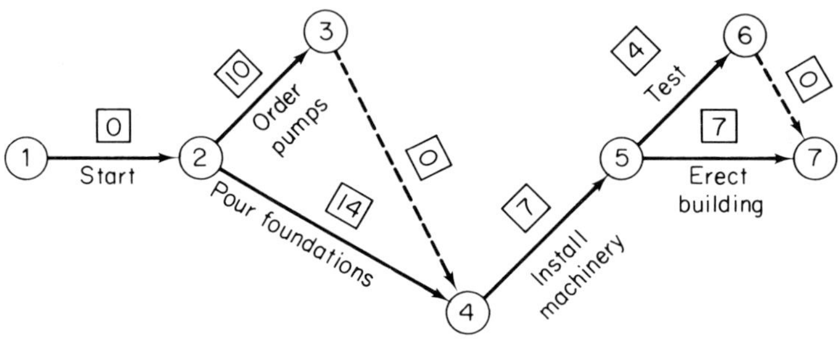

Figure 10.3. Activity on arrow diagram. Use of dummy arrows permits two node numbers to designate only one activity. Estimated time required is shown in the small squares.

for personal inspection. A sequence of activities can be traced through the network by virtue of the arrangement wherein the number that designates the end of one activity also designates the beginning of the next activity.

In the activity-on-arrow diagram the time required to perform the activity is considered to be allocated to the arrow. Conversely, the events are instantaneous and require no time. An event can be said to occur when all activities that lead into the event circle have been completed. Only after completion of all entering activities can any of the one or more leaving activities be started. This expresses the idea that some particular activities or jobs must be completed before beginning certain other ones.

An activity is designated by two numbers and is said to start at i and end at j. The same i and j values also designate two particular events. Events are designated by a single number. Both the earliest and latest times at which events can occur, without disturbing the starting and finishing times of the entire network, are calculated. If these coincide for a particular event, that event is on the critical path.

Example: *Installing a Water Injection Plant* An example will be considered where an oil-producing company desires to install a small water injection plant. The first planning regarding installation of the plant might be very much as shown in Fig. 10.1. This indicates that, given the decision to go ahead, the pumps will be ordered and the concrete foundations can be poured and given time to harden while waiting for the pumps to arrive. The pumps can then be installed. When this is done, a building can be built over the pumps and testing of the machinery can proceed concurrently, which would complete the project. So an outline has been formulated which gives the main things that must be done and the sequence for doing them.

The same information is shown in Fig. 10.2. In this figure the activities, or jobs, are printed along the arrows. There may have been a loss in clarity in the diagram and little gain. Note that two arrows go from 2 to 3, one designating the ordering of the pumps and the other designating the pouring of concrete foundations. Both for manual computations and particularly for computer calculations it is desirable to have the (i, j) number, such as (2, 3), mean one thing only. This has been achieved in Fig. 10.3 by adding dummy arrows (3, 4) and (6, 7) and assigning zero time for the performance of these dummy activities.

The critical path on Fig. 10.3 can be found by visual analysis. There is only one way to go from event 1 to event 2, which here is used only to indicate that the decision to start must be made. A branching, or burst, occurs at event 2. Event 4 can be reached directly by (2, 4), which requires 14 days. Event 4 must also be reached by (2, 3), which requires 10 days, plus (3, 4) which requires 0 days, a total of 10 days. Of these, path (2, 4) requires the longer time and is therefore on the critical path. The activity arrow (4, 5) is obviously on the critical path since it is the only way to reach event 5. Finally, activity (5, 7) requires a longer time than (5, 6) plus (6, 7). The critical path, therefore, goes in the order 1, 2, 4, 5, and 7. This is indicated in Table 10.1, which lists the activities and their individual duration times required for their completions.

Table 10.1
List of Activities in Figure 10.3*

i	j	T_E	*Critical*	*Activity*
1	2	0	*	(start)
2	3	10		Order and obtain pumps
2	4	14	*	Pour concrete foundations
3	4	0		(dummy)
4	5	7	*	Install machinery
5	6	4		Test machinery
5	7	7	*	Erect building
6	7	0		(dummy)

*Activity arrows start at i and end at j. T_E is the estimated time to complete the individual activity. *indicates that the activity is on the critical path.

10.3 CALCULATION OF EARLIEST START AND LATEST FINISH TIMES

Figure 10.4 illustrates how the duration times can be arranged in a tableau. The spaces are empty except where an activity should be indicated. In the first line the 0 indicates that there is an activity that begins at event 1 and ends at event 2 which requires 0 time for its completion. In the second line there are indicated two activities that begin at event 2. One of these ends at event 3 and requires 10 days; the other ends at event 4 and requires 14 days. In this manner all the individual activities in the entire project can be recorded in the tableau.

i \ j	1	2	3	4	5	6	7
1		0					
2			10	14			
3				0			
4					7		
5						4	7
6							0
7							

Figure 10.4. Tableau showing activity duration times, TE_{ij}. Data from Fig. 10.3.

Figure 10.5 illustrates how the tableau of Fig. 10.4 can be used to calculate the earliest start and the latest finish times. The earliest start times are shown in the column on the left margin. The first activity is assumed to start at time 0. Beginning at the first 0 in the left margin, add the duration time (also 0 in this case) to obtain the earliest when event 2 can be reached, 0 in this case, which is entered in the second line of the earliest start column. Since there is only one activity in the first line, proceed to the second line. Starting with the 0 in the left margin, add to it the 10 in column 3, which is the earliest that event 3 could be reached through activity (2, 3), and enter 10 in the left margin of line 3. Again in line 2, start with the 0 in the left margin and add the 14 in column 4, and enter 14 in the left margin of line 4, which is the earliest that event 4 can be reached through activity (2, 4). When more than one

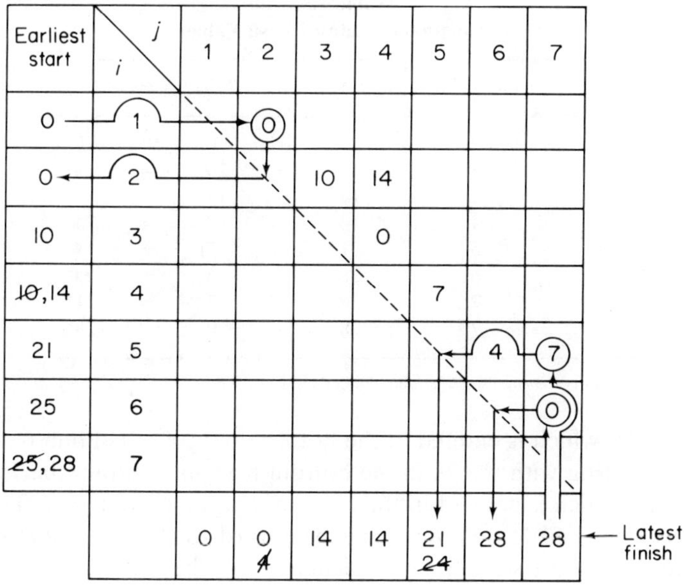

Figure 10.5. Tableau showing calculation of earliest and latest finish times. Data from Fig. 10.3.

earliest start time is calculated for the same *j*-value, only the greatest is retained. Proceeding in this manner, it is found that event 7 can be reached in 28 days. These same calculations are exhibited in Table 10.2.

Table 10.2
Calculation of Earliest Start Values*

i	j	ES_i	+	TE_{ij}	=	ES_j
1	2	0	+	0	=	0
2	3	0	+	10	=	10
2	4	0	+	14	=	14
3	4	10	+	0	=	~~10~~
4	5	14	+	7	=	21
5	6	21	+	4	=	25
5	7	21	+	7	=	28
6	7	25	+	0	=	~~25~~

*Retain only the largest ES_j value. Data from Fig. 10-3.

The bottom margin of Fig. 10.5 shows the latest finish times. These are calculated by making a backward pass (see Table 10.3). The final event was reached in 28 days, which is entered in the bottom margin in the last column. In column 7, starting with the 28 in the bottom margin, subtract the 0 in line

Table 10.3
Calculation of Latest Finish Values*

i	j	LF_j	$-$	TE_{ij}	$=$	LF_i
1	2	0	$-$	0	$=$	0
2	3	14	$-$	10	$=$	4
2	4	14	$-$	14	$=$	0
3	4	14	$-$	0	$=$	14
4	5	21	$-$	7	$=$	14
5	6	28	$-$	4	$=$	24
5	7	28	$-$	7	$=$	21
6	7	28	$-$	0	$=$	28

*Retain only the smallest LF_i value. Data from Fig. 10.3.

6 to obtain 28, which is entered in the bottom margin of column 6. Again in column 7, starting with the 28 in the bottom margin, subtract the 7 in line 5 to obtain 21, which is entered in the bottom margin of column 5. The calculations next proceed to column 6 and backward to the first column. When more than one latest start time is calculated for the same i-value, only the smallest is retained.

Earliest start times are calculated by the relation

$$ES_j = ES_i + TE_{ij}, \tag{10-1}$$

where

TE_{ij} = estimated time required for the completion of activity (i, j)

ES_i = earliest starting time for activities that start at event (i)

ES_j = earliest starting time for activities that start at event (j); retain only the largest value.

Latest finish times can be calculated as follows:

$$LF_i = LF_j - TE_{ij}, \tag{10-2}$$

where

LF_i = latest finish time for activities that end at event (i); retain only the smallest value,

LF_j = latest finish time for activities that end at event (j).

Two other relations, which were not used in this example, but which are rather obvious, are as follows:

$$EF_{ij} = ES_i + TE_{ij}, \tag{10-3}$$

where

$$EF_{ij} = \text{earliest finish of activity } (ij),$$

and

$$LS_{ij} = LF_j - TE_{ij}, \tag{10-4}$$

where

$$LS_{ij} = \text{latest start of activity } (ij).$$

10.4 CALCULATION OF SLACK TIMES, *TF* AND *FF*

Activities that are not on the critical path have slack time; the job does not require all the time available. Final completion of the entire project will not be delayed if an activity containing slack is not started as soon as possible. Of course, if it is delayed as long as possible, it then becomes critical and must then be diligently completed. Slack time is commonly referred to as *float*.

Two types of float are recognized. First is *total float* (*TF*). This is the amount of time which an activity, or job, can be delayed, without affecting the final completion time. If its total float is zero, an activity is on a critical path. This is how critical activities are detected in a computer solution, by the lack of positive total float. Two or more consecutive activities may share the same total float, so that if it is used up by one such activity it may automatically disappear for the others.

Total float can be calculated as the latest finish of an activity minus the earliest finish. Thus

$$\begin{aligned} TF_{ij} &= LF_j - EF_{ij}, \\ &= LF_j - TE_{ij} - ES_i, \end{aligned} \tag{10-5}$$

where

$$TF_{ij} = \text{total float (slack time) of activity } (ij)$$

Free float (*FF*) is the time by which an actual completion can exceed the expected completion without affecting any other activity or event in the network. It cannot exceed total float. It can be calculated as the earliest start from the following node (event) minus the earliest finish of the activity. Thus

$$\begin{aligned} FF_{ij} &= ES_j - EF_{ij} \\ &= ES_j - TE_{ij} - ES_i. \end{aligned} \tag{10-6}$$

Tables 10.4 and 10.5 illustrate the calculations of total float and free float, respectively.

Table 10.4
Calculation of Total Float (Slack Time)*

i	j	LF_j	$-$	TE_{ij}	$-$	ES_i	$=$	TF_{ij}	
1	2	0	−	0	−	0	=	0	*
2	3	14	−	10	−	0	=	4	
2	4	14	−	14	−	0	=	0	*
3	4	14	−	0	−	10	=	4	
4	5	21	−	7	−	14	=	0	*
5	6	28	−	4	−	21	=	3	
5	7	28	−	7	−	21	=	0	*
6	7	28	−	0	−	25	=	3	

*Data from Fig. 10.3.

Table 10.5
Calculation of Free Float (Slack Time of an Activity)*

i	j	ES_j	$-$	TE_{ij}	$-$	ES_i	$=$	FF_{ij}
1	2	0	−	0	−	0	=	0
2	3	10	−	10	−	0	=	0
2	4	14	−	14	−	0	=	0
3	4	14	−	0	−	10	=	4
4	5	21	−	7	−	14	=	0
5	6	25	−	4	−	21	=	0
5	7	28	−	7	−	21	=	0
6	7	28	−	0	−	25	=	3

*Data from Fig. 10.3.

10.5 AN EXPANDED VERSION OF THE WATER-INJECTION-PLANT EXAMPLE

The activity diagrams shown in Fig. 10.1 and 10.3 can be considered as preliminary sketches that list only the major phases of the project. Such diagrams would have very limited usefulness in the detailed planning and control of installing the water injection plant. They can serve, however, as a basis for an expanded and more detailed diagram. A detailed diagram can be converted into a detailed list. Greater detail will permit the realization of the advantages claimed for the CPM method.

Figure 10.6 is an arrow diagram in which has been achieved a greater detail and breakdown of the individual activities necessary for installing the water injection plant. The tableau method for analyzing the network is shown in Fig. 10.7. In general, the people who will be directing and controlling the actual work should be consulted during the preparation of the

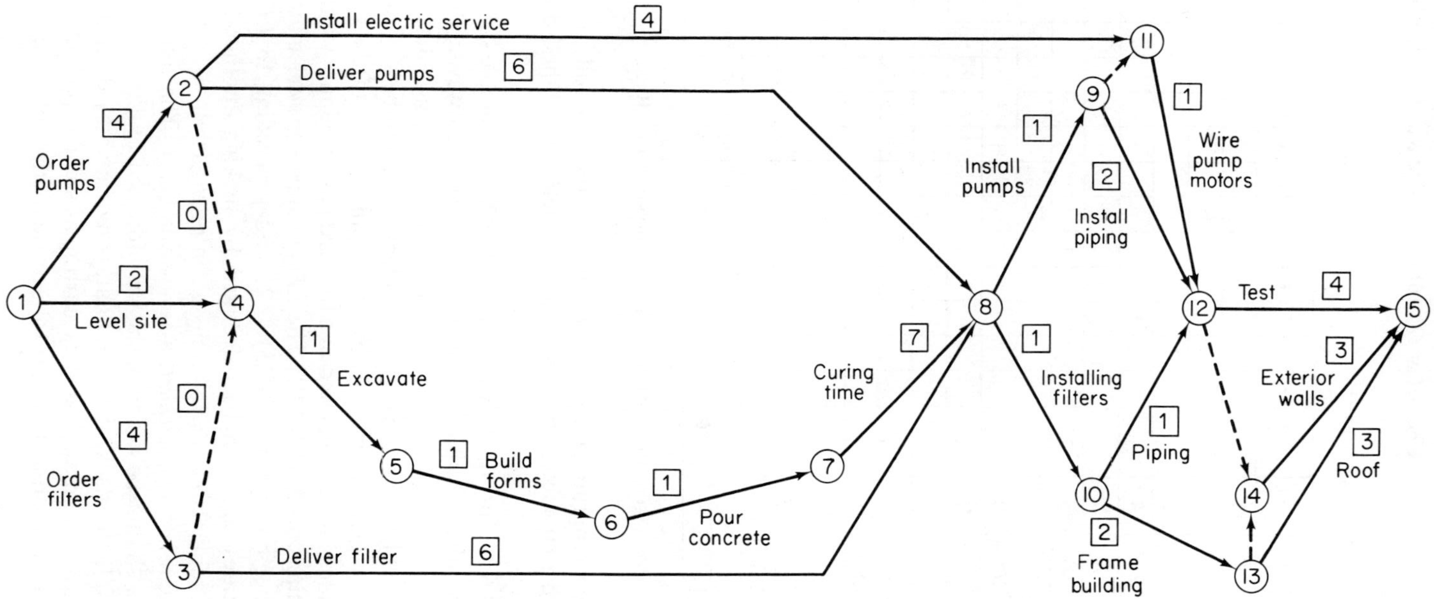

Figure 10.6. Expanded diagram of activities for installing a water-injection plant.

Earliest start	i \ j	1	2	3	4	5	6	7	8	9	10	11	12	13	14	15
0	1		4	4	2											
4	2				0				6							
4	3				0				6							
~~2~~,4	4					1										
5	5						1									
6	6							1								
7	7								7							
~~10~~,14	8									1	1					
15	9											0	2			
15	10												1	2		
15	11												1			
~~16~~,17	12														0	4
17	13														0	3
17	14															3
~~20~~,21	15															
Latest finish		0 ~~2~~	4 ~~8~~	4 ~~8~~	4	5	6	7	14 ~~15~~	15 ~~16~~	16 ~~18~~	16	17 ~~18~~	18	18	21

Figure 10.7. Tableau of data shown in Fig. 10.6.

detailed arrow diagram and list of activities. It must be fully appreciated by all concerned that successful application of the CPM method depends upon team effort.

Table 10.6 is a list of activities in the expanded example. In this case the list was made after completing the arrow diagram. However, the list can be used as the starting point. When this method is used, it is desirable to have a column in which can be written the identifying numbers of activities that must precede a given activity. Table 10.7 is an analysis of the time network. The $ES(i)$ values were calculated in a forward pass, beginning with $ES(1) = 0$ and proceeding to $ES(15) = 21$. The $LF(j)$ values were calculated in a backward pass beginning with $LF(15) = 21$ and proceeding to $LF(1) = 0$. All other quantities can then be calculated by making another forward pass through the table using the relations given previously.

Figure 10.8 is a *Gantt chart*, a time-scale chart. The time during which the activities should be accomplished is indicated by the heavy black lines. In this example the slack, the total float, is indicated by dashed lines. In this figure the activities are indicated as starting as soon as possible. Where there is float, it is obvious that the start for such activities could be delayed one or more days.

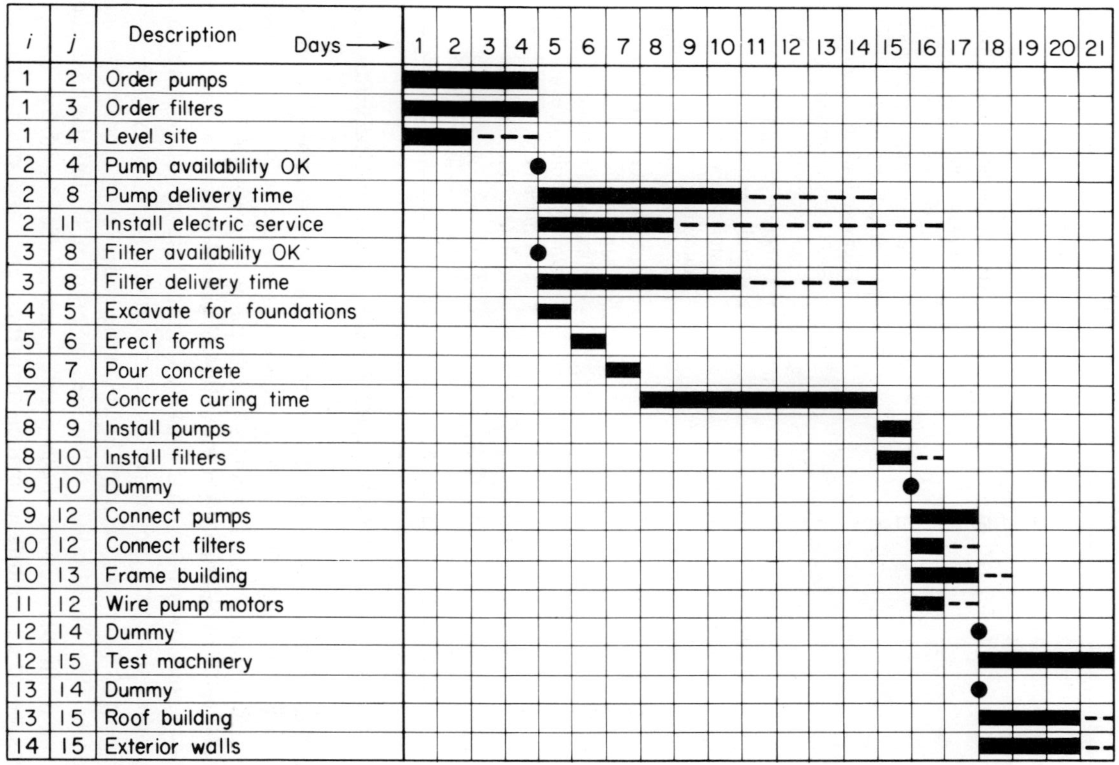

Figure 10.8. Gantt or bar chart. Time-scale presentation of data shown in Fig. 10.6.

Table 10.6
Expanded List of Activities as Shown in Figure 10.6

i	j	T_E	*Critical*	*Activity*
1	2	4	*	Order pumps
1	3	4	*	Order filters
1	4	2		Level site
2	4	0	*	Pump availability confirmed
2	8	6		Pump delivery time
2	11	4		Install electric service
3	4	0	*	Filter availability confirmed
3	8	6		Filter delivery time
4	5	1	*	Excavate for foundations
5	6	1	*	Erect forms and templates
6	7	1	*	Pour concrete
7	8	7	*	Concrete curing time
8	9	1	*	Install pumps
8	10	1		Install filters
9	11	0		(dummy)
9	12	2	*	Pipe connections to pumps
10	12	1		Pipe connections to filters
10	13	2		Frame building
11	12	1		Wire pump motors
12	14	0		(dummy, hold on exterior walls)
12	15	4	*	Test machinery
13	14	0		(dummy, hold on exterior walls)
13	15	3		Roof
14	15	3		Exterior walls

10.6 SORTING UNORDERED NODE NUMBERS

In the process of developing an arrow diagram it invariably happens that the node numbers get out of order. This occurs where activities are added or broken into smaller components somewhere in the network, after the event nodes have been numbered. Changing all the following node numbers is a time-consuming process that is likely to introduce errors. Sorting on some basis other than strict numerical order is more satisfactory.

Figure 10.9 shows a network in which the nodes (events) are not in numerical order. The initial node is identified by number 12 and the final node by number 8. In between are both greater and smaller numbers. Thus some arrows begin at a number smaller than their end, and other arrows begin with a number larger than their end event number. Listing the activities in this network according to natural numerical order would be misleading.

Table 10.8 illustrates two ways of sorting. It can be observed in previous examples, such as Table 10.7, where the (i, j) activity numbers are in natural numerical order throughout, that the j-value of any activity in the table is not

Table 10.7
Analysis of Data from Figure 10.6 and Table 10.6

i	j	TE_{ij}	ES_i	ES_j	LF_j	FF_{ij}	TF_{ij}	*Critical*
1	2	4	0	4	4	0	0	*
1	3	4	0	4	4	0	0	*
1	4	2	0	4	4	2	2	
2	4	0	4	4	4	0	0	*
2	8	6	4	14	14	4	4	
2	11	4	4	15	16	7	8	
3	4	0	4	4	4	0	0	*
3	8	6	4	14	14	4	4	
4	5	1	4	5	5	0	0	*
5	6	1	5	6	6	0	0	*
6	7	1	6	7	7	0	0	*
7	8	7	7	14	14	0	0	*
8	9	1	14	15	15	0	0	*
8	10	1	14	15	16	0	1	
9	11	0	15	15	16	0	1	
9	12	2	15	17	17	0	0	*
10	12	1	15	17	17	1	1	
10	13	2	15	17	18	0	1	
11	12	1	15	17	17	1	1	
12	14	0	17	17	18	0	1	
12	15	4	17	21	21	0	0	*
13	14	0	17	17	18	0	1	
13	15	3	17	21	21	1	1	
14	15	3	17	21	21	1	1	

matched by the i-value of any preceding activity. The same principle can be observed in Table 10.8. In either of these lists, if the TE values of the activities were given, it would be possible to make the usual forward pass to determine the earliest start (ES) values and then make the backward pass to determine the latest finish (LF) values. All other network quantities could then be calculated.

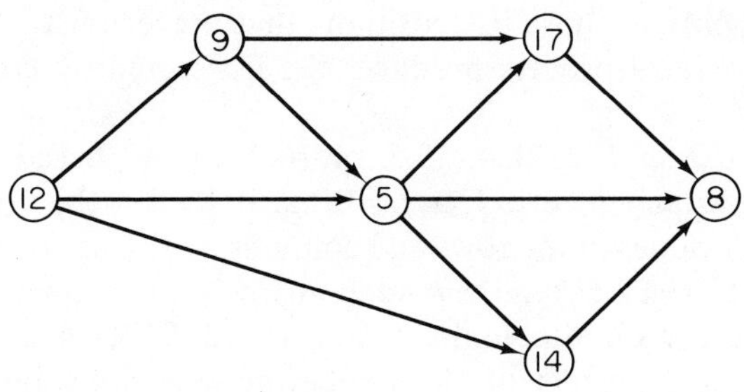

Figure 10.9. CPM network with unordered node numbers.

Table 10.8
Sorting Unordered Node Numbers

List Position	Sorted by I-Values		Sorted by J-Values	
	I	J	I	J
1	12	5	12	9
2	12	9	9	5
3	12	14	12	5
4	9	5	5	14
5	9	17	12	14
6	5	8	5	17
7	5	14	9	17
8	5	17	5	8
9	14	8	14	8
10	17	8	17	8

Where it is desirable to sort by groups having the same i-value, one starts at list position 1. Various (i, j) activity values are tried in that position and interchanged until one is found such that no following activity has a j-value that is the same as the i-value in position 1. Then activities having the same i-value are moved into adjacent following positions, positions 2 and 3 in this case. Then the entire process is repeated, starting at position 4 and checking only those activities in the following list positions. Repetitions of this process are continued until the end of the list is reached.

Where it is desirable to sort by groups having the same j-values, one starts at the highest list position, list position 10 in this example. Various (i, j) activity values are tried in that position, and the activities are interchanged in the list until one is found with a j-value that is not matched by any i-value in the preceding list positions. Then other activities having that j-value are interchanged into the immediately preceding list positions. In this example that process determined the activities in list positions 10, 9, and 8. Next the process was repeated, starting with list position 7, and searching and comparing i-values only in those list positions that precede list position 7. Continuing this backward process produces the list sorted by groups that have similar j-values.

A tableau such as Fig. 10.4, 10.5, or 10.7, can be based upon either of the two sorting arrangements. That is, Table 10.8 gives the order of arrangement of the (i, j) values of the rows and columns, which are of course identical within any one tableau. From the i-value sorting arrangement, the numbering sequence of the rows and columns would be 12, 9, 5, 14, 17, and 8. From the j-value sorting arrangement, the numbering sequence would be identically the same, in this example. With this sequence of (i, j) values, the activity duration (TE) values appear only in the upper-right half of the tableau, in the portion above the main diagonal, as in Fig. 10.4, for example. This is a desirable arrangement for manual calculations.

10.7 PERT

Activity-duration times cannot always be estimated precisely. Sources of uncertainty include weather, uncertain delivery of supplies, equipment failure, lack of knowledge concerning procedures and methods to be used, interactions among persons, and personnel problems. External conditions can induce uncertainty into a familiar procedure. New developments rather naturally contain uncertainty.

The PERT method consists of making estimates from which the expected duration times as well as their variance can be obtained. Adding the expected duration times for individual activities along a path gives the expected time for completion of the path. Adding their variances gives the total variance. Where there are three or more activities along a path, the central limit theorem of statistics supports the finding that the combined variance of the path corresponds to a normal distribution. This result is independent of the probability characteristics of the individual activity durations. Use of the normal distribution then offers a very convenient way to calculate the probability of achieving various completion dates.

Three time estimates are used in the PERT method. These estimates are generally based upon normal operating procedures, as opposed to crash programs. The amounts of resources, such as the number of men and machines available, form the basis of any estimates made. The amounts of resources should be recorded, unless understood fully by all concerned. The following duration-time estimates are generally used.

Optimistic: An estimate of the minimum time required for completing an activity, if there is unusually good luck and all goes well without any trouble.

Most likely: An estimate of the normal duration time, which would be repeated most often if a large number of trials were made.

Pessimistic: An estimate of the maximum time an activity will require for its completion if unusually bad luck is encountered. It should, however, be based upon the activity itself and not upon unlikely external occurrences.

Reduction of the estimates to a mean and variance is based upon the beta distribution.[2] The following equations are customarily used:

$$TE = \frac{L + 4M + H}{6}, \tag{10-7}$$

$$\sigma^2 = \left(\frac{H - L}{6}\right)^2, \tag{10-8}$$

where

TE = expected activity-duration time,

L = optimistic duration-time estimate,

M = mode, the most likely duration-time estimate,

H = pessimistic duration-time estimate,

σ^2 = variance of the time duration of an activity.

An example network is shown in Fig. 10.10 and supporting data are given in Table 10.9. The TE values are calculated according to Eq. (10-7), and the variances are calculated according to Eq. (10-8). In Table 10.10 the means and variances of the individual activities along the critical path are added together. In a similar manner, Table 10.11 contains the means and variances

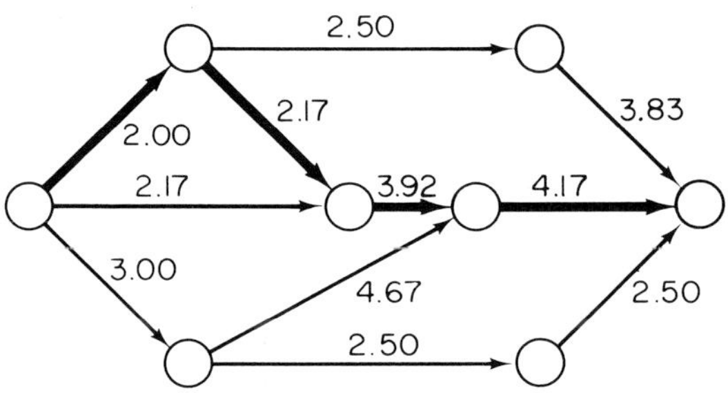

Figure 10.10. Network of PERT example. Critical path is shown by heavy lines.

Table 10.9
Pert Time Estimates for Figure 10-10

		Time Estimates					
i	*j*	*L*	*M*	*H*	*TE*	*Variance,* σ^2	*Critical*
1	2	1	2	3	2.00	0.111	*
1	3	1	2	4	2.17	0.250	
1	4	1.5	2.5	6.5	3.00	0.695	
2	3	1.5	2	3.5	2.17	0.111	*
2	5	1.5	2.5	3.5	2.50	0.111	
3	6	3	4	4.5	3.92	0.063	*
4	6	2	3	8	4.67	1.000	
4	7	1.5	2.5	3.5	2.50	0.111	
5	8	2	4	5	3.83	0.250	
6	8	2.5	4	6.5	4.17	0.444	*
7	8	2	2.5	3	2.50	0.028	

Table 10.10
Estimated Time and Variance Along Critical Path 1, 2, 3, 6, 8

i	j	TE	σ^2
1	2	2.00	0.111
2	3	2.17	0.111
3	6	3.92	0.063
6	8	4.17	0.444
		12.26	0.729

Table 10.11
Estimated Time and Variance Along Near-Critical Path 1, 4, 6, 8

i	j	TE	σ^2
1	4	3.00	0.695
4	6	4.67	1.000
6	8	4.17	0.444
		11.34	2.139

along a near-critical path. These quantities are used to estimate the probabilities of attaining various completion dates.

Figure 10.11 shows the basis for calculating probabilities of meeting various completion dates. Consider first the critical path. The standard deviation of the critical path is calculated as follows:

$$\sigma = \sqrt{\sigma^2} = \sqrt{0.729} = 0.858 \text{ week.}$$

The probability of completing the project 1 week early is arrived at as follows:

$$z = \frac{11.26 - 12.26}{0.858} = -1.17.$$

From a normal probability table one finds the corresponding probability to be 0.121.

The probability of completing the project prior to the end of 1 additional week beyond the estimated completion date is calculated as follows:

$$z = \frac{13.26 - 12.26}{0.858} = 1.17.$$

The corresponding probability, obtained in a normal table, is 0.879.

Consider next the near-critical path of Table 10.11. The standard deviation of this path is

$$\sigma = \sqrt{\sigma^2} = \sqrt{2.139} = 1.461.$$

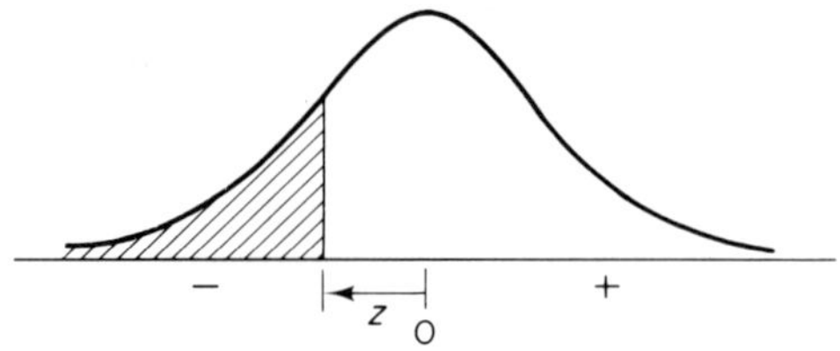

(a) Probability of completing project early is less than 50%

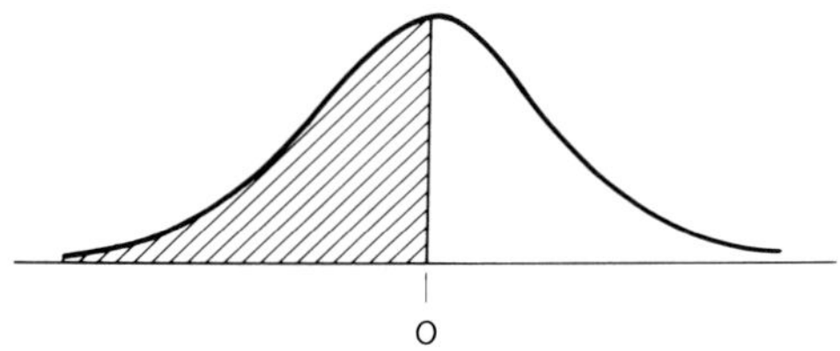

(b) Probability of completing project by estimated completion date is 50%.

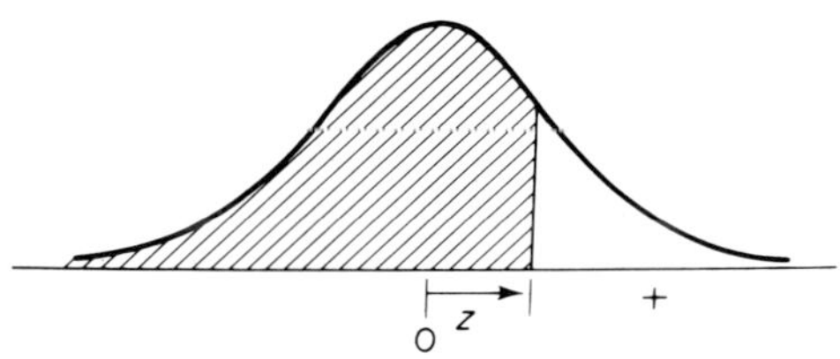

(c) Probability of completing project on or before some date after estimated completion date is greater than 50%

Figure 10.11. Probabilities of meeting completion dates.

The probability of completing the project by 11.26 weeks, according to the activities along the near-critical path, is calculated as follows:

$$z = \frac{11.26 - 11.84}{1.461} = -0.397.$$

From a normal probability table the corresponding probability is 0.154. The probability of completing the project by 13.26 weeks, on the same basis, is

$$z = \frac{13.26 - 11.84}{1.461} = 0.970.$$

From a normal probability table the corresponding probability is 0.834.

The more pessimistic view in both cases results in the following interpretation. Based on the critical path, there is a probability of only 12.1% of completing the project as much as 1 week ahead of time. Based on a near-critical path, there is a $1 - 0.834$, or 16.6%, chance that the project will extend 1 week or more beyond the estimated completion date.

10.8 TIME-COST TRADEOFF

Each activity when completed within its normal time will result in a certain cost, called the *normal cost*. The time for an activity to be completed can be reduced, in most cases, but this results in higher direct operating costs. This statement can be justified in the language of economics, where it would be said that if a workman or machine could be used more efficiently someplace else it would be moved there. Thus there is a tendency for normal time to correspond to the most efficient local conditions and amount of resources applied to each job.

It is usually possible to arrive at an estimate of how much it would cost to reduce an activity duration by one time unit. Such time-reduction cost may or may not be linear. Usually the cost tends to increase more as additional time compression is required. Some lower time limit will be reached, and at that point one either sets an arbitrary limit or in effect assigns an infinite cost to further time reduction. However, as a practical matter, although it is not necessary, it is usual to assign a constant cost per unit time reduction and use it between prescribed limits of an activity duration. This convention is used in the accompanying example.

A project network is shown in Fig. 10.12. The activity duration times are shown on each arrow, and the first number in each case is the normal duration

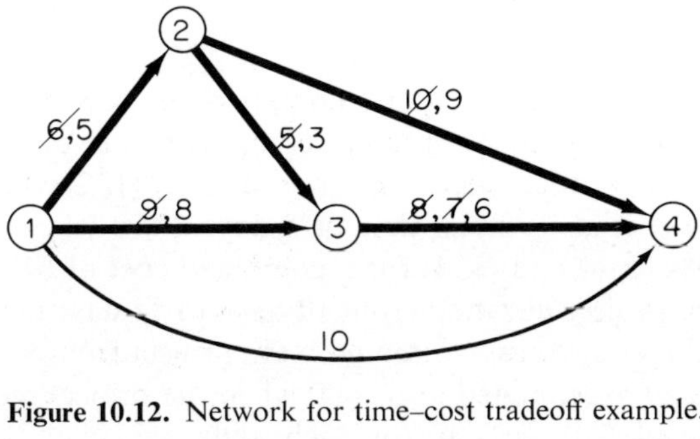

Figure 10.12. Network for time–cost tradeoff example.

time. Only in activity (1, 4) is the normal time still in effect; in the others it has been replaced by a smaller number.

The time reductions were made in several steps, as shown in Table 10.12. The objective, of course, is to achieve a time reduction at the smallest possible cost. Only activities on the critical path or paths when compressed will result in reduction of time for the total project. Activity (2, 3) is on the original critical path and has the smallest cost slope. It can be reduced by 2 days without affecting any other path. This results in plan 1. However, in plan 1, activity (1, 3) becomes critical since path 1, 2, 3 and path 1, 3 both take 9 days, at this stage of the development. Plan 2 is reached by reducing activity (3, 4) by 1 day, and this makes activity (2, 4) become critical, since path 2, 3, 4 and path 2, 4 both take 10 days at this second stage of the development.

Table 10.12
Time–Cost Tradeoff for Network in Figure 10.12

				*Plan Number**				
Activity	*Normal Time*	*Crash Time*	*Cost Slope*	*0*	*1*	*2*	*3*	*4*
1, 2	6	5	60	*	*	* →	1*	*
1, 3	9	5	40		*	* →	1*	*
2, 3	5	3	30	* →	2*	*	*	*
1, 4	10	6	25					
2, 4	10	5	70			*	* →	1*
3, 4	8	6	60	*	* →	1*	* →	1*
Project duration (days)				19	17	16	15	14
Time compression (days)				0	2	1	1	1
Total cost/day to compress ($)				0	30	60	100	130
Increase in cost ($)				0	60	60	100	130
Project cost ($)				1,300	1,360	1,420	1,520	1,650
Indirect costs ($)				1,750	1,610	1,540	1,470	1,400
Total cost ($)				3,050	2,970	2,960	2,990	3,050

* Denotes activities on critical paths.

Further time reductions require compressing two activities simultaneously. Activities (1, 2) and (1, 3) can be compressed one day for a combined cost of $100, which produces plan 3. But now activity (1, 2) can be compressed no further, since it is at its crash time of 5 days. Plan 4 is reached by compressing activities (2, 4) and (3, 4) for a combined cost of $130.

Reducing the project duration from 19 days to 14 days has resulted in an increase in the direct costs associated with the project from $1,300 to $1,650. The indirect costs are estimated at $1,400 when the project requires 14 days, plus an increase of $70 per day for each additional day. This additional information permits calculating the total costs and determining which plan will result in the smallest total cost. In Table 10.12 it may be seen that plan 2, with a project duration of 16 days, is most advantageous. The same information is shown on Fig. 10.13, where the costs are plotted.

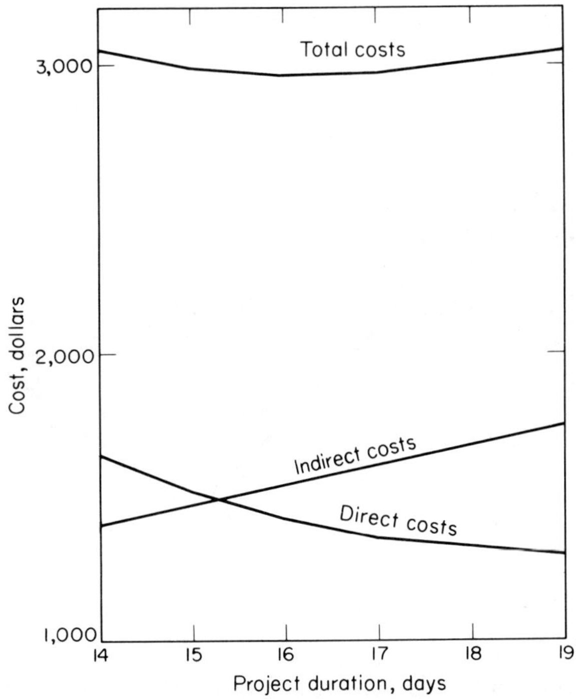

Figure 10.13. Time–cost tradeoff.

10.9 RESOURCE CONSTRAINTS, LEVELING

In many cases it is highly desirable to have the least possible variation in the resources committed to a project. In the case of manpower, bringing new men into a job may require hiring costs, extra overhead costs, and lessened efficiency and safety during a training phase. In the case of machinery, there is either the cost of investment in idle facilities or the cost of transportation from one job to another. The problem considered here is that of reducing the fluctuations without increasing the total time required to complete the project. In general, this can be done by adjusting the starting times of activities which are not on the critical path. That is, we must adjust starting times so that concurrent activities contain minimum fluctuation in the total quanity of a resource used in the project.

Figure 10.14 is a project network where a key resource is the welders used on the various individual jobs. The total project requires 14 days. Activities on the critical path are 1-2, 2-4, 4-6, and 6-7. The starting times of these individual jobs will remain fixed and not be subject to any shifting, since to do so would lengthen the completion time of the entire project.

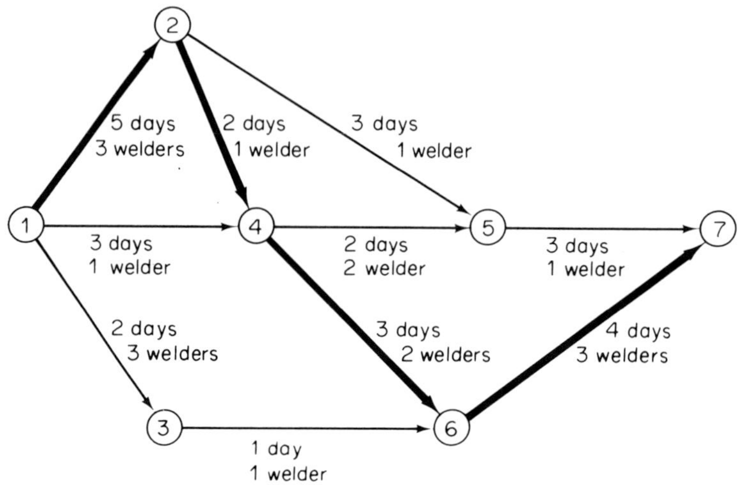

Figure 10.14. Number of welders used on individual jobs.

Figure 10.15 is a Gantt chart which shows the requirement for welders by days. The upper part of the figure shows the case where all activities are started as soon as possible. Those activities not on the critical path have slack, and this is indicated as following the time when the activity is performed. The total slack is shown, which is a quantity generally assigned to more than one activity. For example, delaying the start of activity 1-3 would advance the time when activity 3-6 could be started. For the upper chart the total number of welders required varies from a maximum of 7 to a minimum of 2, with many fluctuations throughout the 14-day period. The chart in the lower part of the figure shows how the starting times of the activities can be adjusted to greatly reduce the fluctuations in the total number of welders required on the project. In some cases, certain activities can be interrupted in order to achieve maximum leveling, but this was not possible in this particular case.

In the chart in the upper part of Fig. 10.15, counting the initial moving of the welding machines onto the job, the daily fluctuations, and the final moving of the machines off the job, there are a total of 22 machine moves. On the same basis, there are 10 machine moves in the lower chart. If the cost of moving the welding machines back and forth between projects averages $40 per move, the 12 fewer moves would amount to a savings of $480.

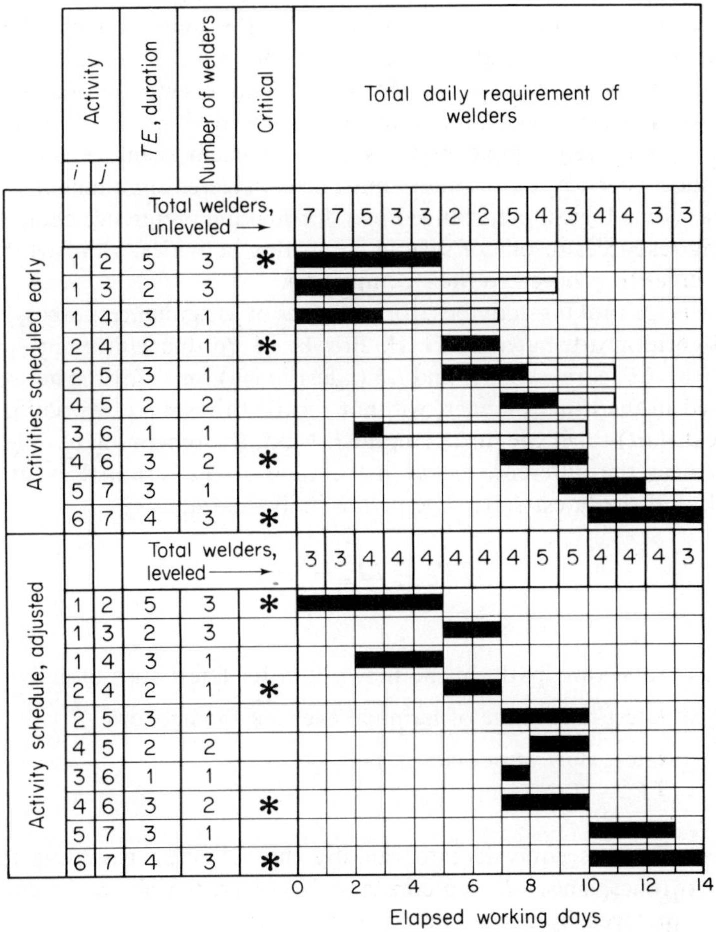

Figure 10.15. Gantt chart showing the leveling of resource requirements.

10.10 RESOURCE CONSTRAINTS, LIMITED RESOURCES

In all cases the resources, such as men, machines, and materials, that can be allocated to a particular job or project, are limited. Although the foregoing discussion has apparently presumed the availability of unlimited resources, this is not the case. Rather, the time estimates used were implicitly based upon either common practice or an understood total pool of available resources. It is now time to consider more explicitly the case where different jobs compete for the same resources, and a choice must be made on which jobs to complete first and which to stay for the time being.

Obtaining an optimum solution to this problem, and proving the solution to be optimal, is generally conceded to involve the computing of all possible network permutations. That is, for each particular initial choice, there would be several secondary choices, and for each of these several tertiary choices, and so on, so that a very large number of computations would be required for even a relatively simple network. The problem then becomes greatly confounded when several resources must be considered simultaneously. Thus one seeks some logical prearranged plan for making preference choices when awarding resources to one of several competing activities, all of which need the resources to proceed to their completion.

A basis for ranking activities, for purposes of assigning resources, is given in an algorithm attributed to G. H. Brooks of Purdue University.[3] In this method the *ES* (earliest start) and *LF* (latest finish) times for the network are computed in the usual manner, without regard to resource restrictions. The *LS* (latest start) times are also computed. Next, the remaining path length is computed as the difference between the time of the terminal event of the network and the latest start time of the individual activity:

$$TP_{ij} = LF_n - LS_{ij}, \tag{10-9}$$

where

TP_{ij} = total time path, in the network, after latest start of activity (i, j)

LF_n = latest finish time of terminal event of the network

LS_{ij} = latest start of activity (i, j)
$= LF_j - TE_{ij}$.

This method is equivalent to ranking the activities according to their latest start times, since LF_n is a constant. However, the physical significance is better portrayed by the method as described.

A network problem in which the total resources consist of a maximum of seven crews is shown in Fig. 10.16. This has been analyzed, without regard to resource requirements and constraints, in the customary CPM method, and the results are shown in Table 10.13. The earliest start and latest finish times were calculated by the methods shown in Fig. 10.5 and Tables 10.2 and 10.3. Next, the *TP*-values were calculated from Eq. (10-9). The *TP*-values were then used for the sole purpose of ordering the activities in Table 10.14. In this method, whenever two or more activities can be started, the necessary resources will first be assigned to the activity nearest the top of the table.

The *decision set* is defined as all activities, not yet started, whose leading events have been completed. With unlimited resources, these would all be started as soon as their leading events were completed. However, where

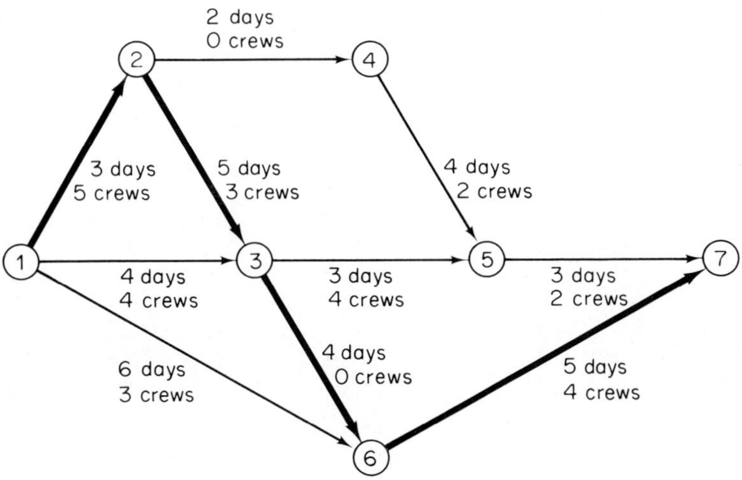

Figure 10.16. Project network showing resource requirements.

Table 10.13
Analysis of Network in Figure 10.16 Without Regard to Resource Constraints

i	*j*	*TE*	*ES*	*LF*	*TP*
1	2	3	0	3	17
1	3	4	0	8	13
1	6	6	0	12	11
2	3	5	3	8	14
2	4	2	3	10	9
3	5	3	8	14	6
3	6	4	8	12	9
4	5	4	5	14	7
5	7	3	11	17	3
6	7	5	12	17	5

resources are limited, the activity in the decision set which is nearest the top of the list will be examined first. If there are sufficient unconsigned resources, that activity will be started. Then the member of the decision set next to the top of the list will be examined and started if possible. All activities within a decision set must be examined, since some may not require any of the strategic resource.

The plan evolved by the algorithm for scheduling the activities is shown in Tables 10.14 and 10.15. These two tables must be developed simultaneously. In Table 10.14 the *TP*-values have already been calculated and

Table 10.14
Application of Algorithm*

i	*j*	*Total path, TP*	*Duration, TE*	*Number of crews, C*	*Time of Entering Decision Set (days)*	*Time of Starting (days)*	*Time of Finishing (days)*
1	2	17	3	5	0	0	3
2	3	14	5	3	3	3	8
1	3	13	4	4	0	3	7
1	6	11	6	3	0	7	13
2	4	9	2	0	3	3	5
3	6	9	4	0	8	8	12
4	5	7	4	2	5	8	12
3	5	6	3	4	8	12	15
6	7	5	5	4	13	15	20
5	7	3	3	2	15	15	18

*A maximum of 7 crews is available.

Table 10.15
Application of Algorithm*

Time (days)	*0*	*3*	*5*	*7*	*8*	*12*	*13*	*15*	*18*	*20*
Decision set	1,2 1,3 1,6	2,3 1,3 1,6 2,4	1,6 4,5	1,6 4,5	3,6 4,5 3,5	3,5	6,7	6,7 5,7		
Working set	1,2	2,3 1,3 2,4	2,3 1,3	2,3 1,6	1,6 3,6 4,5	1,6 3,5	3,5	6,7 5,7	6,7	
Crews	5	3 4 0	3 4	3 3	3 0 2	3 4	4	4 2	4	
Total crews	5	7	7	6	5	7	4	6	4	

*A maximum of 7 crews is available.

serve only as the basis for arranging the activities in the table. The *TE* (duration) and *C* (resources required) are obtained from Fig. 10.16. This leaves the last three columns to be completed. Tables 10.14 and 10.15 can be completed according to the following algorithm.

1. Note the number of resources available and check to see that the requirements of any individual activity do not exceed the maximum availability. Set current time, $T = 0$. Enter all initial activities into the decision set.
2. Examine all the activities in the decision set according to their order in the list. If sufficient resources are available, assign them to the activity, reduce the pool of available resources by that amount, and assign the activity to the working set. Calculate the completion time of the activity.
3. Examine the completion times of all activities in the working set. If T is already as great as the largest of these, terminate the algorithm. Otherwise, advance T to the smallest of these completion times.
4. Terminate the activities whose completion times are equal to current time T. Remove them from the working set and return the resources assigned to them to the available pool.
5. Determine if any events have occurred. An event occurs when all activities, which terminate at a particular node, have been completed. When this happens, enter the activities, which start with the same node, into the decision set. Return to step 2.

In Table 10.14 it may be seen that the algorithm arrives at a project completion time of 20 days for the case where there is a maximum of seven crews available. This compares to a completion time of 17 days when resources are unrestricted.

The problem considered here has been restricted to one project and only one resource. The general problem includes restriction on several resources. At least one development includes a multiproject approach, the RAMPS (resource allocation and multiproject scheduling) algorithm.[4] A good discussion of the problems inherent in developing algorithms for resource allocation is given by Davis.[5]

REFERENCES

1. LEVY, F. K., G. L. THOMPSON, and J. D. WEST, "The ABC's of Critical Path Planning." *Harvard Business Review*, Sept.–Oct. 1963.
2. PERT Summary Report, Phase I. Washington, D.C.: Special Projects Office, Department of the Navy, July 1958.
3. MODER, J. J., and C. R. PHILLIPS, *Project Management with CPM and PERT.* New York: Van Nostrand Reinhold Company, 1964.
4. LAMBOURN, S., "RAMPS—A New Tool in Planning and Control." The Computer Journal, January 1963.
5. DAVIS, E. W., "Resource Allocation in Project Network Models—A Survey." *Journal of Industrial Engineering*, Vol. 27, No. 4, p. 177.

PROBLEMS

10-1. A worn-out pump must be replaced with a new one. Following is a list of activities.

No.	*Preceded by*	*Activity Description*
1	None	Disconnect electric power
2	None	Tear out old piping
3	1, 2	Remove old pump
4	3	Prepare foundation
5	4	Set new pump
6	2	Fabricate new pipe
7	1	Install new control box
8	5, 6	Connect new piping
9	8	Pressure-test new piping
10	9	Insulate new piping
11	5, 7	Connect electric power
12	10, 11	Test pump performance

(a) Draw an activity-on-node network of the project.
(b) Draw an activity-on-arrow network.

10-2. A gas gathering system is to be installed that will connect a small group of gas wells to a main trunk line. Following is a breakdown of the major activities.

No.	*Preceded by*	*Activity*	*Duration (days)*
1	–	Design and cost estimate	4
2	1	Field routing, survey	3
3	2	Purchase right-of-way	3
4	1	Order pipe	1
5	1	Prepare pipe-handling dock	2
6	3, 4, 5	Lay pipelines	10
7	3	Install meter runs	5
8	3	Install dehydrator	3
9	6, 7, 8	Inspect welds	1
10	9	Connect to wells	1
11	10	Test	1

(a) Make an activity-on-node network of the project.
(b) Make an activity-on-arrow network.
(c) Calculate the estimated time of completion of the project. Calculate the earliest start, latest finish, and slack times for each activity. Identify the critical path.

10-3. The network (see figure) indicates the sequence of major activities on a construction project. The time required for each activity is given in the table. Also shown is the number of trucks required for each activity.

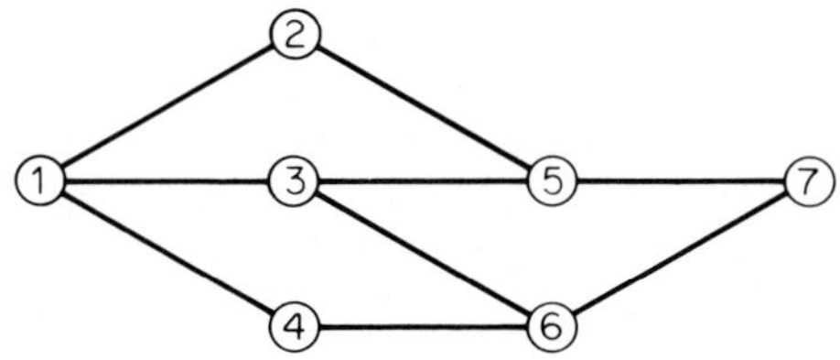

Activity *i*	Activity *j*	Duration (days)	Trucks Required
1	2	4	3
1	3	3	2
1	4	4	4
2	5	6	3
3	5	3	3
3	6	5	3
4	6	3	3
5	7	2	2
6	7	7	4

(a) If the number of trucks is unlimited, what would be the number of days required for completion?

(b) It has been calculated that it will require a total of 34 days to complete the project if no more than 5 trucks are made available. Calculate the completion times for the cases where totals of 6, 7, and 8 trucks are made available. Calculate the number of idle truck days for each case. Also calculate the number of times trucks would have to be shifted back to the project if the idle days were utilized elsewhere.

10-4. A booster compressor must be installed in a gas delivery line. The table gives the major activities and the minimum, most likely, and maximum duration times (in days) for each activity.

No.	Preceded by	Activity	Duration Times T_a	T_m	T_b
1	–	Design and cost estimate	2	3	4
2	1	Obtain compressor	5	6	9
3	1	Grade location	1	1	2
4	3	Install foundation	2	3	5
5	2, 4	Install compressor	2	2	3
6	1	Install electric supply	2	2	3
7	6	Install electric control lines	1	1	2
8	5, 7	Install control panel	1	1	2
9	5	Connect into gas line	2	3	4
10	8, 9	Test	1	1	2
11	10	Fence location	1	1	1

(a) Draw the activity-on-node network of the project.
(b) Draw the activity-on-arrow network.
(c) Calculate the estimated completion time for the project, assuming that the most likely estimate holds true in each case.
(d) Calculate the probabilities that the project will be completed (1) 2 days early, (2) 1 day early, (3) by the best estimate calculated in part (c), (4) by the end of 1 additional day, and (5) by the end of 2 additional days.

11

EVALUATION OF FUTURE PRODUCTION BY PERFORMANCE TRENDS

11.1 INTRODUCTION

One of the necessary functions in petroleum producing operations is the evaluation of properties. The value of the oil, when produced at the surface, depends upon both economic and political forces. Petroleum is a mobile substance, subject to control. Only a stable worldwide petroleum industry would permit accurate forecasts of future prices. Nevertheless, an estimate of future prices and trends is necessary for evaluation. Also necessary is the determination of the required investments and operating expenses. Another part, which is discussed in this chapter, concerns estimates of the total future recovery of petroleum and its time pattern of production.

Present-value discount factors for declining oil production are presented in Appendix Tables 8, 9, 10, and 11. They cover the range from uniform production, or a series of equal payments, to cases of sharply declining production, arranged to facilitate interpolation; a special interpolation table is included. Several examples of the estimation of present values are presented, as are examples of actual production decline data.

The complete evaluation of a petroleum reservoir involves the march of time. In early stages, estimates of value must be based largely upon comparisons with similar reservoirs, formations, or areas. As additional information is obtained, particularly through drilling wells, the principles of reservoir engineering can be utilized. When sufficient information becomes available, estimates can be made of reservoir performance under different producing conditions. Later, as producing patterns become established, these can be extended to narrow down the final evaluation. These ideas are expressed in Fig. 11.1.[2]

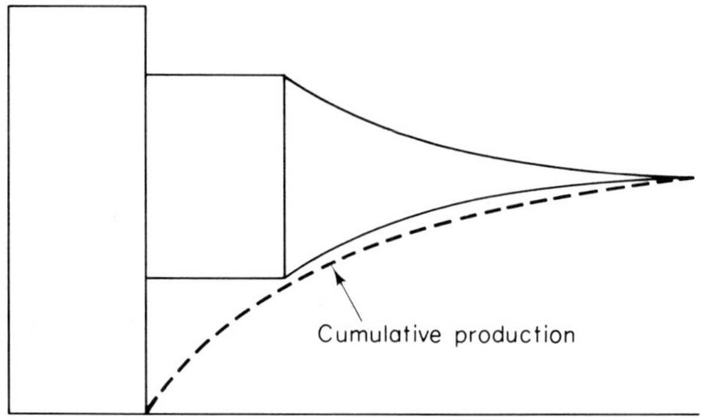

Figure 11.1. Schematic presentation of ranges of recovery estimates made during the life of a producing property. (After Arps[2])

Where the ultimate recovery and the time pattern of production for a property are known, its present value can be calculated. Particularly in the early stages of development it is often necessary to estimate ultimate recovery. Table 11.1 shows values that have been observed. Table 11.2 lists some types of production curves that have been found useful for the prediction of production patterns. This chapter is particularly devoted to the study of production decline curves, which occur often in the later life of producing units.

Table 11.1
Ultimate Recoveries of Oil

Predominant Drive Mechanism	*Sand and Sandstone (bbl/acre-ft)*			*Limestone, Dolomite, and Others (bbl/acre-ft)*		
	Min.	*Med.*	*Max.*	*Min.*	*Med.*	*Max.*
Water drive	155	571	1641	6	172	1422
Solution gas drive	47	154	534	20	88	187
Solution gas + supplemental	109	227	820	32	120	464
Gas cap drive	68	289	864	(Combined with SS)		
Gravity drainage	250	696	1124	(Not analyzed)		

Source: Data from API BUL D14, 1st ed., Oct. 1967, "A Statistical Study of Recovery Efficiency."[1]

Table 11.2
Types of Production Curves

Rate versus Time

1. *Conventional production decline curve.* Raw production amounts for equal time intervals are plotted versus time on any type of coordinate paper.

2. *Percentage decline curve.* The productions of successive equal time units are plotted as a percentage of the production of the first time unit versus time. Superimposing such curves for several wells, all starting at 100%, provides a comparison of production characteristics of the wells.

3. *Persistence curve.* The ratios of successive equal time units are plotted versus time. For a well under proration or for a well producing under an efficient water drive, the ratio is 1. For a well producing without restriction, the ratio of the second year to the first is often about 0.5, and thereafter increases asymptotically toward 1.

Cumulative versus Time

4. *Cumulative production curve.* The total or accumulated production of the well or field is plotted versus time. Similar curves have been constructed on a percentage basis, with the production of the first year taken as the unit value.

5. *Future production curve.* The estimated future production of a well, tract, or field is plotted versus the remaining time units. The estimated remaining reserves can be read directly from such a curve.

Rate versus Cumulative

6. *Rate versus cumulative.* Raw production amounts for equal time units are plotted versus cumulative production on any type of coordinate paper.

7. *Ultimate production or appraisal curve.* For any given field or set of similar conditions, the ultimate production of wells to a specified economic limit may be plotted versus the production of the first year.

Average or Composite Curves

8. *Average or composite performance* of wells, tracts, fields, groups of pools, and whole states have been used for control and appraisal purposes. On a statewide basis, these have been known as "rate of withdrawal curves."

Curves Focusing on Special Relations

9. *Reservoir pressure versus cumulative production.* Such curves plotted on semilog paper have been useful in limestone reservoirs without water drives. The log of pressure tends to be roughly proportional to the remaining energy, for fields producing under gas drive.

10. *Well potential versus cumulative production.* Changes in slope have sometimes indicated when water encroachment became effective.

11. *Gas–oil and/or water–oil ratio* versus daily, monthly, or cumulative productions.

11.2 DECLINE CURVES, THEORETICAL RELATIONS

Estimates of the ultimate recovery of oil and of the producing life are commonly made by extrapolating a performance trend. This technique is most useful after initial development when production patterns have become established. Where production is unrestricted, the production rate is often the most convenient variable to use in such extrapolations. The most common experience is for the productivity of the well or wells to decrease with time and with the total cumulative amount produced. As the wells are produced, the energy causing oil to flow into the wells is depleted, or becomes less effective as part of the resources are removed, so that production rate gradually decreases. Figure 11.2 shows three ways to illustrate the general relationship.

The method of extrapolating a trend, for purposes of estimating future performance, depends upon the assumption that the factors which caused changes in the past will operate in the same way in the future. Thus it must be

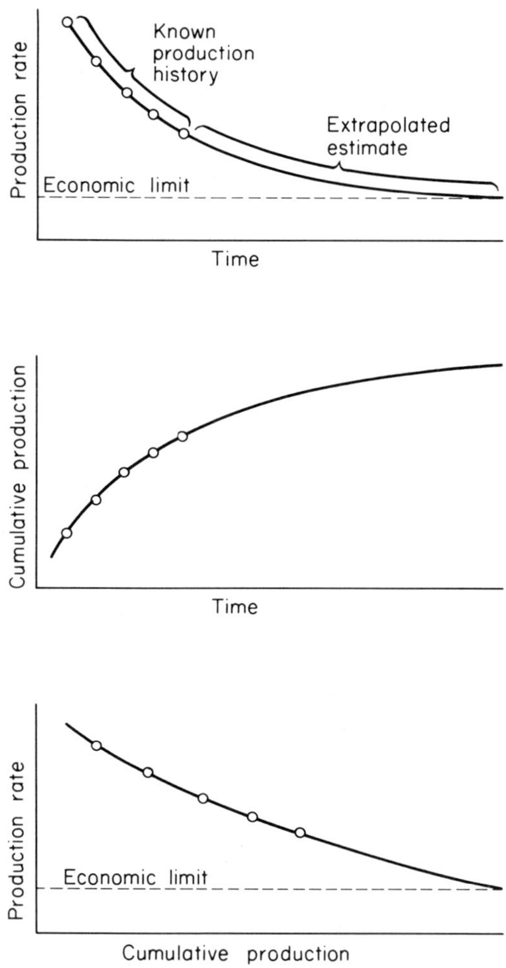

Figure 11.2. Common forms of production curves.

assumed that the producing conditions that prevailed in the past will continue into the future. At some point the producing rate will become so small and net revenue reduced so that it merely equals operating expenses. Beyond such a limit, further production is uneconomical. Therefore, the final limit of production is generally defined as an economic limit. The remaining reserves of oil and the time required to produce them are estimated empirically. This is done either by fitting equations to the known production history or by graphically plotting the data so that it can be extrapolated into the future to the economic limit.

The decline curves most commonly used for the extrapolation of the production rate are members of a hyperbolic family which stems from the following differential equation[3]:

$$\frac{d\left(\frac{q}{dq/dt}\right)}{dt} = -b. \tag{11-1}$$

Integration of Eq. (11-1) gives

$$\frac{q}{dq/dt} = -bt - a, \tag{11-2}$$

where b is normally zero or a positive constant and a is a positive constant. Various possible values of b are considered in the following cases. Figure 11.3 illustrates the general shapes of the curves.

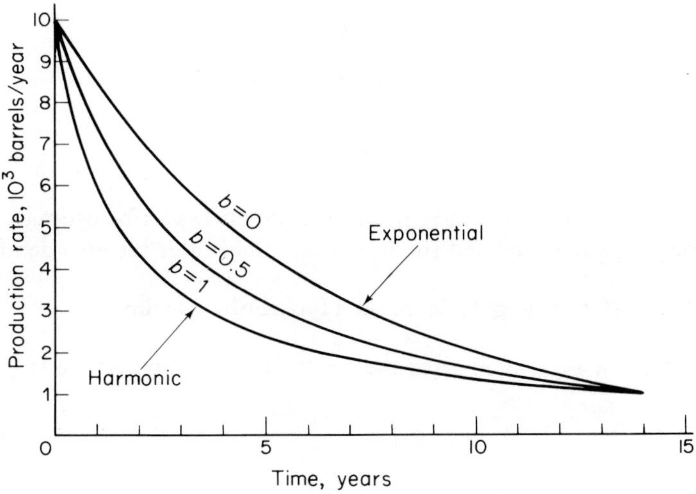

Figure 11.3. Hyperbolic decline curves.

11.2-1 Case of $b = 0$, Exponential Decline

When $b = 0$, the exponential decline results. Integrating Eq. (11-2) for $b = 0$ results in

$$\ln\left(\frac{q}{q_0}\right) = -\frac{1}{a}t$$

or

$$q = q_0 e^{-t/a}. \tag{11-3}$$

In obtaining this rate–time relation, the constant of integration is resolved by letting $q = q_0$ when $t = 0$. The value of a can be found by setting $q = q_F$ when $t = t_F$. The following results:

$$a = \frac{t_F}{\ln(q_0/q_F)}. \tag{11-4}$$

The time-cumulative relation can be determined as follows:

$$N_{PD} = \int_0^t q\,dt. \tag{11-5}$$

Using the expression for q that is contained in Eq. (11-3) results in the following time-cumulative equation:

$$N_{PD} = q_0 a(1 - e^{-t/a}). \tag{11-6}$$

The rate-cumulative relation is obtained by using Eq. (11-3) to replace t, giving

$$N_{PD} = a(q_0 - q). \tag{11-7}$$

Solving for time on decline results in the following:

$$t_F = \frac{N_{PD}}{q_0 - q_F}\ln\left(\frac{q_0}{q_F}\right). \tag{11-8}^*$$

This relation is useful where the producible reserves can be estimated from volumetric considerations and the initial and final rates are also defined.

11.2-2 Case of $0 < b < 1$, General Hyperbolic Decline

Integration of Eq. (11-2) gives the following rate–time relation:

$$q = q_0\left(1 + \frac{bt}{a}\right)^{-1/b}. \tag{11-23}$$

*Equations (11-9) through (11-22) and other equation numbers may be found in the Appendix, Sections 11.A-1, 11.A-2, and 11.A-3, of this chapter.

The value of a is given by

$$a = \frac{bt_F}{(q_0/q_F)^b - 1}. \qquad (11\text{-}24)$$

The cumulative production between $t = 0$ and t is

$$N_{PD} = \frac{aq_0}{1-b}\left[1 - \left(1 + \frac{bt}{a}\right)^{(b-1)/b}\right]. \qquad (11\text{-}25)$$

The rate-cumulative relation is

$$N_{PD} = \frac{aq_0^b}{1-b}(q_0^{1-b} - q^{1-b}). \qquad (11\text{-}26)$$

Time on decline is given by

$$t_F = \frac{1-b}{b}\frac{N_{PD}}{q_0}U^{1-b}\frac{U^b - 1}{U^{1-b} - 1}, \qquad (11\text{-}27)$$

where $U = q_0/q_F$.

For the special case of $b = 0.5$, Eq. (11-27) reduces to

$$t_F = \frac{N_{PD}}{q_0}\sqrt{\frac{q_0}{q_F}}. \qquad (11\text{-}27a)$$

11.2-3 Case of $b = 1$, Harmonic Decline

The rate–time relation of Eq. (11-23) reduces to

$$q = q_0\left(1 + \frac{t}{a}\right)^{-1}. \qquad (11\text{-}39)$$

The value of a is given by

$$a = \frac{t_F}{(q_0/q_F) - 1}. \qquad (11\text{-}40)$$

Integration of the time-cumulative relation, between $t = 0$ and t, results in

$$N_{PD} = aq_0 \ln\left(1 + \frac{t}{a}\right). \qquad (11\text{-}41)$$

Replacing t from Eq. (11-39) gives the rate-cumulative relation

$$N_{PD} = aq_0 \ln\left(\frac{q_0}{q}\right). \qquad (11\text{-}42)$$

Time on decline is given by

$$t_F = \frac{N_{PD}}{q_0} \frac{(q_0/q_F) - 1}{\ln (q_0/q_F)}. \tag{11-43}$$

11.3 CONVERSION TO YEARLY PRODUCTION CALCULATIONS

In the preceding section, sets of equations were presented which are useful in describing declining oil production. In many cases these can be used directly. They presume, however, a knowledge of the instantaneous production rates. Where the data are weekly or even monthly production figures, no great error is introduced by assuming them equal to the instantaneous production rates. However, where the data are yearly production figures, inaccuracies can be introduced by such assumptions. In general, the rate at the beginning of the time period is higher and at the end it is lower, so the available data represent an average over the time period. This is illustrated in Fig. 11.4.

Quite often it is desirable to estimate analytically the position of the

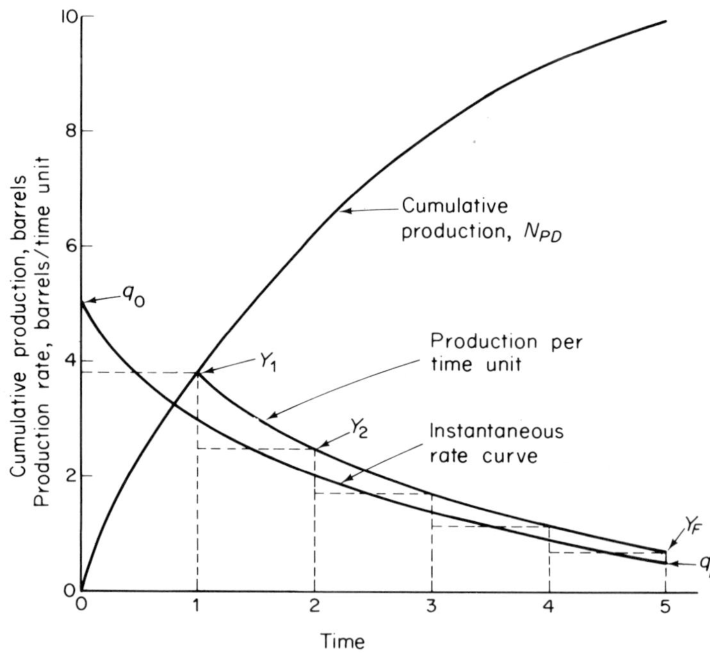

Figure 11.4. Relations between instantaneous rate curves and plots of production per time unit.

instantaneous curve from the available production data. Relations between the two can be developed from the expressions given in the previous section. The method consists of evaluating the time-cumulative relation over the first period, which would normally be the first year. It is then evaluated over the last yearly period. This procedure permits developing relations between the productions of the first and last years, relating them to the instantaneous rates. These relations are given in the appendixes at the end of the chapter.

Another relation of interest becomes apparent from observing Fig. 11.4. The cumulative production can, of course, be obtained by integration of the instantaneous rate curve,

$$N_{PD} = \int_0^{t_F} q\, dt. \tag{11-52}$$

An alternative procedure, which gives good results, utilizes the data that are usually available. The cumulative production, considered as the whole, is equal to the sum of its parts. Hence

$$N_{PD} = \sum_i^f Y, \tag{11-53}$$

where the Y-values represent productions recorded for successive periods such as years. The area under the Y-curve can be obtained by direct integration and also by the trapezoidal rule. Equating these gives

$$\int_{t_i}^{t_f} Y\, dt = \sum_i^f Y - \frac{Y_i + Y_f}{2}.$$

Hence the following relation is obtained:

$$N_{PD} = \int_{t_i}^{t_f} Y\, dt + \frac{Y_i + Y_f}{2}. \tag{11-54}$$

It should be noted that the integral in the above equation is evaluated between the ordinates. For example, in calculating the production from $t = 0$ to t_F, the above integral should be evaluated between $t = 1$ and $t = t_F$.

It must be emphasized that no entirely new relations have been developed here. All that is necessary to convert from the familiar instantaneous rate equations is as follows:

1. Replace N_{PD} in the instantaneous rate equation with $N_{PD} - (Y_i + Y_f)/2$.
2. Replace t, or $(t_F - t_0)$, in the instantaneous rate equation with $(t_F - t_0 - 1)$. (Note that we have used $t_i = t_1 = t_0 + 1$.)
3. Replace q_0 and q_F in the instantaneous rate equation with Y_1 and Y_F.

This produces equations that contain correction terms. These equations will handle yearly, semiannual, quarterly, or monthly production data. In those cases where no such correction terms are really needed, as with monthly production data, their effect actually becomes negligible, so no harm is evidenced from including them.

11.4 ESTIMATING TIME BY EXTRAPOLATING RATE–TIME CURVES AND CALCULATING RESERVES

Production decline data are analyzed directly by graphical methods or by equivalent mathematical procedures. Solutions for total recovery and for the producing life can both be made graphically, or they can both be made analytically. Or, one of these quantities can be estimated graphically, and the corresponding value of the other can be obtained by a convenient mathematical solution.

Actual plots of the data serve two good purposes. First, the value of b, which characterizes the production history, can be determined visually. When plots of the data are made on several types of paper, with different scales, the plot on which the data form a straight line can be used for extrapolation. This plot also indicates the proper value of b. Second, interruptions in the production history tend to prolong the producing life. Such shutdowns can be caused by mechanical troubles. Interpretation of the effects of such factors is aided by a plot of the data.

The three examples that follow are for exponential, hyperbolic, and harmonic production histories. In each of these cases, a production rate versus time curve has been plotted and extrapolated to the economic limit. In this procedure the value of b, which characterizes the data, is determined by the graph which produces a straight-line plot of the data.

The ultimate production has been calculated in each case according to the data and the b-value and time obtained from the rate–time plot. In these examples the data are yearly production figures, and the correction terms for this type of data cannot be neglected. As a check on these calculations, a rate-cumulative plot has also been constructed and extrapolated to the economic limit. In each case the type of rate-cumulative plot corresponds to the b-value obtained from the rate–time graphs.

11.4-1 Exponential Decline Example ($b = 0$)

Production data are given in Table 11.3. The economic limit is 2,000 bbl for the production of the last year. The objective is to estimate the remaining reserves and producing life.

The rate–time plot in Fig. 11.5a shows a straight-line relation on semilog

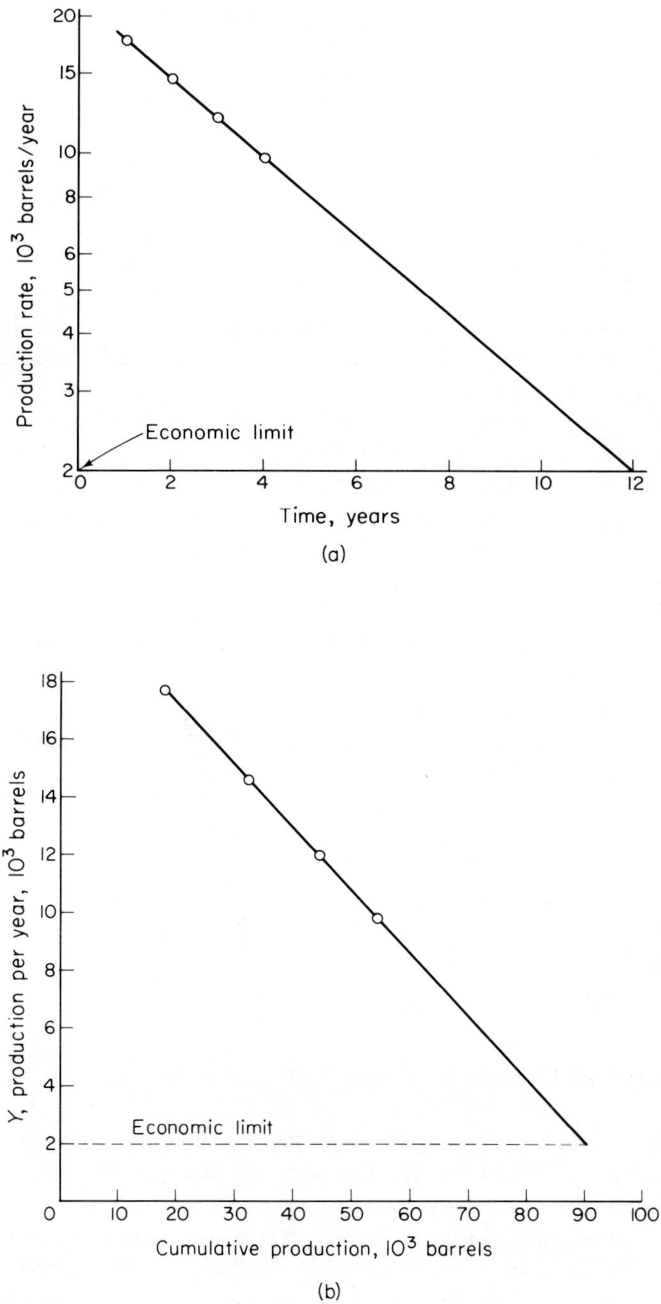

Figure 11.5. Exponential decline example. (a) Rate versus time on semilog plot. (b) Rate versus cumulative on arithmetic plot.

Table 11.3
Yearly Production Data

Year	*Production (bbl)*	*Cumulative (bbl)*
1	17,700	17,700
2	14,600	32,300
3	12,000	44,300
4	9,800	54,100

paper. This indicates that the production is exponential, $b = 0$. The straight line drawn through the yearly productions, when extended to a rate of 2,000 bbl, indicates a life of 12 years.

The rate-cumulative plot shown in Fig. 11.5b is a straight-line relation on regular, or arithmetic, graph paper. This indicates again that the production is exponential, $b = 0$, for this is the only case which plots as a straight line on regular graph paper. Extending the line through the yearly productions down to a rate of 2,000 bbl/year indicates a total ultimate production of approximately 90,000 bbl. The remaining reserves would be 90,000 − 54,100 or 35,900 bbl.

The rate-cumulative data would also form a straight line on a plot such as is shown in Fig. 11.6b, where the value of the slope would indicate that the decline was exponential.

The total cumulative reserves, for years 1 through 12 inclusive, can be calculated, using the yearly production figures, as follows:

$$N_{PD} = (t_F - t_1)\frac{Y_1 - Y_F}{\ln (Y_1/Y_F)} + \frac{Y_1 + Y_F}{2}$$

$$= (12 - 1)\frac{17{,}700 - 2{,}000}{\ln (17{,}700/2{,}000)} + \frac{17{,}700 + 2{,}000}{2} \tag{11-21}$$

$$= 89{,}300 \text{ bbl} \qquad \text{(compare to 90,000 bbl by graphical solution).}$$

11.4-2 General Hyperbolic Example ($0 < b < 1$)

The data in Table 11.4 prove to be intermediate in character. The economic limit is 4,750 bbl for the last year on decline. The objective is to estimate the remaining reserves and producing life.

The rate–time plot in Fig. 11.6a, which is a log–log plot, will straighten if shifted to the right by 5 years ($A = 5$). Shifting the data more than this amount produces a slight curvature opposite to that shown by the original unshifted data. The slope of the straight line can be measured directly on the graph, or it can be calculated as follows, from the available production data.

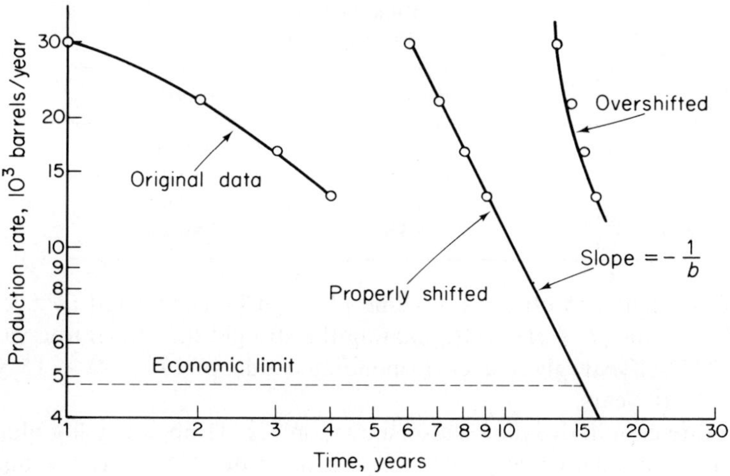

(a)

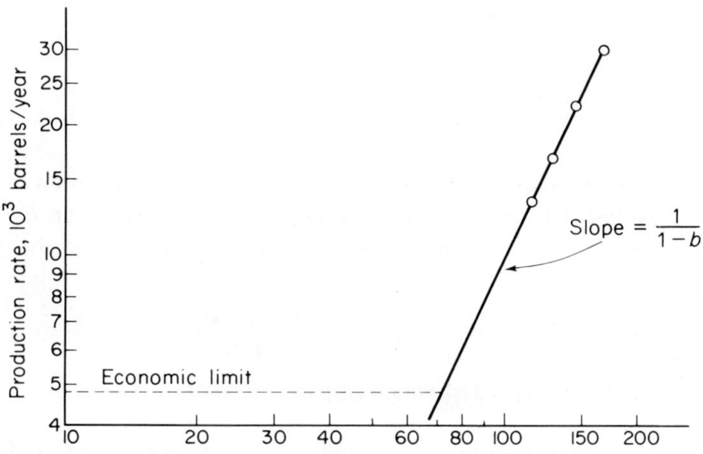

(b)

Figure 11.6. General hyperbolic decline example. (a) Rate versus time on log-log plot. (b) Rate versus cumulative on log-log plot.

$$\text{slope} = \frac{\log (Y_F/Y_1)}{\log \left(\dfrac{T_F + A}{T_1 + A}\right)} = \frac{\log (13{,}330/30{,}000)}{\log \left(\dfrac{4+5}{1+5}\right)} = -2.0$$

$$b = -\frac{1}{\text{slope}} = -\frac{1}{-2.0} = 0.5.$$

Table 11.4
Yearly Production Data

Year	*Production (bbl)*	*Cumulative (bbl)*
1	30,000	30,000
2	22,050	52,050
3	16,900	68,950
4	13,330	82,280

This straight line on the rate–time plot can be considered as a log–log plot of Y versus $(T + A)$. Extrapolating the straight line down to a value of $Y = 4{,}750$ bbl/year gives a corresponding value of $(T + A) = 15$ years; hence $T = 10$ years.

The rate-cumulative plot is also shown on Fig. 11.6b, a log–log plot. Log of the production rate is plotted on the ordinate. However, the quantity plotted on the abscissa is log $(k_2 - N_{PD})$. In order to justify this procedure, it is noted that Eq. (11-26) can be arranged in the following form:

$$\frac{aq_0^b}{1-b} q^{1-b} = \frac{aq_0}{1-b} - N_{PD} \tag{11-26a}$$

or

$$k_1 q^{1-b} = k_2 - N_{PD}. \tag{11-28}$$

The value of k_2 can be found by trial and error, using different values until one is found that gives a straight-line plot on log–log paper. Or, where values of a, b, and q_0 are known, k_2 can be calculated. For the present problem,

$$k_2 = \frac{aq_0}{1-b} = \frac{(2.74)(35{,}400)}{1-0.5}$$
$$= 197{,}000 \text{ bbl.}$$

Table 11.5 contains values of $197{,}000 - N_{PD}$, which are plotted in Fig. 11.6b. Extrapolating the straight line down to the economic limit of 4,750

Table 11.5
Data for Rate-Cumulative Log–Log Plot

Y (bbl)	N_{PD} *(bbl)*	*197,000* − N_{PD} *(bbl)*
30,000	30,000	167,000
22,050	52,050	144,950
16,900	68,950	128,050
13,330	82,280	114,720

gives a corresponding value on the abscissa of 72,000. The corresponding value of N_{PD} is

$$N_{PD} = 197{,}000 - 72{,}000$$
$$= 125{,}000 \text{ bbl.}$$

The remaining reserves would then the 125,000 − 82,280 bbl = 42,700 bbl, approximately.

The total production for years 1 through 10 inclusive, using the yearly production figures, can be calculated as follows:

$$N_{PD} = \frac{b}{1-b} Y_1(T_1 + A)\left[1 - \left(\frac{Y_F}{Y_1}\right)^{1-b}\right] + \frac{Y_1 + Y_F}{2}$$

$$= \frac{0.5}{1-0.5}(30{,}000)(1+5)\left[1 - \left(\frac{4{,}750}{30{,}000}\right)^{1-0.5}\right] + \frac{30{,}000 + 4{,}750}{2}$$

$$= 125{,}400 \text{ bbl} \qquad \text{(compare to 125,000 bbl from graphical solution).} \tag{11-37}$$

11.4-3 Harmonic Decline Example ($b = 1$)

The data in Table 11.6 exhibits harmonic decline characteristics. The economic limit is 3,700 bbl for the last year of production. The objective is to estimate the remaining reserves and producing life.

Table 11.6
Yearly Production Data

Year	*Production (bbl)*	*Cumulative Production (bbl)*
1	44,800	44,800
2	25,820	70,620
3	18,150	88,770
4	14,000	102,770

The rate–time plot is shown in Fig. 11.7a. On this log–log plot, the data straightened after shifting to the right by 0.35 year. Hence $A = 0.35$ year. The slope of the straight line is -1; hence $b = -1/\text{slope} = -1/(-1) = 1$. The shifted straight line can be considered as a plot of Y versus $(t + A)$. Extending the straight line on Fig. 11.7a to the economic limit of 3,700 bbl/year indicates that $t_F + A \simeq 16.5$ years, so $t_F = 16.5 - 0.35 \simeq 16$ years.

The rate-cumulative plot is shown in Fig. 11.7b, which gives a straight line on a semilog plot for the case of $b = 1$. Extending the line to the economic

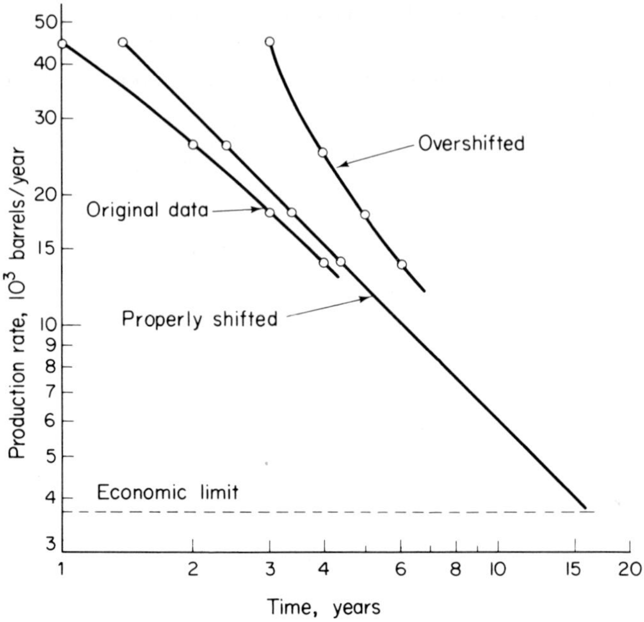

(a)

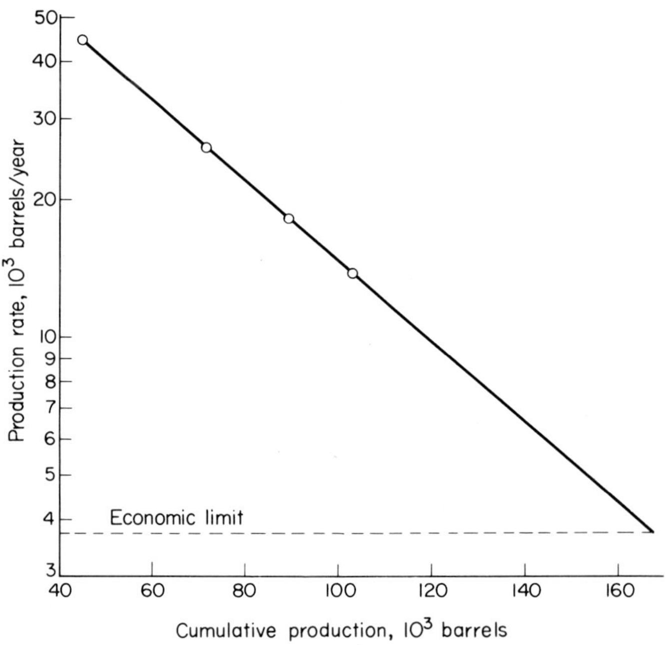

(b)

Figure 11.7. Harmonic decline example. (a) Rate versus time on log-log plot. (b) Rate versus cumulative on semilog plot.

limit of 3,700 bbl/year indicates a total cumulative production of 176,000 bbl. Therefore, the remaining reserves are 176,000 − 102,770 ≃ 73,200 bbl.

The total production for years 1 through 16 inclusive, using yearly production figures, can be calculated as follows:

$$\begin{aligned} N_{PD} &= Y_1(t_1 + A)\ln\left(\frac{Y_1}{Y_F}\right) + \frac{Y_1 + Y_F}{2} \\ &= (44{,}800)(1 + 0.35)\ln\left(\frac{44{,}800}{3{,}700}\right) + \frac{44{,}800 + 3{,}700}{2} \\ &= 175{,}300 \text{ bbl} \qquad \text{(compare to 176,000 bbl from graphical solution),} \end{aligned} \tag{11-50}$$

reserves = 175,300 bbl − 102,770 bbl, produced ≃ 72,500 bbl.

11.5 ESTIMATING RESERVES BY EXTRAPOLATING RATE-CUMULATIVE CURVES AND CALCULATING TIME ON DECLINE

This section is devoted to the use of rate-cumulative plots for estimating the ultimate recovery of oil from a well, or producing unit. Plots of this type have an advantage over rate–time plots, for time interruptions due to mechanical or other failures do not affect the plotting coordinates. Also, there is less curvature in the natural data, so plotting scales or shifting techniques used to straighten the data are much less severe than for corresponding rate–time data. Finally, the evaluation of workovers and remedial treatments tends to be directly evident on such graphs, and any additional recovery can be read easily by extrapolating the plotted data.

The time that will be required to produce the projected reserves is readily calculated. Accuracy of such estimates depends in part upon there being no unforeseen interruptions in the producing operations. The particular formula used for calculating time on decline depends on the value of b corresponding to the data, which is determined by plotting. For extrapolation purposes, the data for the past production should form a straight line.

Where the rate-cumulative data form a straight line when plotted on regular arithmetic graph paper, this indicates that production is exponential and $b = 0$. Where rate-cumulative data straighten on semilog paper, with rate plotted on the log scale and cumulative production on the arithmetic scale, this indicates that production is harmonic and $b = 1$. Other cases are usually intermediate. The best procedure is to use special paper with scales marked off according to q^{1-b}. Such an example is included.[3] Another procedure is to shift the data on a log–log plot, such as Fig. 11.6b, where the slope of the line is $1/(1 - b)$. Thus the procedure is to plot the data according to one or more assumed values of b, and inspect the resulting alignment of the data points.

Three examples of production histories are given which illustrate exponential ($b = 0$), hyperbolic ($b = 0.5$), and harmonic ($b = 1$) relationships. The rate-cumulative graph was plotted in each case and extrapolated into the future, to the rate corresponding to the economic limit. The time required to reach the economic limit was then calculated.

11.5-1 Case of Exponential Decline ($b = 0$)

Production data for a well that was treated with an acid wash are shown in Table 11.7. The rate-cumulative plot is shown in Fig. 11.8. Since the straight-line relationship develops on regular (Cartesian) graph paper, it is evident that the production history is exponential ($b = 0$), and if the rate–time plot were made it would straighten on semilog paper. There is, of course, some question regarding the proper location of the line drawn through the early production history. As the lines are drawn, however, the interpretation would be that the workover would not result in an increased recovery if the well could be produced to a very low rate. However, at the economic limit, there is an increase in ultimate recovery of about 5,000 bbl.

Table 11.7
Production History of a Well on Exponential Decline

	First Year		*Second Year*		*Third Year*	
Month	Y (*bbl/month*)	N_{PD} (*bbl*)	Y (*bbl/month*)	N_{PD} (*bbl*)	Y (*bbl/month*)	N_{PD} (*bbl*)
Jan.	1,240	1,240	991	14,373	1,023	26,259
Feb.	1,036	2,276	1,073	15,446	1,008	27,267
Mar.	1,147	3,423	930	16,376	961	28,228
Apr.	1,350	4,773	870	17,246	960	29,188
May	1,395	6,168	930	18,176	930	30,118
June	1,170	7,338	780	18,956	810	30,928
July	1,054	8,392	775	19,731	806	31,734
Aug.	1,085	9,477	1,240	20,971	775	32,509
Sept.	990	10,467	1,170	22,141	780	33,289
Oct.	1,085	11,522	1,147	23,288	775	34,064
Nov.	900	12,452	1,080	24,368	480	34,544
Dec.	930	13,382	868	25,236	837	35,381

On Fig. 11.8, the solid line drawn after the workover was used for extrapolation purposes. The rate, on this line, is 740 bbl/month, where the cumulative production is 35,381 bbl. The economic limit is taken as 360 bbl/month, where the cumulative production will be approximately 46,000 bbl.

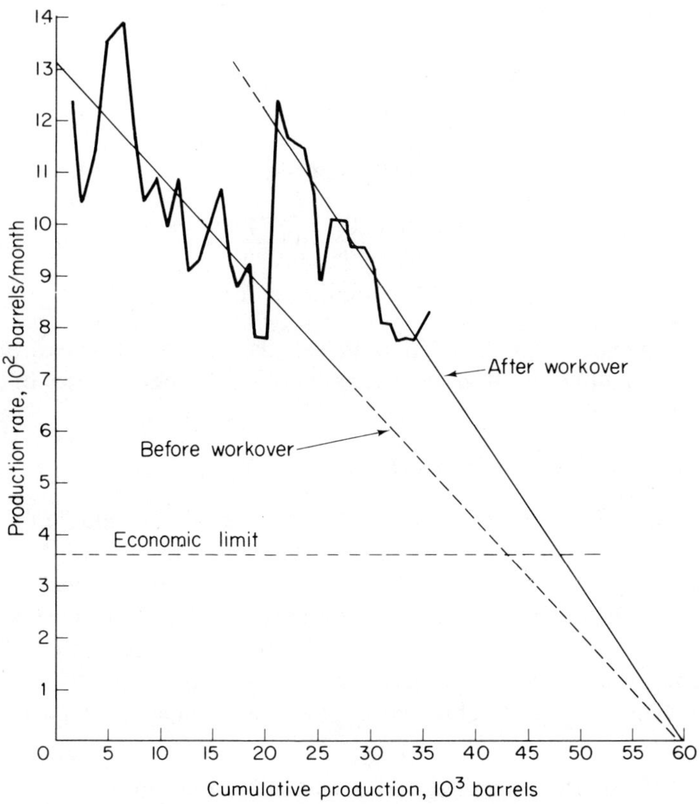

Figure 11.8. Exponential production history, arithmetic plot.

The following calculation is for the case of discrete time intervals. Let the month where (according to the line) 740 bbl were already produced be included as t_i in the projected series. The reserves produced in this series will then be

$$\begin{aligned} N_{PD} &= 46{,}000 \text{ bbl} - (35{,}381 \text{ bbl} - 740 \text{ bbl}) \\ &\simeq 11{,}360 \text{ bbl}. \end{aligned}$$

The time is then calculated:

$$\begin{aligned} t_F - t_i &= \frac{\ln (Y_i/Y_F)}{Y_i - Y_F}\left(N_{PD} - \frac{Y_i + Y_F}{2}\right) \\ &= \frac{\ln (740/360)}{740 - 360}\left(11{,}360 - \frac{740 + 360}{2}\right) \qquad (11\text{-}22) \\ &= 20.4 \text{ months, in future.} \end{aligned}$$

If one takes the monthly productions as equivalent to instantaneous rates, the time is calculated as follows. First, the remaining reserves to be produced are

$$N_{PD} = 46{,}000 \text{ bbl} - 35{,}381$$
$$\simeq 10{,}620 \text{ bbl},$$
$$t = \frac{N_{PD}}{q_0 - q_F} \ln\left(\frac{q_0}{q_F}\right)$$
$$= \frac{10{,}620}{740 - 360} \ln\left(\frac{740}{360}\right) \quad (11\text{-}8)$$
$$= 20.3 \text{ months, in future.}$$

The answers are the same in both calculations, as expected, and one would have to allow for at least 21 months of operating expenses in making an evaluation.

11.5-2 Case of Hyperbolic Decline ($b = 0.5$)

In solving this example, the rate-cumulative plot was made on a special graph.[3] The rate scale is marked off so that it is linear in q^{1-b}. That the cumulative scale will then be linear can be inferred from inspection of Eq. (11-28), and it can be marked off in arithmetic fashion in the desired range. The data for a well are given in Table 11.8 and are plotted in Fig. 11.9. For convenience, the rate values were multiplied by 10 before plotting so that a more convenient range is obtained on the graph paper. For an economic

Table 11.8
Production Decline Data, Hyperbolic Case ($b = 0.5$)*

Time (quarter-year)	*Production (bbl/quarter)*	*Cumulative Production (bbl)*
1	8,247	8,247
2	7,820	16,067
3	7,985	24,052
4	7,544	31,596
5	6,384	37,980
6	6,281	44,261
7	5,589	49,850
8	5,836	55,686
9	5,629	61,315
10	5,327	66,642
11	4,989	71,631
12	4,808	76,439
13	4,407	80,846
14	4,346	85,192
15	4,269	89,461

*The original monthly productions were combined into productions per quarter year for analysis.

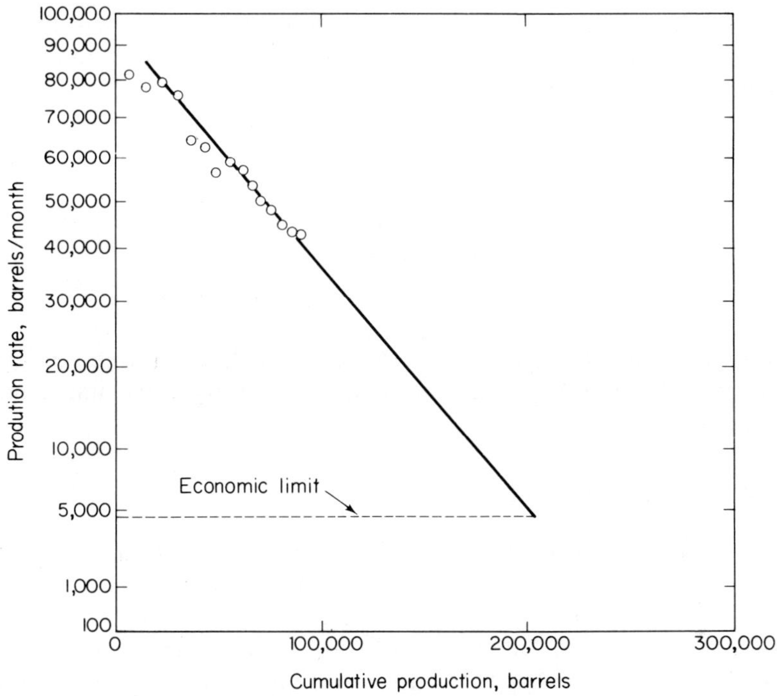

Figure 11.9. Hyperbolic production history.

limit of 450 bbl per quarter, the total ultimate production is read from the graph as 203,000 bbl. Since 89,461 bbl have already been produced, the remaining reserves are 203,000 bbl − 89,461 bbl = 113,540 bbl.

The following calculation uses productions of discrete time intervals. As before, it will be convenient to take the last known period as the initial period for a series extending into the future. The production rate for this last period is read from the line on the graph as 4,200 bbl. The reserves for the entire series is then

$$N_{PD} = 203{,}000 \text{ bbl} - (89{,}461 \text{ bbl} - 4{,}200 \text{ bbl})$$
$$\simeq 117{,}740 \text{ bbl}.$$

The remaining time on decline is then calculated as follows, for the special case where $b = 0.5$:

$$t_F - t_i = \left(\frac{Y_i}{Y_F}\right)^{0.5} \frac{N_{PD} - \dfrac{Y_i + Y_F}{2}}{Y_i}$$

$$= \left(\frac{4{,}200}{450}\right)^{0.5} \frac{117{,}740 - \dfrac{4{,}200 + 450}{2}}{4{,}200} \tag{11-38a}$$

$$= 85.5 \text{ quarters, or 22 years, in future.}$$

Where the quarterly productions are assumed to be equivalent to instantaneous rates, the remaining time on decline is calculated as follows. This is for the special case of $b = 0.5$.

$$t = \frac{N_{PD}}{q_0}\sqrt{\frac{q_0}{q_F}}$$

$$= \frac{113{,}540 \text{ bbl}}{4{,}200 \text{ bbl/qtr}}\sqrt{\frac{4{,}200}{450}} \qquad (11\text{-}27a)$$

$$= 84.0 \text{ quarters, or 21 years, in future.}$$

The small difference in time between the two methods has little significance in this example. However, as the time units become larger, the first method is to be preferred.

11.5-3 Case of Harmonic Decline ($b = 1$)

The production history data for a well are given in Table 11.9 and are shown plotted in Fig. 11.10. This rate-cumulative plot is linear on semilog paper, which is an indication that the production history is harmonic ($b = 1$). At an economic limit of 270 bbl/quarter, the total ultimate production is read from the graph as 152,000 bbl. The remaining reserves are 152,000 bbl − 39,350 bbl = 112,650 bbl, approximately.

Where productions are assumed to occur within discrete time units, the

Table 11.9
Production Decline Data, Harmonic Case ($b = 1$)*

Time (quarter-year)	*Production, (bbl/quarter)*	*Cumulative Production (bbl)*
1	5,050	5,050
2	4,530	9,580
3	4,010	13,590
4	3,590	17,180
5	3,410	20,590
6	3,070	23,660
7	2,860	26,520
8	2,860	29,380
9	2,680	32,060
10	2,520	34,580
11	2,430	37,010
12	2,340	39,350

*The original monthly productions were combined into productions per quarter-year for analysis.

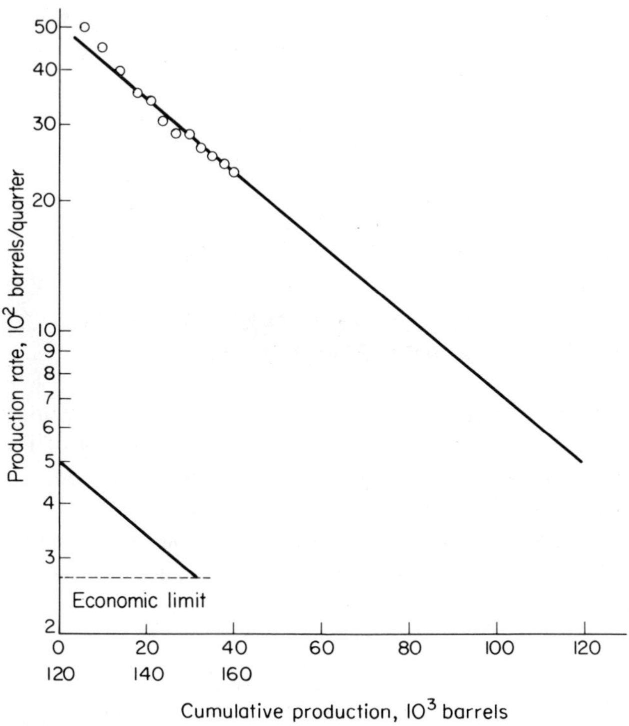

Figure 11.10. Harmonic production history, semilog plot.

remaining time on decline can be calculated as follows. In this example the last recorded period is exactly on the line drawn on the graph. If this period is taken as the initial period for a series extending into the future, the total reserves for that future series would be

$$\begin{aligned} N_{PD} &= 152{,}000 \text{ bbl} - (39{,}350 \text{ bbl} - 2{,}340 \text{ bbl}) \\ &= 114{,}990 \text{ bbl}. \end{aligned}$$

The remaining time on decline can then be calculated:

$$\begin{aligned} t_F - t_i &= \frac{(Y_i/Y_F) - 1}{\ln (Y_i/Y_F)} \frac{N_{PD} - \dfrac{Y_i + Y_F}{2}}{Y_i} \\ &= \frac{(2{,}340/270) - 1}{\ln (2{,}340/270)} \frac{114{,}990 - \dfrac{2{,}340 + 270}{2}}{2{,}340} \qquad (11\text{-}51) \\ &= 172 \text{ quarters, or 43 years.} \end{aligned}$$

Where the productions per quarter are taken as equivalent to instantaneous rates, the calculation for remaining time on decline is as follows:

$$
\begin{aligned}
t &= \frac{N_{PD}}{q_0}\frac{(q_0/q_F) - 1}{\ln (q_0/q_F)} \\
&= \frac{112{,}650 \text{ bbl}}{2{,}340 \text{ bbl/qtr}}\frac{(2{,}340/270) - 1}{\ln (2{,}340/270)} \qquad (11\text{-}43) \\
&= 171 \text{ quarters, or 43 years.}
\end{aligned}
$$

11.6 FRACTION OF RESERVES PRODUCED AT A RESTRICTED RATE

In many cases wells are not produced at their capacity during their early life. The reasons may be physical, such as limited capacity of flow lines or transportation facilities, or the restrictions may be from orders imposed by regulatory bodies or governing agencies.

Figure 11.11 illustrates a model for estimating the time pattern of production where the rate is restricted. The total production that would have occurred under decline from the initial capacity down to the restricted rate

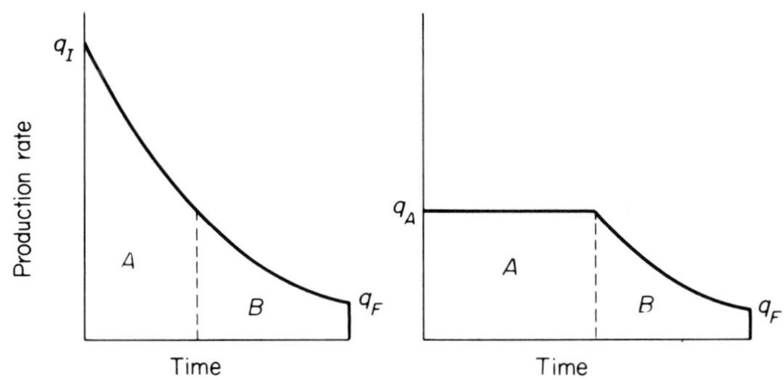

Figure 11.11. Estimating the effect of restricting maximum production rate.

is set equal to the total amount produced under the restricted rate. The total amount produced under decline from the restricted rate down to the final rate is assumed to be the same in both cases. Although some objections to such a procedure can be suggested, it is often accepted as sufficiently accurate for evaluation purposes. The greatest uncertainty is associated with predicting the time pattern in the later stages of production. In turn, these later stages are generally heavily discounted in arriving at a present value, so that inaccuracies in this portion are relatively unimportant.

Relations that give the value of N_{PA}/N_P, the fraction of oil produced under a restricted or allowable rate, are readily derived from rate-cumulative relations. They are listed in the Appendixes, for exponential, hyperbolic, and harmonic declines.

11.7 AN EVALUATION EXAMPLE USING EXPONENTIAL DECLINE

11.7-1 Statement of the Problem

The problem considered in this section is that of estimating the present value of future production according to a specified discount rate. In the example given, the discount rate used is 10% per year. The subsequent use of several different interest rates, in the manner illustrated, would permit defining a present-value profile if such were desired. Income taxes are not included in this example, for it is desired to keep the example uncomplicated. Thus the final answer obtained would be misleading in some instances. The objective here is to focus on the primary aspects of the method.

Example: Calculate the present value, before income taxes, of the following prospect. Use 10% discount factors; assume all receipts and expenditures over the year to occur at the midpoint of the year. Assume that exponential decline will follow after the period of restricted production. The basic date of evaluation is the beginning of the allowable production period, which is the date of well completion.

The data are as follows:

Well spacing:	40 acres
Net pay thickness:	26.5 ft
Recovery:	150 bbl/acre-ft
Initial flow capacity:	110 bbl/day
Maximum allowable production rate:	50 bbl/day
Working interest:	7/8 working interest
Cost of well:	\$100,000
Operating cost, flowing (2 years):	\$100 per month
Operating cost, pumping:	\$280 per month
Pumping equipment and installation:	\$6,000

Assume \$3 per barrel, less 5% state (severence) tax. From the working interest, allow 3% for ad valorem taxes and overhead.

11.7-2 Calculation of Production History

$$\text{Net income per barrel} = \tfrac{7}{8}(1 - 0.05)(1 - 0.03)(\$3.00) = \$2.42.$$

$$\begin{aligned}\text{Total recovery} &= (40 \text{ acres})(26.5 \text{ ft})(150 \text{ bbl/acre-ft})\\ &= 159{,}000 \text{ bbl.}\end{aligned}$$

$$\text{Economic limit} = \frac{\$280 \text{ expenses per month}}{(\$2.42 \text{ per bbl})(30.4 \text{ days/month})} = 3.81 \text{ bbl/day}.$$

$$\text{Yearly production on allowable} = (365 \text{ days/yr})(50 \text{ bbl/day}) = 18{,}250 \text{ bbl/yr}.$$

$$\text{Fraction produced on allowable} = \frac{q_1 - q_A}{q_1 - q_F} = \frac{110 - 50}{110 - 3.81} = 0.565.$$

$$\text{Years producing on allowable} = \frac{(0.565)(159{,}000 \text{ bbl})}{(18{,}250 \text{ bbl/yr.})} \simeq 5 \text{ years}.$$

$$\text{Production on allowable} = (5 \text{ years})(18{,}250 \text{ bbl/yr}) = 91{,}250 \text{ bbl}.$$

$$\text{Production on decline} = 159{,}000 \text{ bbl} - 91{,}250 \text{ bbl} = 67{,}750 \text{ bbl}.$$

$$T = \frac{N_{PD}}{q_0 - q_F} \ln\left(\frac{q_0}{q_F}\right) = \frac{67{,}750 \text{ bbl}}{(365)(50 - 3.81) \text{ bbl/year}} \ln\left(\frac{50}{3.81}\right) = 10.3 \text{ years on decline}.$$

$$R = \left(\frac{q_F}{q_0}\right)^{1/T} = \left(\frac{3.81}{50}\right)^{1/10.3} = 0.78.$$

$$\frac{Y_1}{q_0} = \frac{1 - R}{\ln\left(\frac{1}{R}\right)} = \frac{1 - 0.78}{\ln\left(\frac{1}{0.78}\right)} = 0.888.$$

$$Y_1 = (50 \text{ bbl/day})(365 \text{ days/year})(0.888) = 16{,}210 \text{ bbl, first year on decline}.$$

$$Y_2 = RY_1 = (0.78)(16{,}210 \text{ bbl}) = 12{,}640 \text{ bbl, second year on decline}.$$

$$U = \frac{q_0}{q_F} = \frac{50 \text{ bbl/day}}{3.81 \text{ bbl/day}} = 13.12.$$

In the foregoing calculations, all the quantities needed to describe the production history have been calculated. The production of the third year is calculated as $Y_3 = RY_2$, for the fourth year $Y_4 = RY_3$, and so on down to the economic limit. If one desires to tabulate the production by years, this method is followed. However, it is generally not necessary to exhibit such detail in the estimates. The next section explains how to use present-value factors which transform a whole series of payments into a single present value. This is the more efficient method, particularly when such factors can be found in tables. First, however, their manner of calculation will be shown.

11.7-3 Present Value of Gross Income from Production

The present value of the oil production can be done by parts. The first part consists of evaluating the series of five equal payments received during the allowable (restricted) period of production. The present-value factor is calculated as follows:

$$D\text{, allowable} = (1+i)^{0.5}\frac{1-(1+i)^{-m}}{mi} = (1.10)^{0.5}\frac{1-(1.10)^{-5}}{(5)(0.10)}$$

$$= 0.79516,$$

$$\text{P.V. of allowable production} = (0.79516)(91{,}250\text{ bbl})(\$2.42\text{ per bbl})$$

$$= \$175{,}590.$$

The second part of the oil production consists of an exponential decline. The customary decline factor will reduce the decline production to a lump sum "present value" as of the beginning of the decline. It is then further necessary to apply a single-payment factor to reduce the value to a present value as of the beginning of the allowable production period.

To reduce the decline production to a value as of the beginning of the decline, the time on decline will be rounded off to 10 years. This is not necessary for calculation purposes, particularly if one is very certain of the estimated values, but without introducing appreciable error it will facilitate later comparison with equivalent calculation methods. The factor is calculated as follows:

$$D = (1+i)^{0.5}\frac{1-R}{1-R^T}\frac{1-(R/(1+i))^T}{1+i-R} \qquad (11\text{-}20)$$

$$= (1.10)^{0.5}\frac{1-0.78}{1-(0.78)^{10}}\frac{1-(0.78/1.10)^{10}}{1.10-0.78}$$

$$= 0.76135.$$

The above factor reduces the decline production to a lump-sum value as of the beginning of the decline, which is 5 years after the basic date of evaluation. Accordingly, a single-payment factor must be applied which will reduce the amount to a present value as of the beginning of the allowable period, which is the basic date of evaluation. The single-payment factor is

$$\text{single-payment factor} = \left(\frac{1}{1+i}\right)^m$$

$$= \left(\frac{1}{1.10}\right)^5$$

$$= 0.62092.$$

Combining these two factors then gives the deferment factor for the decline production:

$$D\text{, decline} = (\text{single-payment factor})(\text{P.V. factor, as of beginning of decline})$$

$$= (0.62092)(0.76135)$$

$$= 0.47274,$$

$$\text{P.V. of decline production} = (0.47274)(67{,}750 \text{ bbl})(\$2.42) = \$77{,}510.$$

The total present value of the gross income from the production is then the sum of two parts, or $175,590 + $77,510 = $253,100.

11.7-4 Present Value of Expenses

The present value of the production expenses can also be done by parts. Involved are two series of equal payments. One of these begins immediately after the basic date of evaluation and lasts for 2 years. The second series begins when the first ends and extends over the remaining 13 years of the total production history.

For reducing the 2 years of flowing expenses to a present value, where the disbursements are presumed to be made at the midpoint of each year, the following factor is used:

$$D\text{, flowing} = (1 + i)^{0.5} \frac{1 - (1 + i)^{-n}}{ni}$$

$$= (1.10)^{0.5} \frac{1 - (1.10)^{-2}}{(2)(0.10)}$$

$$= 0.91012,$$

$$\text{P.V. of flowing expenses} = (0.91012)(2 \text{ years})(12 \text{ months/year})(\$100) = \$2{,}180.$$

For reducing the pumping expenses to a present value as of the basic date of evaluation, one first obtains a factor which reduces the series of equal payments to a lump sum value as of the beginning of the series.

$$D = (1 + i)^{0.5} \frac{1 - (1 + i)^{-n}}{ni}$$

$$= (1.10)^{0.5} \frac{1 - (1.10)^{-13}}{(13)(0.10)}$$

$$= 0.57308.$$

Next, the lump-sum value must be reduced by a single-payment factor which will make it correspond to the basic data of evaluation. The single-payment factor is

$$D = \left(\frac{1}{1 + i}\right)^n$$

$$= \left(\frac{1}{1.10}\right)^2$$

$$= 0.82645.$$

The deferment factor for the pumping expenses then becomes

$$\begin{aligned} D\text{, pumping} &= (\text{single-payment factor})(\text{P.V. factor, as of beginning of series}) \\ &= (0.82645)(0.57308) \\ &= 0.47362, \end{aligned}$$

$$\begin{aligned} \text{P.V. of pumping expenses} &= (0.47362)(13\text{ years})(12\text{ months})(\$280\text{ per month}) \\ &= \$20{,}690. \end{aligned}$$

11.7-5 Summary of Present-Value Calculations

So far, the present values of series of payments for both gross income and operating expenses have been considered and suitable calculations formulated. In some cases, although it is omitted here, one might wish to take into account a salvage value, which would be discounted as a single payment received at the end of 15 years. One item that will be included here is the cost of the pumping installation at the end of 2 years. Its present value will be calculated as a single payment at the end of 2 years.

$$\begin{aligned} \text{P.V. of pumping equipment} &= \left(\frac{1}{1+i}\right)^n (\text{undeferred value}) \\ &= \left(\frac{1}{1.10}\right)^2 (\$6{,}000) \\ &= (0.82645)(\$6{,}000) \\ &= \$4{,}960. \end{aligned}$$

The cost of drilling and completing the well is given as \$100,000. This amount of money would be paid by the working interest at the time of well completion, which is the basic date of evaluation. Thus it is entered as a present value. All these items are entered into Table 11.10 and combined to give the net present value before income taxes.

Table 11.10
Summary of Present-Value Calculations

Income from production			
Allowable period (0.79516)(91,250 bbl)(\$2.42)	=	\$175,590	
Decline period (0.47274)(67,750 bbl)(\$2.42)	=	77,510	
			253,100
Expenses			
Operating, flowing (0.91012)(2 yr)(\$1,200)	=	2,180	
Operating, pumping (0.47362)(13 yr)(\$3,360)	=	20,690	
Pumping equipment (0.82645)(\$6,000)	=	4,960	
Drill and comp.	=	100,000	
			127,830
Net present value before income tax			\$125,270

11.7-6 Solution in Table Form

The most explicit way of obtaining the present value of the proposed well is to construct a table. Items included in such a table are the expected yearly productions, the yearly gross incomes, and the yearly expenses. The difference between the latter usually gives the operating profit for each year, and that figure can be discounted by the appropriate single-payment factor for the year. Thus, after obvious additions, one obtains the total present value.

Table 11.11 exhibits the results of calculations of present values for each year. The yearly productions can be seen to consist of one series of 5 equal amounts followed by another series of declining amounts over the next 10 years. The operating expenses can be seen to consist of one series of 2 equal amounts followed by another series of 13 equal amounts. These are the several series which were evaluated as such in the foregoing section. The present-value factors applied to each yearly amount are identical in the table with those incorporated into the calculations in the foregoing section. The slight difference of \$125,270 − \$125,150, or \$120, is a discrepancy of only $\frac{1}{10}\%$, which is the result of roundoff errors.

The last column shows the cumulative net present value. Here it may be observed that if one uses 10% discount, the payout will occur late in the third year. It is again pointed out that income taxes are not included in this evaluation. A chief purpose herein is to show the equivalence of the method of using deferment factors which incorporate whole series of payments to the method of constructing a detailed table.

11.8 USE OF PRESENT-VALUE FACTOR TABLES

11.8-1 Types of Present-Value Factors

Three types of present-value factors can be found in the Appendixes following this section. A fourth type is formed from the product of two of the types listed in the tables.

1. *Single-payment factors.* These factors reduce the value of a single amount of money, which is to be received n years hence, to its present value, according to a given interest rate. The factors are values of $(1 + i)^{-n}$ for conventional calculations. For continuously compounding interest calculations, there are tables containing values of e^{-nj}. These factors are applied to a lump-sum value, as of a specified time.

2. *Factors for amounts distributed over 1 year.* Where money is received or disbursed more or less regularly throughout a yearly period, a convention that has been widely used in the petroleum industry is to assume that the entire transaction occurs at the midpoint of the year. Equivalent factors

Table 11.11
Calculation of Present Values by Years

Year	*Production (bbl)*	*Gross Income at $2.42*	*Operating Expenses ($)*	*Net Operating Profit ($)*	*Present-Value Factor*	*Present Value of Operating Profits ($)*	*Present Value of Investment ($)*	*Cumulative Net Present Value ($)*
0					1.00000		100,000	−100,000
1	18,250	44,165	1,200	42,965	0.95346	40,970		−59,030
2	18,250	44,165	1,200	42,965	0.86678	37,240	4,960	−26,750
3	18,250	44,165	3,360	40,805	0.78799	32,150		5,400
4	18,250	44,165	3,360	40,805	0.71635	29,230		34,630
5	18,250	44,165	3,360	40,805	0.65123	26,570		61,200
6	16,210	39,230	3,360	35,870	0.59203	21,240		82,440
7	12,640	30,590	3,360	27,230	0.53820	14,660		97,100
8	9,860	23,860	3,360	20,500	0.48928	10,030		107,130
9	7,690	18,610	3,360	15,250	0.44480	6,780		113,910
10	6,000	14,520	3,360	11,160	0.40436	4,510		118,420
11	4,680	11,330	3,360	7,970	0.36760	2,930		121,350
12	3,650	8,830	3,360	5,470	0.33418	1,830		123,180
13	2,850	6,900	3,360	3,540	0.30380	1,080		124,260
14	2,220	5,370	3,360	2,010	0.27618	550		124,810
15	1,950	4,720	3,360	1,360	0.25108	340		125,150
	159,000	384,785	46,080	338,705		230,110 −	104,960 =	125,150

based upon continuous interest are listed. The latter are calculated by assuming a continuous uniform flow throughout the nth year, which is then reduced to a present value, by the appropriate e^{-nj} factor.[4]

3. *Factors for reducing a series of payments to a single value as of the beginning of the series.* An example of this type of factor applies to the case of a series of uniform payments. Consider the case of finding the present value, according to 6% yearly interest, of 10 equal yearly payments of \$100 each received at the end of successive years. Factors listed in this book are to be applied to the entire amount of \$1,000. To find the proper factor, look in the 6% table, for exponential decline, payments received at the end of each year; in the column for $U = 1$, opposite 10 years, find $D = 0.73601$. Multiplying the entire amount by this factor gives a present value of \$736.01. In some tables this information is listed as a present value of annuity factor, with a value of 7.3601 for the case cited. Multiplying the value of one payment by this factor again gives the same present value of \$736.01. For the cases of declining productions, it is much simpler to use factors that are applied to the total undeferred value. These factors reduce the value of monies to be received over the life of a series, or over a production decline, to a single "present value" as of the beginning of the series.

4. *Factors for reducing a series of payments to a present value, general case.* For the general case where a series of payments may not start immediately, the product of two factors is required. One of these is of the type described under 3, which reduces the series to a single amount value as of the beginning of the series. The other is described under 1, which brings the value from the beginning of the series to the present. More strictly, one should bring the values to the basic date of evaluation. An example of this would be the "date of well completion." Sometimes an evaluation is made and then not acted upon for a year or more. Thus it often is expedient to specify a date other than a calendar date.

11.8-2 Interpolation in Decline Present-Value Factor Tables

Present-value factors for declining production depend upon four factors: (1) the interest rate used for discounting, (2) the total length of the time period from the beginning to the end of the decline, (3) the time pattern of successive yearly productions, and (4) the ratio of initial to the final production rates. Interpolations involving the first two categories are commonplace; they are generally resolved by linear interpolation between the closest conveniently obtained values. In the third category, as far as hyperbolic decline curves are concerned, one often finds surprisingly little difference in present-value factors for the exponential, the midhyperbolic, and the harmonic declines. The chief purpose of this section is to discuss interpolations relating to the fourth category, the ratio of initial to final production rate.

Each page in the decline deferment factor tables contains values that correspond to a given interest discount rate and a particular type of decline. For example, consider the page containing 8% present-value factors for exponential decline. On this page are listed columns headed $U = 1, 2, 4, 8, 16, 32$, and 64. These are the ratios of initial production rate to the final production rate. For example, if the production declines from 32 bbl/day to 4 bbl/day, $U = 8$, and the present-value factor can be found directly in the column for $U = 8$.

It is an observed fact that plotting the present-value factors versus log U yields lines of moderate curvature. The assumption of a straight-line relation on such a semilog plot serves as a basis for interpolation. This permits the following relationship:

$$\frac{D - D_1}{D_2 - D_1} = \frac{\log (U/U_1)}{\log (U_2/U_1)} = \text{interpolation fraction.}$$

Then, for the present-value factor corresponding to U, one has

$$D = D_1 + (\text{interpolation fraction})(D_2 - D_1).$$

For example, if $i = 8\%$, $T = 10$ years, exponential decline, monies received at the ends of the years, one finds the following corresponding values:

U	D
1	0.67101
2	0.70026
4	0.72864
8	0.75498
16	0.77856
32	0.79906
64	0.81653

For the case where the decline starts at 40 bbl/day and decreases to 4 bbl/day, $U = 40/4 = 10$. The interpolation fraction would then be calculated as follows:

$$\begin{aligned}\text{interpolation fraction} &= \frac{\log (U/U_8)}{\log (U_{16}/U_8)} \\ &= \frac{\log (10/8)}{\log (16/8)} \\ &= 0.322.\end{aligned}$$

The desired deferment factor then is

$$\begin{aligned} D, \text{ for } U = 10 &= D_{U=8} + (0.322)(D_{U=16} - D_{U=8}) \\ &= 0.75498 + (0.322)(0.77856 - 0.75498) \\ &= 0.76257 \\ &= 0.763. \end{aligned}$$

For convenience, values of the interpolation fraction are listed in an interpolation table in Appendix Table 7. Further use of this table is incorporated in the following examples.

Example: Determine the present-value factor for exponentially declining production, where the initial rate is 24 bbl/day, the final rate is 3 bbl/day, and the length of the decline is 10 years. Use the 8% annual discount, and assume that all monies are received at the end of the year.
Solution:

$$U = \frac{q_0}{q_F} = \frac{24 \text{ bbl/day}}{3 \text{ bbl/day}} = 8, \qquad T = 10 \text{ years.}$$

The required table is for exponential decline, sums received at the end of year, $i = 0.08$. Under the column headed $U = 8$, in the line for time $= 10$ years, find $D = 0.75498$.

Example: Determine the 8% discount factors, monies received continuously, for $U = 8$, $T = 20$ years, for both exponential decline and for hyperbolic, $b = 0.5$.
Answer: For exponential decline, $D = 0.62959$.
For hyperbolic, $b = 0.5$, decline, $D = 0.63761$.

Example: Determine the 12% present-value factor, monies received continuously, hyperbolic decline $b = 0.5$, $T = 15$ years, $U = 12$.
Solution: Interpolation is needed between the columns $U = 8$ and $U = 16$. For $U = 8$, $D = 0.60877$, and for $U = 16$, $D = 0.65734$. In the interpolation table, the fourth column contains U-values from 8 to 16. Find $U = 12$ in that column, and in the same line find the interpolation fraction 0.585 in the right-hand column. The present-value factor for $U = 12$ is found by interpolation as follows:

$$\begin{aligned} D &= 0.60877 + (0.585)(0.65734 - 0.60877) \\ &= 0.63718, \text{ or } 0.637. \end{aligned}$$

Example: Determine the 12% present-value factor, monies received continuously, hyperbolic decline $b = 0.5$, $T = 20$ years, $U = 24$.

Solution: Interpolation is needed between the columns $U = 16$ and $U = 32$. For $U = 16$, $D = 0.58982$, and for $U = 32$, $D = 0.64286$. In the interpolation table, the fifth column contains U-values from 16 to 32. Find $U = 24$ in that column, and in the same line find the interpolation fraction 0.585 in the right-hand column. The present-value factor for $U = 24$ is found by interpolation as follows:

$$D = 0.58982 + (0.585)(0.64286 - 0.58982)$$
$$= 0.62085, \text{ or } 0.621.$$

Example: Determine the composite present-value factor and the present value of the gross operating income of the following described production. Assume a gross operating income of \$1.80 per barrel and a total ultimate recovery of 292,000 bbl. Use 10% annual interest, end-of-year convention.

Duration (years)	*Production Rate (bbl/day)*	*Fraction Produced*
2	80, constant	0.2
4	60, constant	0.3
17	60–6, decline	0.5

Solution: The present-value factor for the first 2 years is found in the tables, under $U = 1$, as 0.86777. The present-value factor for the next 4 years is calculated as

$$D = (\text{single-payment factor, 2 years})(\text{P.V. factor as of beginning uniform series})$$
$$= (0.82645)(0.79247)$$
$$= 0.65494.$$

The present-value factor for the decline where $U = (60 \text{ bbl/day})/(6 \text{ bbl/day}) = 10$ and $T = 17$ years, is calculated as follows. First, the interpolation factor corresponding to $U = 10$ is found to be 0.322, in the interpolation table. For 17 years, interpolation is made between $U = 8$, $D = 0.59650$, and $U = 16$, $D = 0.63291$.

$$D = 0.59650 + (0.322)(0.63291 - 0.59650)$$
$$= 0.60822.$$

It is further necessary to reduce this to the basic date of evaluation.

$$D, \text{ decline} = (\text{single-payment factor, 6 years})(\text{P.V. factor as of beginning of decline})$$
$$= (0.56447)(0.60822)$$
$$= 0.34332.$$

For the entire oil production, the present-value factor is

$$D\text{, composite} = \sum (\text{fractional weighting factor})(D \text{ for the fraction})$$
$$= (0.2)(0.86777) + (0.3)(0.65494) + (0.5)(0.34332)$$
$$= 0.54170.$$

The present value of the gross income from the entire production is then calculated as follows:

$$\text{P.V. of gross income} = (0.54170)(292{,}000 \text{ bbl})(\$1.80 \text{ per bbl})$$
$$= \$285{,}000.$$

NOMENCLATURE

a = constant used to describe decline relations,

A = time shift for straightening rate–time log–log plot,

b = constant characterizing decline relations,

N_P = total oil produced, under set allowable plus later decline,

N_{PA} = total oil produced under restricted, allowable rate,

N_{PD} = total oil produced during the decline,

N_{PA}/N_P = fraction produced under allowable rate,

q = instantaneous production rate, barrels/year,

q_A = restricted or set allowable production rate,

q_F = final production rate, usually the economic limit,

q_1 = initial producing capacity, where production is restricted,

q_0 = initial production rate at beginning of decline,

R = ratio, less than 1, of successive yearly productions,

t = time, reckoned from the beginning of the decline,

t_F = total time required to reach the economic limit,

$T = t_F$, used interchangeably,

$U = q_0/q_F$ = ratio of initial to final rate,

Y = production during a time period (such as barrels/year),

Y_1 = production during the first year on decline,

Y_F = production during the last year on decline.

APPENDIX

11.A-1 Decline Equations, Exponential Decline, $b = 0$

Relations based on instantaneous rates:

$$q = q_0 e^{-t/a}, \tag{11-3}$$

$$a = \frac{t_F}{\ln (q_0/q_F)} = \frac{1}{\ln (1/R)}, \tag{11-4}$$

$$N_{PD} = q_0 a(1 - e^{-t/a}), \tag{11-6}$$

$$= a(q_0 - q), \tag{11-7}$$

$$t_F = \frac{N_{PD}}{q_0 - q_F} \ln \left(\frac{q_0}{q_F}\right). \tag{11-8}$$

Fraction of reserves produced at allowable rate:

$$\frac{N_{PA}}{N_P} = \frac{q_1 - q_A}{q_1 - q_F}. \tag{11-9}$$

Relations between instantaneous rates and yearly productions:

$$\frac{Y_1}{Y_F} = U^{(T-1)/T} \tag{11-10}$$

$$\frac{Y_1}{q_0} = \frac{T}{\ln U} \frac{U^{1/T} - 1}{U^{1/T}} = \frac{1 - R}{\ln (1/R)}, \tag{11-11}$$

$$\frac{Y_F}{q_F} = \frac{T}{\ln U}(U^{1/T} - 1) = \frac{1 - R}{R \ln (1/R)}. \tag{11-12}$$

Relations from a geometric series:

$$R = \frac{Y_2}{Y_1} = \frac{Y_3}{Y_2} = \cdots = \frac{Y_F}{Y_{F-1}}, \tag{11-13}$$

$$Y_F = Y_1 R^{T-1}, \tag{11-14}$$

$$R = \left(\frac{Y_F}{Y_1}\right)^{1/(T-1)} = \left(\frac{q_F}{q_0}\right)^{1/T} = \left(\frac{1}{U}\right)^{1/T}, \tag{11-15}$$

$$N_{PD} = Y_1 \frac{1 - R^T}{1 - R}. \tag{11-16}$$

Relations based on yearly productions:

$$\ln Y = \ln Y_0 - kt, \tag{11-17}$$

$$Y = Y_0 e^{-kt}, \tag{11-18}$$

$$k = \frac{\ln (Y_1/Y_F)}{t_F - t_1}, \tag{11-19}$$

$$N_{PD} = (t_F - t_1)\frac{Y_1 - Y_F}{\ln (Y_1/Y_F)} + \frac{Y_1 + Y_F}{2}, \tag{11-21}$$

$$t_F - t_1 = \frac{\ln (Y_1/Y_F)}{Y_1 - Y_F}\left(N_{PD} - \frac{Y_1 + Y_F}{2}\right). \tag{11-22}$$

11.A-2 Decline Equations, General Hyperbolic Decline, $0 < b < 1$

Relations based on instantaneous rates:

$$q = q_0\left(1 + \frac{bt}{a}\right)^{-1/b}, \tag{11-23}$$

$$a = \frac{bt_F}{(q_0/q_F)^b - 1}, \tag{11-24}$$

$$N_{PD} = \frac{aq_0}{1 - b}\left[1 - \left(1 + \frac{bt}{a}\right)^{(b-1)/b}\right], \tag{11-25}$$

$$= \frac{aq_0^b}{1 - b}(q_0^{1-b} - q^{1-b}), \tag{11-26}$$

$$t_F = \frac{1 - b}{b}\frac{N_{PD}}{q_0}U^{1-b}\frac{U^b - 1}{U^{1-b} - 1}, \tag{11-27}$$

$$= \frac{N_{PD}}{q_0}\sqrt{\frac{q_0}{q_F}}, \qquad \text{for } b = 0.5, \tag{11-27a}$$

$$N_{PD} = k_2 - k_1 q^{1-b}, \tag{11-28}$$

$$k_1 = \frac{aq_0^b}{1 - b}, \tag{11-29}$$

$$k_2 = \frac{aq_0}{1 - b}. \tag{11-30}$$

Fraction of reserves produced at allowable rate:

$$\frac{N_{PA}}{N_P} = \frac{1 - (q_A/q_1)^{1-b}}{1 - (q_F/q_1)^{1-b}}. \tag{11-31}$$

Relations between instantaneous rates and yearly productions:

$$\frac{Y_1}{q_0} = \frac{b}{1 - b}\frac{T}{U^b - 1}\left[1 - \left(\frac{T}{T - 1 + U^b}\right)^{(1-b)/b}\right], \tag{11-32}$$

$$\frac{Y_F}{q_F} = \frac{b}{1 - b}\frac{T}{1 - U^{-b}}\left[\left(\frac{T}{T - 1 + U^{-b}}\right)^{(1-b)/b} - 1\right], \tag{11-33}$$

$$\frac{Y_1}{Y_F} = U^{1-b} \frac{1 - \left(\dfrac{T}{T - 1 + U^b}\right)^{(1-b)/b}}{\left(\dfrac{T}{T - 1 + U^{-b}}\right)^{(1-b)/b} - 1} \tag{11-34}$$

Relations based on yearly productions:

$$Y = Y_0(t + A)^{-1/b}, \tag{11-35}$$

$$A = \frac{t_F - t_1(Y_1/Y_F)^b}{(Y_1/Y_F)^b - 1}, \tag{11-36}$$

$$N_{PD} = \frac{b}{1-b} Y_1(t_1 + A)\left[1 - \left(\frac{Y_F}{Y_1}\right)^{1-b}\right] + \frac{Y_1 + Y_F}{2}, \tag{11-37}$$

$$t_F - t_1 = \frac{1-b}{b}\left(\frac{Y_1}{Y_F}\right)^b \left[\frac{1 - (Y_F/Y_1)^b}{1 - (Y_F/Y_1)^{1-b}}\right] \frac{N_{PD} - \dfrac{Y_1 + Y_F}{2}}{Y_1}. \tag{11-38}$$

For $b = 0.5$,

$$t_F - t_1 = \left(\frac{Y_1}{Y_F}\right)^{0.5} \frac{N_{PD} - \dfrac{Y_1 + Y_F}{2}}{Y_1} \tag{11-38a}$$

11.A-3 Decline Equations, Harmonic Decline, $b = 1$

Relations based on instantaneous rates:

$$q = q_0\left(1 + \frac{t}{a}\right)^{-1}, \tag{11-39}$$

$$a = \frac{t_F}{(q_0/q_F) - 1}, \tag{11-40}$$

$$N_{PD} = aq_0 \ln\left(1 + \frac{t}{a}\right), \tag{11-41}$$

$$= aq_0 \ln\left(\frac{q_0}{q}\right), \tag{11-42}$$

$$t_F = \frac{N_{PD}}{q_0} \frac{(q_0/q_F - 1)}{\ln(q_0/q_F)}. \tag{11-43}$$

Fraction of reserves produced at allowable rate:

$$\frac{N_{PA}}{N_P} = \frac{\ln(q_1/q_A)}{\ln(q_1/q_F)}. \tag{11-44}$$

Relations between instantaneous rates and yearly productions:

$$\frac{Y_1}{q_0} = \frac{T}{U-1} \ln\left(\frac{T+U-1}{T}\right), \tag{11-45}$$

$$\frac{Y_F}{q_F} = \frac{TU}{U-1} \ln\left[\frac{TU}{U(T-1)+1}\right], \tag{11-46}$$

$$\frac{Y_1}{Y_F} = \frac{\ln\left(\frac{T+U-1}{T}\right)}{\ln\left[\frac{TU}{U(T-1)+1}\right]}. \tag{11-47}$$

Relations based on yearly productions:

$$Y = Y_0(t+A)^{-1}, \tag{11-48}$$

$$A = \frac{t_F - t_1(Y_1/Y_F)}{(Y_1/Y_F)-1}, \tag{11-49}$$

$$N_{PD} = Y_1(t_1 + A)\ln\left(\frac{Y_1}{Y_F}\right) + \frac{Y_1 + Y_F}{2}, \tag{11-50}$$

$$t_F - t_1 = \frac{(Y_1/Y_F)-1}{\ln(Y_1/Y_F)} \frac{N_{PD} - \frac{Y_1+Y_F}{2}}{Y_1}. \tag{11-51}$$

REFERENCES

1. "A Statistical Study of Recovery Efficiency" (API BUL D14). Dallas: American Petroleum Institute, Division of Production, Oct. 1967.
2. ARPS, J. J., "Estimation of Primary Oil Reserves." *Transactions of the AIME*, Vol. 207, 1956, p. 182.
3. ARPS, J. J., "Analysis of Decline Curves." *Transactions of the AIME*, Vol. 160, 1945, p. 228.
4. BRONS, F., and M. W. MCGARRY, Jr., "Method for Calculating Profitabilities" (Petroleum Transactions Reprint Series No. 3, Oil and Gas Property Evaluation and Reserve Estimates). Dallas: Society of Petroleum Engineers of AIME, 1970.

PROBLEMS

11-1. An oil property intially produced at the rate of 300 bbl/day. Six years later it was producing at the rate of 100 bbl/day. Calculate the value of parameter

a, which is introduced in Eq. (11-2), and also calculate N_{PD}, the total oil produced during the 6-year period, by assuming successively that parameter b has the following values: (a) $b = 0$, (b) $b = 0.5$, and (c) $b = 1.0$.

11-2. Referring to Problem 11-1, calculate the production rate at the end of the next 6-year period, and also calculate the total oil that will be produced during this second 6-year period, again for the following three values of b: (a) $b = 0$, (b) $b = 0.5$, and (c) $b = 1.0$.

11-3. An oil property is estimated to have total reserves amounting to 800,000 bbl. It will produce unrestricted from an initial rate of 400 bbl/day to an economic limit of 30 bbl/day. Calculate the total time on production by assuming successively the following values of parameter b: (a) $b = 0$, (b) $b = 0.5$, and (c) $b = 1.0$.

11-4. When the five yearly productions given below are plotted versus time on log–log paper, and shifted by adding 4 years to each time value, the points fall on a straight line with a slope of -2.5. (Hence $b = -1/-2.5 = 0.4$.) Determine the value of parameter K in the equation $Y = K(t + A)^n$, which describes this line.

(a) Calculate the cumulative production by integrating the equation determined above between the limits $t = 0$ and $t = 5$. For this purpose presume that the original plot of the data had been made with the production amounts plotted at the midpoints of the years, in order to approximate the instantaneous rate. Thus for this calculation one uses $A = 4.5$, while K retains its same value and $n = -2.5$, as before.

(b) Calculate the cumulative production by integrating the equation between the limits $t = 1$ and $t = 5$ and using the correction derived from the trapezoidal rule. For this purpose $A = 4$, since the production of the first year is presumed to be plotted at $t = 1$, etc. Equation (11-37) is a convenient form of the integration and correction indicated.

Year	*Production (bbl)*
1	60,000
2	38,200
3	25,800
4	18,600
5	13,800
	156,400

11-5. After shifting monthly production data to the right on log–log paper by 10 months, the rate–time production data all fell along a straight line. The production of the first month was 20,000 bbl, production of the 24th (last) month was 12,000 bbl .Calculate the equation of the straight line on log–log paper. Calculate the production of the next 5 years.

11-6. A well has produced a total of 5 years. During its second year it produced 24,000 bbl. The rate–time data was plotted on log–log paper. Originally the production data for the first year was plotted above 1 on the time scale. After shifting the data 6 years, all the data points were observed to fall along a straight line with a slope of −4.0. Estimate the reserves (not including the first 5 years) to an economic limit of 1,000 bbl/year. Also, estimate the remaining time required to reach the economic limit.

11-7. A plot of yearly productions vs time was extrapolated and it was determined that production would continue for 15 more years before reaching the economic limit of 1,800 bbl/year for the particular property. The best line drawn through the data, which was used for the extrapolation, passed through the value of 15,000 bbl for the previous year. However, production records show that the actual production was 14,400 bbl and that some downtime was experienced. Calculate the reserves expected to be produced during the next 15 years for the following three cases: (a) where the complete plot of the production data indicates that parameter $b = 0$, (b) where $b = 0.5$, and (c) where $b = 1.0$.

11-8. A plot of yearly productions versus cumulative production was extrapolated and it was indicated that the ultimate recovery would be 205,200 bbl at the economic limit of 2,000 bbl/year. Total recovery to date has been 90,500 bbl, 14,400 of which was produced during the previous year. However, for this last year, the best line used for extrapolating the data passed through the 14,700-bbl value on the graph. Calculate the remaining time on production for the following three cases: (a) where the complete plot of the data indicated that parameter $b = 0$, (b) where $b = 0.5$, and (c) where $b = 1.0$.

11-9. Find the reserves for the well whose production history is given in the following table. The economic limit is 1,000 bbl/year. Use a graphical solution.

Year	*Production (bbl)*
1	31,175
2	20,041
3	15,105
4	10,892
5	9,038

11-10. A Hunton Limestone well in McLain County since completion has produced oil as shown in the table. Assume that the well is producing at a wide-open rate and extrapolate a decline curve on suitable paper to the economic limit. The total well cost was \$65,000. Total depth is 6,675 ft. Oil gravity, 40°API, no gas sales. Lifting cost is estimated at \$380 per month, including taxes. Use the latest posted price throughout for oil value.

Bbl/Month	*Bbl/Month*	*Bbl/Month*
5,400	1,700	980
5,000	1,620	960
4,800	1,550	940
4,100	1,500	910
3,900	1,410	900
3,600	1,370	890
3,300	1,300	870
3,100	1,280	850
2,900	1,230	840
2,650	1,200	815
2,400	1,160	806
2,350	1,120	800
2,160	1,100	790
2,050	1,060	780
1,910	1,030	
1,790	1,000	

Using a $\frac{7}{8}$ working interest, calculate
(a) Remaining reserves.
(b) Net present value using 8% discount rate, assuming that the basic date is at the beginning of production.
(c) Net present value, assuming that the basic date is the end of the above past production history.

11-11. A development well is planned in a field where spacing is 40 acres per well. Net pay thickness is 20 ft. A recovery of 200 bbl/acre-ft is expected. Initial potential of offset wells is about 80 bbl/day. The economic limit is about 6 bbl/day. List the expected productions by years, assuming exponential decline.

11-12. Calculate the present value of the production in Problem 11-11, assuming a net of $3 per barrel, and 8% yearly discount factors, and assuming that the income for each year is received at the midpoint of each year. The basic date is the date of well completion.

11-13. Check the answer to Problem 11-12 by determining a present-value factor that applies directly to the entire production. For this particular case, the factor can be calculated by Eq. (11-20), p. 335.
Note: Present-value factors in Appendix Table 8 can be multiplied by the appropriate $(1 + i)^{0.5}$ value to convert them to the corresponding mid-year factors.

11-14. It is estimated that a well will produce as follows:
(a) 4 years at a constant rate of 40 bbl/day.
(b) 6 years at a constant rate of 20 bbl/day.
(c) 15 years on exponential decline to an economic limit of 4 bbl/day. Calculate the composite discount factor that would give the present value of the entire reserves at the discount rate of 8% per year continuous interest ($j = 0.08$).

11-15. Calculate the value as of January 1, 1970, of an exponential decline beginning January 1, 1975. The initial rate of production is 30 bbl/day and the final rate is 2 bbl/day. The decline will extend over a period of 15 years. Assume that all payments are received at the end of each 6-month period, and use an interest rate of 5% for each semiannual period. Note that the number of time periods is $T = 30$. The value of R for semiannual periods is the square root of R for annual periods. Where the payments are received at the ends of the periods, the factor $(1 + i)^{0.5}$ is omitted from Equation (11-20).

11-16. A well has reserves of 71,300 bbl. It will produce 45% of the reserves at an allowable rate of 30 bbl/day and then go on exponential decline to an economic limit of 3 bbl/day. The well is in Oklahoma and produces 41°API oil. The cost of drilling and completing the well is \$40,000. An additional expense of \$8,000 must be made at the end of 3 years for artificial lift equipment. Operating expenses will be \$100 per month. Deduct 7% state tax and $\frac{1}{8}$ royalty. Salvage is estimated at \$2,000 at well abandonment.

(a) Calculate the net present value (before income taxes) by calculating the appropriate deferment factor to apply to each expense and income. Use 8% per year and assume that all incomes and expenses are lumped at the midpoint of each year. Exceptions to this are the initial expense and the additional expense for artificial lift, both of which are presumed to be single payments at the time indicated. The evaluation date is the date of well completion.

(b) Calculate the present value by preparing a table that shows the production for each successive year together with the undeferred income, undeferred expenses, net undeferred operating income, deferred net operating income, deferred investments (which include the initial investment, the artificial lift expense, and the salvage as a negative quantity), and the overall net cumulative present worth. Again use 8% per year, midyear calculations except in the cases of the investments, which are treated as single payments.

Note: The results of parts (a) and (b) should be in agreement, with perhaps 1% deviation due to numerical errors.

11-17. One well is to be drilled on a 40-acre lease. It is estimated that the total cost of the completed well and necessary surface equipment will be \$70,000. The average sand thickness is 15 ft, the average porosity is 18%, connate water averages 22%, and the initial formation volume factor is 1.15. Assume 15% recovery of the initial stock-tank oil in place to an economic limit of 3 bbl/day and that 50% of the reserves will be produced at an allowable rate of 20 bbl/day. Further assume that the 7/8 working interest will realize an operating profit of \$2 per gross (total) barrel produced. Calculate the maximum bonus that you would be willing to pay to acquire this lease. Indicate clearly all the assumptions that you must make to obtain your solution.

11-18. How much would you pay for an oil lease in the center of proved production, where the recovery is estimated at 80 bbl/acre-ft with a net pay of 25 ft?

Spacing is 40 acres. Estimate the deferment factor by exponential relations based on a potential of 30 bbl/day but restricted to a maximum allowable of 20 bbl/day declining to an economic limit of 2.5 bbl/day. Development cost is estimated at $70,000 per well and producing expenses will average $100 per month.

11-19. How much would you pay for a 10% overriding royalty on the well described in Problem 11-18,

(a) If the well has already been drilled?

(b) If the well has not yet been drilled?

List several considerations that you investigate in such cases.

11-20. There are two partners; *A* owns 60%, *B* owns 40%. Together they paid $100 per acre for a 40-acre lease. It now appears that there will be 15 ft of net pay yielding a recovery of 100 bbl/acre-ft, with a deferment factor of 0.8. It will cost $70,000 to drill and equip one well. Partner *B* has other demands on his capital and prefers to take a $\frac{1}{16}$ override. Prepare calculations so that you can advise partner *A* whether or not to give *B* the $\frac{1}{16}$ override. (The partnership now has $\frac{7}{8}$, the landowner $\frac{1}{8}$.) What should be *A*'s maximum offer?

APPENDIX TABLES

BINOMIAL PROBABILITIES

P, PROBABILITY OF OCCURRENCE

N	X	0.01	0.05	0.10	0.15	0.20	0.25	0.30	0.40	0.50
2	0	0.9801	0.9025	0.8100	0.7225	0.6400	0.5625	0.4900	0.3600	0.2500
	1	0.0198	0.0950	0.1800	0.2550	0.3200	0.3750	0.4200	0.4800	0.5000
	2	0.0001	0.0025	0.0100	0.0225	0.0400	0.0625	0.0900	0.1600	0.2500
3	0	0.9703	0.8574	0.7290	0.6141	0.5120	0.4219	0.3430	0.2160	0.1250
	1	0.0294	0.1354	0.2430	0.3251	0.3840	0.4219	0.4410	0.4320	0.3750
	2	0.0003	0.0071	0.0270	0.0574	0.0960	0.1406	0.1890	0.2880	0.3750
	3	0.0000	0.0001	0.0010	0.0034	0.0080	0.0156	0.0270	0.0640	0.1250
4	0	0.9606	0.8145	0.6561	0.5220	0.4096	0.3164	0.2401	0.1296	0.0625
	1	0.0388	0.1715	0.2916	0.3685	0.4096	0.4219	0.4116	0.3456	0.2500
	2	0.0006	0.0135	0.0486	0.0975	0.1536	0.2109	0.2646	0.3456	0.3750
	3	0.0000	0.0005	0.0036	0.0115	0.0256	0.0469	0.0756	0.1536	0.2500
	4	0.0000	0.0000	0.0001	0.0005	0.0016	0.0039	0.0081	0.0256	0.0625
5	0	0.9510	0.7738	0.5905	0.4437	0.3277	0.2373	0.1681	0.0778	0.0312
	1	0.0480	0.2036	0.3281	0.3915	0.4096	0.3955	0.3601	0.2592	0.1562
	2	0.0010	0.0215	0.0729	0.1382	0.2048	0.2637	0.3087	0.3456	0.3125
	3	0.0000	0.0011	0.0081	0.0244	0.0512	0.0879	0.1323	0.2304	0.3125
	4	0.0000	0.0000	0.0004	0.0021	0.0064	0.0146	0.0284	0.0768	0.1563
	5	0.0000	0.0000	0.0000	0.0001	0.0003	0.0010	0.0024	0.0102	0.0313
6	0	0.9415	0.7351	0.5314	0.3771	0.2621	0.1780	0.1177	0.0467	0.0156
	1	0.0571	0.2321	0.3543	0.3993	0.3932	0.3560	0.3025	0.1866	0.0937
	2	0.0014	0.0306	0.0984	0.1762	0.2458	0.2966	0.3241	0.3110	0.2344
	3	0.0000	0.0021	0.0146	0.0415	0.0819	0.1318	0.1852	0.2765	0.3125
	4	0.0000	0.0001	0.0012	0.0055	0.0154	0.0330	0.0596	0.1382	0.2344
	5	0.0000	0.0000	0.0001	0.0004	0.0015	0.0044	0.0102	0.0369	0.0938
	6	0.0000	0.0000	0.0000	0.0000	0.0001	0.0002	0.0007	0.0041	0.0156
7	0	0.9321	0.6983	0.4783	0.3206	0.2097	0.1335	0.0824	0.0280	0.0078
	1	0.0659	0.2573	0.3720	0.3960	0.3670	0.3115	0.2471	0.1306	0.0547
	2	0.0020	0.0406	0.1240	0.2096	0.2752	0.3115	0.3176	0.2613	0.1641
	3	0.0000	0.0036	0.0230	0.0617	0.1147	0.1730	0.2269	0.2903	0.2734
	4	0.0000	0.0002	0.0025	0.0109	0.0287	0.0577	0.0972	0.1935	0.2734
	5	0.0000	0.0000	0.0002	0.0011	0.0043	0.0115	0.0250	0.0774	0.1641
	6	0.0000	0.0000	0.0000	0.0001	0.0004	0.0013	0.0036	0.0172	0.0547
	7	0.0000	0.0000	0.0000	0.0000	0.0000	0.0000	0.0002	0.0017	0.0078

BINOMIAL PROBABILITIES

P, PROBABILITY OF OCCURRENCE

N	X	0.01	0.05	0.10	0.15	0.20	0.25	0.30	0.40	0.50
8	0	0.9227	0.6634	0.4305	0.2725	0.1678	0.1001	0.0576	0.0168	0.0039
	1	0.0746	0.2793	0.3826	0.3847	0.3355	0.2670	0.1977	0.0896	0.0313
	2	0.0026	0.0515	0.1488	0.2376	0.2936	0.3115	0.2965	0.2090	0.1094
	3	0.0001	0.0054	0.0331	0.0839	0.1468	0.2076	0.2541	0.2787	0.2187
	4	0.0000	0.0004	0.0046	0.0185	0.0459	0.0865	0.1361	0.2322	0.2734
	5	0.0000	0.0000	0.0004	0.0026	0.0092	0.0231	0.0467	0.1239	0.2187
	6	0.0000	0.0000	0.0000	0.0002	0.0011	0.0038	0.0100	0.0413	0.1094
	7	0.0000	0.0000	0.0000	0.0000	0.0001	0.0004	0.0012	0.0079	0.0313
	8	0.0000	0.0000	0.0000	0.0000	0.0000	0.0000	0.0001	0.0006	0.0039
9	0	0.9135	0.6303	0.3874	0.2316	0.1342	0.0751	0.0404	0.0101	0.0019
	1	0.0830	0.2985	0.3874	0.3679	0.3020	0.2253	0.1557	0.0605	0.0176
	2	0.0034	0.0629	0.1722	0.2597	0.3020	0.3003	0.2668	0.1613	0.0703
	3	0.0001	0.0077	0.0447	0.1069	0.1762	0.2336	0.2668	0.2508	0.1641
	4	0.0000	0.0006	0.0074	0.0283	0.0661	0.1168	0.1715	0.2508	0.2461
	5	0.0000	0.0000	0.0008	0.0050	0.0165	0.0389	0.0735	0.1672	0.2461
	6	0.0000	0.0000	0.0001	0.0006	0.0027	0.0087	0.0210	0.0743	0.1641
	7	0.0000	0.0000	0.0000	0.0000	0.0003	0.0012	0.0039	0.0212	0.0703
	8	0.0000	0.0000	0.0000	0.0000	0.0000	0.0001	0.0004	0.0035	0.0176
	9	0.0000	0.0000	0.0000	0.0000	0.0000	0.0000	0.0000	0.0003	0.0019
10	0	0.9044	0.5987	0.3487	0.1969	0.1074	0.0563	0.0283	0.0060	0.0010
	1	0.0913	0.3151	0.3874	0.3474	0.2684	0.1877	0.1211	0.0403	0.0097
	2	0.0042	0.0746	0.1937	0.2759	0.3020	0.2816	0.2335	0.1209	0.0439
	3	0.0001	0.0105	0.0574	0.1298	0.2013	0.2503	0.2668	0.2150	0.1172
	4	0.0000	0.0010	0.0112	0.0401	0.0881	0.1460	0.2001	0.2508	0.2051
	5	0.0000	0.0001	0.0015	0.0085	0.0264	0.0584	0.1029	0.2007	0.2461
	6	0.0000	0.0000	0.0001	0.0013	0.0055	0.0162	0.0368	0.1115	0.2051
	7	0.0000	0.0000	0.0000	0.0001	0.0008	0.0031	0.0090	0.0425	0.1172
	8	0.0000	0.0000	0.0000	0.0000	0.0001	0.0004	0.0014	0.0106	0.0439
	9	0.0000	0.0000	0.0000	0.0000	0.0000	0.0000	0.0001	0.0016	0.0098
	10	0.0000	0.0000	0.0000	0.0000	0.0000	0.0000	0.0000	0.0001	0.0010

BINOMIAL PROBABILITIES

P, PROBABILITY OF OCCURRENCE

N	X	0.01	0.05	0.10	0.15	0.20	0.25	0.30	0.40	0.50
15	0	0.8601	0.4633	0.2059	0.0874	0.0352	0.0134	0.0048	0.0005	0.0000
	1	0.1303	0.3658	0.3432	0.2312	0.1319	0.0668	0.0305	0.0047	0.0005
	2	0.0092	0.1347	0.2669	0.2856	0.2309	0.1559	0.0916	0.0219	0.0032
	3	0.0004	0.0307	0.1285	0.2184	0.2501	0.2252	0.1700	0.0634	0.0139
	4	0.0000	0.0049	0.0428	0.1157	0.1876	0.2252	0.2186	0.1268	0.0417
	5	0.0000	0.0006	0.0105	0.0449	0.1032	0.1651	0.2061	0.1859	0.0916
	6	0.0000	0.0000	0.0019	0.0132	0.0430	0.0918	0.1472	0.2066	0.1527
	7	0.0000	0.0000	0.0003	0.0030	0.0138	0.0393	0.0811	0.1771	0.1964
	8	0.0000	0.0000	0.0000	0.0005	0.0035	0.0131	0.0348	0.1181	0.1964
	9	0.0000	0.0000	0.0000	0.0001	0.0007	0.0034	0.0116	0.0612	0.1527
	10	0.0000	0.0000	0.0000	0.0000	0.0001	0.0007	0.0030	0.0245	0.0916
	11	0.0000	0.0000	0.0000	0.0000	0.0000	0.0001	0.0006	0.0074	0.0417
	12	0.0000	0.0000	0.0000	0.0000	0.0000	0.0000	0.0001	0.0016	0.0139
	13	0.0000	0.0000	0.0000	0.0000	0.0000	0.0000	0.0000	0.0003	0.0032
	14	0.0000	0.0000	0.0000	0.0000	0.0000	0.0000	0.0000	0.0000	0.0005
	15	0.0000	0.0000	0.0000	0.0000	0.0000	0.0000	0.0000	0.0000	0.0000
20	0	0.8179	0.3585	0.1216	0.0388	0.0115	0.0032	0.0008	0.0000	0.0000
	1	0.1652	0.3774	0.2702	0.1368	0.0576	0.0211	0.0068	0.0005	0.0000
	2	0.0159	0.1887	0.2852	0.2293	0.1369	0.0669	0.0279	0.0031	0.0002
	3	0.0010	0.0596	0.1901	0.2428	0.2054	0.1339	0.0716	0.0124	0.0011
	4	0.0000	0.0133	0.0898	0.1821	0.2182	0.1897	0.1304	0.0350	0.0046
	5	0.0000	0.0022	0.0319	0.1029	0.1746	0.2023	0.1789	0.0746	0.0148
	6	0.0000	0.0003	0.0089	0.0454	0.1091	0.1686	0.1916	0.1244	0.0370
	7	0.0000	0.0000	0.0020	0.0160	0.0545	0.1124	0.1643	0.1659	0.0739
	8	0.0000	0.0000	0.0003	0.0046	0.0222	0.0609	0.1144	0.1797	0.1201
	9	0.0000	0.0000	0.0000	0.0011	0.0074	0.0271	0.0654	0.1597	0.1602
	10	0.0000	0.0000	0.0000	0.0002	0.0020	0.0099	0.0308	0.1171	0.1762
	11	0.0000	0.0000	0.0000	0.0000	0.0005	0.0030	0.0120	0.0710	0.1602
	12	0.0000	0.0000	0.0000	0.0000	0.0001	0.0008	0.0039	0.0355	0.1201
	13	0.0000	0.0000	0.0000	0.0000	0.0000	0.0002	0.0010	0.0146	0.0739
	14	0.0000	0.0000	0.0000	0.0000	0.0000	0.0000	0.0002	0.0049	0.0370
	15	0.0000	0.0000	0.0000	0.0000	0.0000	0.0000	0.0000	0.0013	0.0148
	16	0.0000	0.0000	0.0000	0.0000	0.0000	0.0000	0.0000	0.0003	0.0046
	17	0.0000	0.0000	0.0000	0.0000	0.0000	0.0000	0.0000	0.0000	0.0011
	18	0.0000	0.0000	0.0000	0.0000	0.0000	0.0000	0.0000	0.0000	0.0002
	19	0.0000	0.0000	0.0000	0.0000	0.0000	0.0000	0.0000	0.0000	0.0000
	20	0.0000	0.0000	0.0000	0.0000	0.0000	0.0000	0.0000	0.0000	0.0000

BINOMIAL PROBABILITIES

P, PROBABILITY OF OCCURRENCE

N	X	0.01	0.05	0.10	0.15	0.20	0.25	0.30	0.40	0.50
30	0	0.7397	0.2146	0.0424	0.0076	0.0012	0.0002	0.0000	0.0000	0.0000
	1	0.2242	0.3389	0.1413	0.0404	0.0093	0.0018	0.0003	0.0000	0.0000
	2	0.0328	0.2586	0.2277	0.1034	0.0337	0.0086	0.0018	0.0000	0.0000
	3	0.0031	0.1270	0.2361	0.1703	0.0785	0.0269	0.0072	0.0003	0.0000
	4	0.0002	0.0451	0.1771	0.2028	0.1325	0.0604	0.0208	0.0012	0.0000
	5	0.0000	0.0124	0.1023	0.1861	0.1723	0.1047	0.0464	0.0041	0.0001
	6	0.0000	0.0027	0.0474	0.1368	0.1795	0.1455	0.0829	0.0115	0.0006
	7	0.0000	0.0005	0.0180	0.0828	0.1538	0.1662	0.1219	0.0263	0.0019
	8	0.0000	0.0001	0.0058	0.0420	0.1106	0.1593	0.1501	0.0505	0.0055
	9	0.0000	0.0000	0.0016	0.0181	0.0676	0.1298	0.1573	0.0823	0.0133
	10	0.0000	0.0000	0.0004	0.0067	0.0355	0.0909	0.1416	0.1152	0.0280
40	0	0.6690	0.1285	0.0148	0.0015	0.0001	0.0000	0.0000	0.0000	0.0000
	1	0.2703	0.2706	0.0657	0.0106	0.0013	0.0001	0.0000	0.0000	0.0000
	2	0.0532	0.2777	0.1423	0.0365	0.0065	0.0009	0.0001	0.0000	0.0000
	3	0.0068	0.1851	0.2003	0.0816	0.0205	0.0037	0.0005	0.0000	0.0000
	4	0.0006	0.0901	0.2059	0.1332	0.0475	0.0113	0.0020	0.0000	0.0000
	5	0.0000	0.0342	0.1647	0.1692	0.0854	0.0272	0.0061	0.0001	0.0000
	6	0.0000	0.0105	0.1068	0.1742	0.1246	0.0530	0.0151	0.0005	0.0000
	7	0.0000	0.0027	0.0576	0.1493	0.1513	0.0857	0.0315	0.0015	0.0000
	8	0.0000	0.0006	0.0264	0.1087	0.1560	0.1179	0.0557	0.0040	0.0001
	9	0.0000	0.0001	0.0104	0.0682	0.1386	0.1397	0.0849	0.0095	0.0002
	10	0.0000	0.0000	0.0036	0.0373	0.1075	0.1444	0.1128	0.0196	0.0008
50	0	0.6050	0.0769	0.0052	0.0003	0.0000	0.0000	0.0000	0.0000	0.0000
	1	0.3056	0.2025	0.0286	0.0026	0.0002	0.0000	0.0000	0.0000	0.0000
	2	0.0756	0.2611	0.0779	0.0113	0.0011	0.0001	0.0000	0.0000	0.0000
	3	0.0122	0.2199	0.1386	0.0319	0.0044	0.0004	0.0000	0.0000	0.0000
	4	0.0015	0.1360	0.1809	0.0661	0.0128	0.0016	0.0001	0.0000	0.0000
	5	0.0001	0.0658	0.1849	0.1072	0.0295	0.0049	0.0006	0.0000	0.0000
	6	0.0000	0.0260	0.1541	0.1419	0.0554	0.0123	0.0018	0.0000	0.0000
	7	0.0000	0.0086	0.1076	0.1575	0.0870	0.0259	0.0048	0.0000	0.0000
	8	0.0000	0.0024	0.0643	0.1493	0.1169	0.0463	0.0110	0.0002	0.0000
	9	0.0000	0.0006	0.0333	0.1230	0.1364	0.0721	0.0220	0.0005	0.0000
	10	0.0000	0.0001	0.0152	0.0890	0.1398	0.0985	0.0386	0.0014	0.0000

STANDARDIZED NORMAL DISTRIBUTION

CUMULATIVE PROBABILITY

Z	0.00	0.01	0.02	0.03	0.04	0.05	0.06	0.07	0.08	0.09
0.0	0.5000	0.5040	0.5080	0.5120	0.5160	0.5199	0.5239	0.5279	0.5319	0.5359
0.1	0.5398	0.5438	0.5478	0.5517	0.5557	0.5596	0.5636	0.5675	0.5714	0.5753
0.2	0.5793	0.5832	0.5871	0.5910	0.5948	0.5987	0.6026	0.6064	0.6103	0.6141
0.3	0.6179	0.6217	0.6255	0.6293	0.6331	0.6368	0.6406	0.6443	0.6480	0.6517
0.4	0.6554	0.6591	0.6628	0.6664	0.6700	0.6736	0.6772	0.6808	0.6844	0.6879
0.5	0.6915	0.6950	0.6985	0.7019	0.7054	0.7088	0.7123	0.7157	0.7190	0.7224
0.6	0.7257	0.7291	0.7324	0.7357	0.7389	0.7422	0.7454	0.7486	0.7517	0.7549
0.7	0.7580	0.7611	0.7642	0.7673	0.7703	0.7734	0.7764	0.7793	0.7823	0.7852
0.8	0.7881	0.7910	0.7939	0.7967	0.7995	0.8023	0.8051	0.8078	0.8106	0.8133
0.9	0.8159	0.8186	0.8212	0.8238	0.8264	0.8289	0.8315	0.8340	0.8365	0.8389
1.0	0.8413	0.8438	0.8461	0.8485	0.8508	0.8531	0.8554	0.8577	0.8599	0.8621
1.1	0.8643	0.8665	0.8686	0.8708	0.8729	0.8749	0.8770	0.8790	0.8810	0.8830
1.2	0.8849	0.8869	0.8888	0.8907	0.8925	0.8943	0.8962	0.8980	0.8997	0.9015
1.3	0.9032	0.9049	0.9066	0.9082	0.9099	0.9115	0.9131	0.9147	0.9162	0.9177
1.4	0.9192	0.9207	0.9222	0.9236	0.9251	0.9265	0.9279	0.9292	0.9306	0.9319
1.5	0.9332	0.9345	0.9357	0.9370	0.9382	0.9394	0.9406	0.9418	0.9429	0.9441
1.6	0.9452	0.9463	0.9474	0.9484	0.9495	0.9505	0.9515	0.9525	0.9535	0.9545
1.7	0.9554	0.9564	0.9573	0.9582	0.9591	0.9599	0.9608	0.9616	0.9625	0.9633
1.8	0.9641	0.9649	0.9656	0.9664	0.9671	0.9678	0.9686	0.9693	0.9699	0.9706
1.9	0.9713	0.9719	0.9726	0.9732	0.9738	0.9744	0.9750	0.9756	0.9761	0.9767
2.0	0.9772	0.9778	0.9783	0.9788	0.9793	0.9798	0.9803	0.9808	0.9812	0.9817
2.1	0.9821	0.9826	0.9830	0.9834	0.9838	0.9842	0.9846	0.9850	0.9854	0.9857
2.2	0.9861	0.9864	0.9868	0.9871	0.9875	0.9878	0.9881	0.9884	0.9887	0.9890
2.3	0.9893	0.9896	0.9898	0.9901	0.9904	0.9906	0.9909	0.9911	0.9913	0.9916
2.4	0.9918	0.9920	0.9922	0.9924	0.9927	0.9929	0.9931	0.9932	0.9934	0.9936
2.5	0.9938	0.9940	0.9941	0.9943	0.9945	0.9946	0.9948	0.9949	0.9951	0.9952
2.6	0.9953	0.9955	0.9956	0.9957	0.9959	0.9960	0.9961	0.9962	0.9963	0.9964
2.7	0.9965	0.9966	0.9967	0.9968	0.9969	0.9970	0.9971	0.9972	0.9973	0.9974
2.8	0.9974	0.9975	0.9976	0.9977	0.9977	0.9978	0.9979	0.9979	0.9980	0.9981
2.9	0.9981	0.9982	0.9982	0.9983	0.9984	0.9984	0.9985	0.9985	0.9986	0.9986
3.0	0.9986	0.9987	0.9987	0.9988	0.9988	0.9989	0.9989	0.9989	0.9990	0.9990
3.1	0.9990	0.9991	0.9991	0.9991	0.9992	0.9992	0.9992	0.9992	0.9993	0.9993
3.2	0.9993	0.9993	0.9994	0.9994	0.9994	0.9994	0.9994	0.9995	0.9995	0.9995
3.3	0.9995	0.9995	0.9995	0.9996	0.9996	0.9996	0.9996	0.9996	0.9996	0.9996
3.4	0.9997	0.9997	0.9997	0.9997	0.9997	0.9997	0.9997	0.9997	0.9997	0.9998
3.5	0.9998	0.9998	0.9998	0.9998	0.9998	0.9998	0.9998	0.9998	0.9998	0.9998
3.6	0.9998	0.9998	0.9999	0.9999	0.9999	0.9999	0.9999	0.9999	0.9999	0.9999
3.7	0.9999	0.9999	0.9999	0.9999	0.9999	0.9999	0.9999	0.9999	0.9999	0.9999

TABLE OF POISSON PROBABILITIES

LAMBDA	VALUES OF X												
	0	1	2	3	4	5	6	7	8	9	10	11	12
0.1	0.9048	0.0905	0.0045	0.0002	0.0000								
0.2	0.8187	0.1637	0.0164	0.0011	0.0001	0.0000							
0.3	0.7408	0.2222	0.0333	0.0033	0.0003	0.0000							
0.4	0.6703	0.2681	0.0536	0.0072	0.0007	0.0001	0.0000						
0.5	0.6065	0.3033	0.0758	0.0126	0.0016	0.0002	0.0000						
0.6	0.5488	0.3293	0.0988	0.0198	0.0030	0.0004	0.0000						
0.7	0.4966	0.3476	0.1217	0.0284	0.0050	0.0007	0.0001	0.0000					
0.8	0.4493	0.3595	0.1438	0.0383	0.0077	0.0012	0.0002	0.0000					
0.9	0.4066	0.3659	0.1647	0.0494	0.0111	0.0020	0.0003	0.0000					
1.0	0.3679	0.3679	0.1839	0.0613	0.0153	0.0031	0.0005	0.0001	0.0000				
1.1	0.3329	0.3662	0.2014	0.0738	0.0203	0.0045	0.0008	0.0001	0.0000				
1.2	0.3012	0.3614	0.2169	0.0867	0.0260	0.0062	0.0012	0.0002	0.0000				
1.3	0.2725	0.3543	0.2303	0.0998	0.0324	0.0084	0.0018	0.0003	0.0001	0.0000			
1.4	0.2466	0.3452	0.2417	0.1128	0.0395	0.0111	0.0026	0.0005	0.0001	0.0000			
1.5	0.2231	0.3347	0.2510	0.1255	0.0471	0.0141	0.0035	0.0008	0.0001	0.0000			
1.6	0.2019	0.3230	0.2584	0.1378	0.0551	0.0176	0.0047	0.0011	0.0002	0.0000			
1.7	0.1827	0.3106	0.2640	0.1496	0.0636	0.0216	0.0061	0.0015	0.0003	0.0001	0.0000		
1.8	0.1653	0.2975	0.2678	0.1607	0.0723	0.0260	0.0078	0.0020	0.0005	0.0001	0.0000		
1.9	0.1496	0.2842	0.2700	0.1710	0.0812	0.0309	0.0098	0.0027	0.0006	0.0001	0.0000		
2.0	0.1353	0.2707	0.2707	0.1804	0.0902	0.0361	0.0120	0.0034	0.0009	0.0002	0.0000		
2.2	0.1108	0.2438	0.2681	0.1966	0.1082	0.0476	0.0174	0.0055	0.0015	0.0004	0.0001	0.0000	
2.4	0.0907	0.2177	0.2613	0.2090	0.1254	0.0602	0.0241	0.0083	0.0025	0.0007	0.0002	0.0000	
2.6	0.0743	0.1931	0.2510	0.2176	0.1414	0.0735	0.0319	0.0118	0.0038	0.0011	0.0003	0.0001	0.0000
2.8	0.0608	0.1703	0.2384	0.2225	0.1557	0.0872	0.0407	0.0163	0.0057	0.0018	0.0005	0.0001	0.0000
3.0	0.0498	0.1494	0.2240	0.2240	0.1680	0.1008	0.0504	0.0216	0.0081	0.0027	0.0008	0.0002	0.0001

3.2	0.0408	0.1304	0.2087	0.2226	0.1781	0.1140	0.0608	0.0278	0.0111	0.0040	0.0013	0.0004	0.0001
3.4	0.0334	0.1135	0.1929	0.2186	0.1858	0.1264	0.0716	0.0348	0.0148	0.0056	0.0019	0.0005	0.0002
3.6	0.0273	0.0984	0.1771	0.2125	0.1912	0.1377	0.0826	0.0425	0.0191	0.0076	0.0028	0.0009	0.0003
3.8	0.0224	0.0850	0.1615	0.2046	0.1944	0.1477	0.0936	0.0508	0.0241	0.0102	0.0039	0.0013	0.0004
4.0	0.0183	0.0733	0.1465	0.1954	0.1954	0.1563	0.1042	0.0595	0.0298	0.0132	0.0053	0.0019	0.0006
5.0	0.0067	0.0337	0.0842	0.1404	0.1755	0.1755	0.1462	0.1044	0.0653	0.0363	0.0181	0.0082	0.0034
6.0	0.0025	0.0149	0.0446	0.0892	0.1339	0.1606	0.1606	0.1377	0.1033	0.0688	0.0413	0.0225	0.0113
7.0	0.0009	0.0064	0.0223	0.0521	0.0912	0.1277	0.1490	0.1490	0.1304	0.1014	0.0710	0.0452	0.0263
8.0	0.0003	0.0027	0.0107	0.0286	0.0573	0.0916	0.1221	0.1396	0.1396	0.1241	0.0993	0.0722	0.0481
9.0	0.0001	0.0011	0.0050	0.0150	0.0337	0.0607	0.0911	0.1171	0.1318	0.1318	0.1186	0.0970	0.0728
10.0	0.0000	0.0005	0.0023	0.0076	0.0189	0.0378	0.0631	0.0901	0.1126	0.1251	0.1251	0.1137	0.0948

LAMBDA	VALUES OF X											
	13	14	15	16	17	18	19	20	21	22	23	24
5.0	0.0013	0.0005	0.0002									
6.0	0.0052	0.0022	0.0009	0.0003	0.0001							
7.0	0.0142	0.0071	0.0033	0.0014	0.0006	0.0002	0.0001					
8.0	0.0296	0.0169	0.0090	0.0045	0.0021	0.0009	0.0004	0.0002	0.0001			
9.0	0.0504	0.0324	0.0194	0.0109	0.0058	0.0029	0.0014	0.0006	0.0003	0.0001	0.0000	
10.0	0.0729	0.0521	0.0347	0.0217	0.0128	0.0071	0.0037	0.0019	0.0009	0.0004	0.0002	0.0001

SINGLE PAYMENT PRESENT VALUE FACTORS

YEARS	ANNUAL INTEREST RATES						
	0.02	0.03	0.04	0.05	0.06	0.07	0.08
1	0.98039	0.97087	0.96154	0.95238	0.94340	0.93458	0.92593
2	0.96117	0.94260	0.92456	0.90703	0.89000	0.87344	0.85734
3	0.94232	0.91514	0.88900	0.86384	0.83962	0.81630	0.79383
4	0.92385	0.88849	0.85480	0.82270	0.79209	0.76290	0.73503
5	0.90573	0.86261	0.82193	0.78353	0.74726	0.71299	0.68058
6	0.88797	0.83748	0.79031	0.74622	0.70496	0.66634	0.63017
7	0.87056	0.81309	0.75992	0.71068	0.66506	0.62275	0.58349
8	0.85349	0.78941	0.73069	0.67684	0.62741	0.58201	0.54027
9	0.83676	0.76642	0.70259	0.64461	0.59190	0.54393	0.50025
10	0.82035	0.74409	0.67556	0.61391	0.55839	0.50835	0.46319
11	0.80426	0.72242	0.64958	0.58468	0.52679	0.47509	0.42888
12	0.78849	0.70138	0.62460	0.55684	0.49697	0.44401	0.39711
13	0.77303	0.68095	0.60057	0.53032	0.46884	0.41496	0.36770
14	0.75788	0.66112	0.57748	0.50507	0.44230	0.38782	0.34046
15	0.74301	0.64186	0.55526	0.48102	0.41727	0.36245	0.31524
16	0.72845	0.62317	0.53391	0.45811	0.39365	0.33873	0.29189
17	0.71416	0.60502	0.51337	0.43630	0.37136	0.31657	0.27027
18	0.70016	0.58739	0.49363	0.41552	0.35034	0.29586	0.25025
19	0.68643	0.57029	0.47464	0.39573	0.33051	0.27651	0.23171
20	0.67297	0.55368	0.45639	0.37689	0.31180	0.25842	0.21455
21	0.65978	0.53755	0.43883	0.35894	0.29416	0.24151	0.19866
22	0.64684	0.52189	0.42196	0.34185	0.27751	0.22571	0.18394
23	0.63416	0.50669	0.40573	0.32557	0.26180	0.21095	0.17032
24	0.62172	0.49193	0.39012	0.31007	0.24698	0.19715	0.15770
25	0.60953	0.47761	0.37512	0.29530	0.23300	0.18425	0.14602
26	0.59758	0.46369	0.36069	0.28124	0.21981	0.17220	0.13520
27	0.58586	0.45019	0.34682	0.26785	0.20737	0.16093	0.12519
28	0.57437	0.43708	0.33348	0.25509	0.19563	0.15040	0.11591
29	0.56311	0.42435	0.32065	0.24295	0.18456	0.14056	0.10733
30	0.55207	0.41199	0.30832	0.23138	0.17411	0.13137	0.09938
35	0.50003	0.35538	0.25342	0.18129	0.13011	0.09366	0.06763
40	0.45289	0.30656	0.20829	0.14205	0.09722	0.06678	0.04603
45	0.41020	0.26444	0.17120	0.11130	0.07265	0.04761	0.03133
50	0.37153	0.22811	0.14071	0.08720	0.05429	0.03395	0.02132
60	0.30478	0.16973	0.09506	0.05354	0.03031	0.01726	0.00988
70	0.25003	0.12630	0.06422	0.03287	0.01693	0.00877	0.00457
80	0.20511	0.09398	0.04338	0.02018	0.00945	0.00446	0.00212
90	0.16826	0.06993	0.02931	0.01239	0.00528	0.00227	0.00098
100	0.13803	0.05203	0.01980	0.00760	0.00295	0.00115	0.00045

SINGLE PAYMENT PRESENT VALUE FACTORS
ANNUAL INTEREST RATES

YEARS	0.09	0.10	0.11	0.12	0.13	0.14	0.15
1	0.91743	0.90909	0.90090	0.89286	0.88496	0.87719	0.86957
2	0.84168	0.82645	0.81162	0.79719	0.78315	0.76947	0.75614
3	0.77218	0.75131	0.73119	0.71178	0.69305	0.67497	0.65752
4	0.70843	0.68301	0.65873	0.63552	0.61332	0.59208	0.57175
5	0.64993	0.62092	0.59345	0.56743	0.54276	0.51937	0.49718
6	0.59627	0.56447	0.53464	0.50663	0.48032	0.45559	0.43233
7	0.54703	0.51316	0.48166	0.45235	0.42506	0.39964	0.37594
8	0.50187	0.46651	0.43393	0.40388	0.37616	0.35056	0.32690
9	0.46043	0.42410	0.39092	0.36061	0.33288	0.30751	0.28426
10	0.42241	0.38554	0.35218	0.32197	0.29459	0.26974	0.24718
11	0.38753	J.35049	0.31728	0.28748	0.26070	0.23662	0.21494
12	0.35553	0.31863	0.28584	0.25668	0.23071	0.20756	0.18691
13	0.32618	0.28966	0.25751	0.22917	0.20416	0.18207	0.16253
14	0.29925	0.26333	0.23199	0.20462	0.18068	0.15971	0.14133
15	0.27454	0.23939	0.20900	0.18270	0.15989	0.14010	0.12289
16	0.25187	0.21763	0.18829	0.16312	0.14150	0.12289	0.10686
17	0.23107	0.19784	0.16963	0.14564	0.12522	0.10780	0.09293
18	0.21199	0.17986	0.15282	0.13004	0.11081	0.09456	0.08081
19	0.19449	0.16351	0.13768	0.11611	0.09806	0.08295	0.07027
20	0.17843	0.14864	0.12403	0.10367	0.08678	0.07276	0.06110
21	0.16370	0.13513	0.11174	0.09256	0.07680	0.06383	0.05313
22	0.15018	0.12285	0.10067	0.08264	0.06796	0.05599	0.04620
23	0.13778	0.11168	0.09069	0.07379	0.06014	0.04911	0.04017
24	0.12640	0.10153	0.08170	0.06588	0.05323	0.04308	0.03493
25	0.11597	0.09230	0.07361	0.05882	0.04710	0.03779	0.03038
26	0.10639	0.08391	0.06631	0.05252	0.04168	0.03315	0.02642
27	0.09761	0.07628	0.05974	0.04689	0.03689	0.02908	0.02297
28	0.08955	0.06934	0.05382	0.04187	0.03264	0.02551	0.01997
29	0.08215	0.06304	0.04849	0.03738	0.02889	0.02237	0.01737
30	0.07537	0.05731	0.04368	0.03338	0.02557	0.01963	0.01510
35	0.04899	0.03558	0.02592	0.01894	0.01388	0.01019	0.00751
40	0.03184	0.02209	0.01538	0.01075	0.00753	0.00529	0.00373
45	0.02069	0.01372	0.00913	0.00610	0.00409	0.00275	0.00186
50	0.01345	0.00852	0.00542	0.00346	0.00222	0.00143	0.00092
60	0.00568	0.00328	0.00191	0.00111	0.00065	0.00039	0.00023
70	0.00240	0.00127	0.00067	0.00036	0.00019	0.00010	0.00006
80	0.00101	0.00049	0.00024	0.00012	0.00006	0.00003	0.00001
90	0.00043	0.00019	0.00008	0.00004	0.00002	0.00001	0.00000
100	0.00018	0.00007	0.00003	0.00001	0.00000	0.00000	0.00000

SINGLE PAYMENT PRESENT VALUE FACTORS

ANNUAL INTEREST RATES

YEARS	0.16	0.17	0.18	0.19	0.20	0.21	0.22
1	0.86207	0.85470	0.84746	0.84034	0.83333	0.82645	0.81967
2	0.74316	0.73051	0.71818	0.70616	0.69444	0.68301	0.67186
3	0.64066	0.62437	0.60863	0.59342	0.57870	0.56447	0.55071
4	0.55229	0.53365	0.51579	0.49867	0.48225	0.46651	0.45140
5	0.47611	0.45611	0.43711	0.41905	0.40188	0.38554	0.37000
6	0.41044	0.38984	0.37043	0.35214	0.33490	0.31863	0.30328
7	0.35383	0.33320	0.31393	0.29592	0.27908	0.26333	0.24859
8	0.30503	0.28478	0.26604	0.24867	0.23257	0.21763	0.20376
9	0.26295	0.24340	0.22546	0.20897	0.19381	0.17986	0.16702
10	0.22668	0.20804	0.19106	0.17560	0.16151	0.14864	0.13690
11	0.19542	0.17781	0.16192	0.14757	0.13459	0.12285	0.11221
12	0.16846	0.15197	0.13722	0.12400	0.11216	0.10153	0.09198
13	0.14523	0.12989	0.11629	0.10421	0.09346	0.08391	0.07539
14	0.12520	0.11102	0.09855	0.08757	0.07789	0.06934	0.06180
15	0.10793	0.09489	0.08352	0.07359	0.06491	0.05731	0.05065
16	0.09304	0.08110	0.07078	0.06184	0.05409	0.04736	0.04152
17	0.08021	0.06932	0.05998	0.05196	0.04507	0.03914	0.03403
18	0.06914	0.05925	0.05083	0.04367	0.03756	0.03235	0.02789
19	0.05961	0.05064	0.04308	0.03670	0.03130	0.02673	0.02286
20	0.05139	0.04328	0.03651	0.03084	0.02608	0.02209	0.01874
21	0.04430	0.03699	0.03094	0.02591	0.02174	0.01826	0.01536
22	0.03819	0.03162	0.02622	0.02178	0.01811	0.01509	0.01259
23	0.03292	0.02702	0.02222	0.01830	0.01509	0.01247	0.01032
24	0.02838	0.02310	0.01883	0.01538	0.01258	0.01031	0.00846
25	0.02447	0.01974	0.01596	0.01292	0.01048	0.00852	0.00693
26	0.02109	0.01687	0.01352	0.01086	0.00874	0.00704	0.00568
27	0.01818	0.01442	0.01146	0.00912	0.00728	0.00582	0.00466
28	0.01567	0.01233	0.00971	0.00767	0.00607	0.00481	0.00382
29	0.01351	0.01053	0.00823	0.00644	0.00506	0.00397	0.00313
30	0.01165	0.00900	0.00697	0.00541	0.00421	0.00328	0.00257
35	0.00555	0.00411	0.00305	0.00227	0.00169	0.00127	0.00095
40	0.00264	0.00187	0.00133	0.00095	0.00068	0.00049	0.00035
45	0.00126	0.00085	0.00058	0.00040	0.00027	0.00019	0.00013
50	0.00060	0.00039	0.00025	0.00017	0.00011	0.00007	0.00005
60	0.00014	0.00008	0.00005	0.00003	0.00002	0.00001	0.00001
70	0.00003	0.00002	0.00001	0.00001	0.00000	0.00000	0.00000
80	0.00001	0.00000	0.00000	0.00000	0.00000	0.00000	0.00000

SINGLE PAYMENT PRESENT VALUE FACTORS

ANNUAL INTEREST RATES

YEARS	0.23	0.24	0.25	0.26	0.27	0.28	0.29
1	0.81301	0.80645	0.80000	0.79365	0.78740	0.78125	0.77519
2	0.66098	0.65036	0.64000	0.62988	0.62000	0.61035	0.60093
3	0.53738	0.52449	0.51200	0.49991	0.48819	0.47684	0.46583
4	0.43690	0.42297	0.40960	0.39675	0.38440	0.37253	0.36111
5	0.35520	0.34111	0.32768	0.31488	0.30268	0.29104	0.27993
6	0.28878	0.27509	0.26214	0.24991	0.23833	0.22737	0.21700
7	0.23478	0.22184	0.20972	0.19834	0.18766	0.17764	0.16822
8	0.19088	0.17891	0.16777	0.15741	0.14776	0.13878	0.13040
9	0.15519	0.14428	0.13422	0.12493	0.11635	0.10842	0.10109
10	0.12617	0.11635	0.10737	0.09915	0.09161	0.08470	0.07836
11	0.10258	0.09383	0.08590	0.07869	0.07214	0.06617	0.06075
12	0.08339	0.07567	0.06872	0.06245	0.05680	0.05170	0.04709
13	0.06780	0.06103	0.05498	0.04957	0.04473	0.04039	0.03650
14	0.05512	0.04921	0.04398	0.03934	0.03522	0.03155	0.02830
15	0.04481	0.03969	0.03518	0.03122	0.02773	0.02465	0.02194
16	0.03643	0.03201	0.02815	0.02478	0.02183	0.01926	0.01700
17	0.02962	0.02581	0.02252	0.01967	0.01719	0.01505	0.01318
18	0.02408	0.02082	0.01801	0.01561	0.01354	0.01175	0.01022
19	0.01958	0.01679	0.01441	0.01239	0.01066	0.00918	0.00792
20	0.01592	0.01354	0.01153	0.00983	0.00839	0.00717	0.00614
21	0.01294	0.01092	0.00922	0.00780	0.00661	0.00561	0.00476
22	0.01052	0.00880	0.00738	0.00619	0.00520	0.00438	0.00369
23	0.00855	0.00710	0.00590	0.00491	0.00410	0.00342	0.00286
24	0.00695	0.00573	0.00472	0.00390	0.00323	0.00267	0.00222
25	0.00565	0.00462	0.00378	0.00310	0.00254	0.00209	0.00172
26	0.00460	0.00372	0.00302	0.00246	0.00200	0.00163	0.00133
27	0.00374	0.00300	0.00242	0.00195	0.00158	0.00127	0.00103
28	0.00304	0.00242	0.00193	0.00155	0.00124	0.00100	0.00080
29	0.00247	0.00195	0.00155	0.00123	0.00098	0.00078	0.00062
30	0.00201	0.00158	0.00124	0.00097	0.00077	0.00061	0.00048
35	0.00071	0.00054	0.00041	0.00031	0.00023	0.00018	0.00013
40	0.00025	0.00018	0.00013	0.00010	0.00007	0.00005	0.00004
45	0.00009	0.00006	0.00004	0.00003	0.00002	0.00001	0.00001
50	0.00003	0.00002	0.00001	0.00001	0.00001	0.00000	0.00000

SINGLE PAYMENT PRESENT VALUE FACTORS
ANNUAL INTEREST RATES

YEARS	0.30	0.31	0.32	0.33	0.34	0.35	0.36
1	0.76923	0.76336	0.75758	0.75188	0.74627	0.74074	0.73529
2	0.59172	0.58272	0.57392	0.56532	0.55692	0.54870	0.54066
3	0.45517	0.44482	0.43479	0.42505	0.41561	0.40644	0.39754
4	0.35013	0.33956	0.32939	0.31959	0.31016	0.30107	0.29231
5	0.26933	0.25921	0.24953	0.24029	0.23146	0.22301	0.21493
6	0.20718	0.19787	0.18904	0.18067	0.17273	0.16520	0.15804
7	0.15937	0.15104	0.14321	0.13584	0.12890	0.12237	0.11621
8	0.12259	0.11530	0.10849	0.10214	0.09620	0.09064	0.08545
9	0.09430	0.08802	0.08219	0.07680	0.07179	0.06714	0.06283
10	0.07254	0.06719	0.06227	0.05774	0.05357	0.04974	0.04620
11	0.05580	0.05129	0.04717	0.04341	0.03998	0.03684	0.03397
12	0.04292	0.03915	0.03574	0.03264	0.02984	0.02729	0.02498
13	0.03302	0.02989	0.02707	0.02454	0.02227	0.02021	0.01837
14	0.02540	0.02281	0.02051	0.01845	0.01662	0.01497	0.01350
15	0.01954	0.01742	0.01554	0.01387	0.01240	0.01109	0.00993
16	0.01503	0.01329	0.01177	0.01043	0.00925	0.00822	0.00730
17	0.01156	0.01015	0.00892	0.00784	0.00691	0.00609	0.00537
18	0.00889	0.00775	0.00676	0.00590	0.00515	0.00451	0.00395
19	0.00684	0.00591	0.00512	0.00443	0.00385	0.00334	0.00290
20	0.00526	0.00451	0.00388	0.00333	0.00287	0.00247	0.00213
21	0.00405	0.00345	0.00294	0.00251	0.00214	0.00183	0.00157
22	0.00311	0.00263	0.00223	0.00188	0.00160	0.00136	0.00115
23	0.00239	0.00201	0.00169	0.00142	0.00119	0.00101	0.00085
24	0.00184	0.00153	0.00128	0.00107	0.00089	0.00074	0.00062
25	0.00142	0.00117	0.00097	0.00080	0.00066	0.00055	0.00046
26	0.00109	0.00089	0.00073	0.00060	0.00050	0.00041	0.00034
27	0.00084	0.00068	0.00056	0.00045	0.00037	0.00030	0.00025
28	0.00065	0.00052	0.00042	0.00034	0.00028	0.00022	0.00018
29	0.00050	0.00040	0.00032	0.00026	0.00021	0.00017	0.00013
30	0.00038	0.00030	0.00024	0.00019	0.00015	0.00012	0.00010
35	0.00010	0.00008	0.00006	0.00005	0.00004	0.00003	0.00002
40	0.00003	0.00002	0.00002	0.00001	0.00001	0.00001	0.00000
45	0.00001	0.00001	0.00000	0.00000	0.00000	0.00000	0.00000

SINGLE PAYMENT PRESENT VALUE FACTORS

ANNUAL INTEREST RATES

YEARS	0.37	0.38	0.39	0.40	0.41	0.42	0.43
1	0.72993	0.72464	0.71942	0.71429	0.70922	0.70423	0.69930
2	0.53279	0.52510	0.51757	0.51020	0.50299	0.49593	0.48902
3	0.38890	0.38051	0.37235	0.36443	0.35673	0.34925	0.34197
4	0.28387	0.27573	0.26788	0.26031	0.25300	0.24595	0.23914
5	0.20720	0.19980	0.19272	0.18593	0.17943	0.17320	0.16723
6	0.15124	0.14479	0.13865	0.13281	0.12726	0.12197	0.11695
7	0.11040	0.10492	0.09975	0.09486	0.09025	0.08590	0.08178
8	0.08058	0.07603	0.07176	0.06776	0.06401	0.06049	0.05719
9	0.05882	0.05509	0.05163	0.04840	0.04540	0.04260	0.03999
10	0.04293	0.03992	0.03714	0.03457	0.03220	0.03000	0.02797
11	0.03134	0.02893	0.02672	0.02469	0.02283	0.02113	0.01956
12	0.02287	0.02096	0.01922	0.01764	0.01619	0.01488	0.01368
13	0.01670	0.01519	0.01383	0.01260	0.01149	0.01048	0.00956
14	0.01219	0.01101	0.00995	0.00900	0.00815	0.00738	0.00669
15	0.00890	0.00798	0.00716	0.00643	0.00578	0.00520	0.00468
16	0.00649	0.00578	0.00515	0.00459	0.00410	0.00366	0.00327
17	0.00474	0.00419	0.00370	0.00328	0.00291	0.00258	0.00229
18	0.00346	0.00304	0.00267	0.00234	0.00206	0.00181	0.00160
19	0.00253	0.00220	0.00192	0.00167	0.00146	0.00128	0.00112
20	0.00184	0.00159	0.00138	0.00120	0.00104	0.00090	0.00078
21	0.00135	0.00115	0.00099	0.00085	0.00074	0.00063	0.00055
22	0.00098	0.00084	0.00071	0.00061	0.00052	0.00045	0.00038
23	0.00072	0.00061	0.00051	0.00044	0.00037	0.00031	0.00027
24	0.00052	0.00044	0.00037	0.00031	0.00026	0.00022	0.00019
25	0.00038	0.00032	0.00027	0.00022	0.00019	0.00016	0.00013
26	0.00028	0.00023	0.00019	0.00016	0.00013	0.00011	0.00009
27	0.00020	0.00017	0.00014	0.00011	0.00009	0.00008	0.00006
28	0.00015	0.00012	0.00010	0.00008	0.00007	0.00005	0.00004
29	0.00011	0.00009	0.00007	0.00006	0.00005	0.00004	0.00003
30	0.00008	0.00006	0.00005	0.00004	0.00003	0.00003	0.00002
35	0.00002	0.00001	0.00001	0.00001	0.00001	0.00000	0.00000

SINGLE PAYMENT PRESENT VALUE FACTORS
ANNUAL INTEREST RATES

YEARS	0.44	0.45	0.46	0.47	0.48	0.49	0.50
1	0.69444	0.68966	0.68493	0.68027	0.67568	0.67114	0.66667
2	0.48225	0.47562	0.46913	0.46277	0.45654	0.45043	0.44444
3	0.33490	0.32802	0.32132	0.31481	0.30847	0.30230	0.29630
4	0.23257	0.22622	0.22008	0.21416	0.20843	0.20289	0.19753
5	0.16151	0.15601	0.15074	0.14568	0.14083	0.13617	0.13169
6	0.11216	0.10759	0.10325	0.09911	0.09515	0.09139	0.08779
7	0.07789	0.07420	0.07072	0.06742	0.06429	0.06133	0.05853
8	0.05409	0.05117	0.04844	0.04586	0.04344	0.04116	0.03902
9	0.03756	0.03529	0.03318	0.03120	0.02935	0.02763	0.02601
10	0.02608	0.02434	0.02272	0.02122	0.01983	0.01854	0.01734
11	0.01811	0.01679	0.01556	0.01444	0.01340	0.01244	0.01156
12	0.01258	0.01158	0.01066	0.00982	0.00905	0.00835	0.00771
13	0.00874	0.00798	0.00730	0.00668	0.00612	0.00561	0.00514
14	0.00607	0.00551	0.00500	0.00455	0.00413	0.00376	0.00343
15	0.00421	0.00380	0.00343	0.00309	0.00279	0.00252	0.00228
16	0.00293	0.00262	0.00235	0.00210	0.00189	0.00169	0.00152
17	0.00203	0.00181	0.00161	0.00143	0.00128	0.00114	0.00101
18	0.00141	0.00125	0.00110	0.00097	0.00086	0.00076	0.00068
19	0.00098	0.00086	0.00075	0.00066	0.00058	0.00051	0.00045
20	0.00068	0.00059	0.00052	0.00045	0.00039	0.00034	0.00030
21	0.00047	0.00041	0.00035	0.00031	0.00027	0.00023	0.00020
22	0.00033	0.00028	0.00024	0.00021	0.00018	0.00015	0.00013
23	0.00023	0.00019	0.00017	0.00014	0.00012	0.00010	0.00009
24	0.00016	0.00013	0.00011	0.00010	0.00008	0.00007	0.00006
25	0.00011	0.00009	0.00008	0.00007	0.00006	0.00005	0.00004
26	0.00008	0.00006	0.00005	0.00004	0.00004	0.00003	0.00003
27	0.00005	0.00004	0.00004	0.00003	0.00003	0.00002	0.00002
28	0.00004	0.00003	0.00003	0.00002	0.00002	0.00001	0.00001
29	0.00003	0.00002	0.00002	0.00001	0.00001	0.00001	0.00001
30	0.00002	0.00001	0.00001	0.00001	0.00001	0.00001	0.00001

SINGLE PAYMENT PRESENT VALUE FACTORS
CONTINUOUS INTEREST RATES

YEARS	0.02	0.03	0.04	0.05	0.06	0.07	0.08
1	0.98020	0.97045	0.96079	0.95123	0.94176	0.93239	0.92312
2	0.96079	0.94176	0.92312	0.90484	0.88692	0.86936	0.85214
3	0.94176	0.91393	0.88692	0.86071	0.83527	0.81058	0.78663
4	0.92312	0.88692	0.85214	0.81873	0.78663	0.75578	0.72615
5	0.90484	0.86071	0.81873	0.77880	0.74082	0.70469	0.67032
6	0.88692	0.83527	0.78663	0.74082	0.69768	0.65705	0.61878
7	0.86936	0.81058	0.75578	0.70469	0.65705	0.61263	0.57121
8	0.85214	0.78663	0.72615	0.67032	0.61878	0.57121	0.52729
9	0.83527	0.76338	0.69768	0.63763	0.58275	0.53259	0.48675
10	0.81873	0.74082	0.67032	0.60653	0.54881	0.49659	0.44933
11	0.80252	0.71892	0.64404	0.57695	0.51685	0.46301	0.41478
12	0.78663	0.69768	0.61878	0.54881	0.48675	0.43171	0.38289
13	0.77105	0.67706	0.59452	0.52205	0.45841	0.40252	0.35345
14	0.75578	0.65705	0.57121	0.49659	0.43171	0.37531	0.32628
15	0.74082	0.63763	0.54881	0.47237	0.40657	0.34994	0.30119
16	0.72615	0.61878	0.52729	0.44933	0.38289	0.32628	0.27804
17	0.71177	0.60050	0.50662	0.42741	0.36059	0.30422	0.25666
18	0.69768	0.58275	0.48675	0.40657	0.33960	0.28365	0.23693
19	0.68386	0.56553	0.46767	0.38674	0.31982	0.26448	0.21871
20	0.67032	0.54881	0.44933	0.36788	0.30119	0.24660	0.20190
21	0.65705	0.53259	0.43171	0.34994	0.28365	0.22993	0.18637
22	0.64404	0.51685	0.41478	0.33287	0.26714	0.21438	0.17204
23	0.63128	0.50158	0.39852	0.31664	0.25158	0.19989	0.15882
24	0.61878	0.48675	0.38289	0.30119	0.23693	0.18637	0.14661
25	0.60653	0.47237	0.36788	0.28650	0.22313	0.17377	0.13534
26	0.59452	0.45841	0.35345	0.27253	0.21014	0.16203	0.12493
27	0.58275	0.44486	0.33960	0.25924	0.19790	0.15107	0.11533
28	0.57121	0.43171	0.32628	0.24660	0.18637	0.14086	0.10646
29	0.55990	0.41895	0.31349	0.23457	0.17552	0.13134	0.09827
30	0.54881	0.40657	0.30119	0.22313	0.16530	0.12246	0.09072
35	0.49659	0.34994	0.24660	0.17377	0.12246	0.08629	0.06081
40	0.44933	0.30119	0.20190	0.13534	0.09072	0.06081	0.04076
45	0.40657	0.25924	0.16530	0.10540	0.06721	0.04285	0.02732
50	0.36788	0.22313	0.13534	0.08208	0.04979	0.03020	0.01832
60	0.30119	0.16530	0.09072	0.04979	0.02732	0.01500	0.00823
70	0.24660	0.12246	0.06081	0.03020	0.01500	0.00745	0.00370
80	0.20190	0.09072	0.04076	0.01832	0.00823	0.00370	0.00166
90	0.16530	0.06721	0.02732	0.01111	0.00452	0.00184	0.00075
100	0.13534	0.04979	0.01832	0.00674	0.00248	0.00091	0.00034

SINGLE PAYMENT PRESENT VALUE FACTORS
CONTINUOUS INTEREST RATES

YEARS	0.09	0.10	0.11	0.12	0.13	0.14	0.15
1	0.91393	0.90484	0.89583	0.88692	0.87810	0.86936	0.86071
2	0.83527	0.81873	0.80252	0.78663	0.77105	0.75578	0.74082
3	0.76338	0.74082	0.71892	0.69768	0.67706	0.65705	0.63763
4	0.69768	0.67032	0.64404	0.61878	0.59452	0.57121	0.54881
5	0.63763	0.60653	0.57695	0.54881	0.52205	0.49659	0.47237
6	0.58275	0.54881	0.51685	0.48675	0.45841	0.43171	0.40657
7	0.53259	0.49659	0.46301	0.43171	0.40252	0.37531	0.34994
8	0.48675	0.44933	0.41478	0.38289	0.35345	0.32628	0.30119
9	0.44486	0.40657	0.37158	0.33960	0.31037	0.28365	0.25924
10	0.40657	0.36788	0.33287	0.30119	0.27253	0.24660	0.22313
11	0.37158	0.33287	0.29820	0.26714	0.23931	0.21438	0.19205
12	0.33960	0.30119	0.26714	0.23693	0.21014	0.18637	0.16530
13	0.31037	0.27253	0.23931	0.21014	0.18452	0.16203	0.14227
14	0.28365	0.24660	0.21438	0.18637	0.16203	0.14086	0.12246
15	0.25924	0.22313	0.19205	0.16530	0.14227	0.12246	0.10540
16	0.23693	0.20190	0.17204	0.14661	0.12493	0.10646	0.09072
17	0.21654	0.18268	0.15412	0.13003	0.10970	0.09255	0.07808
18	0.19790	0.16530	0.13807	0.11533	0.09633	0.08046	0.06721
19	0.18087	0.14957	0.12369	0.10228	0.08458	0.06995	0.05784
20	0.16530	0.13534	0.11080	0.09072	0.07427	0.06081	0.04979
21	0.15107	0.12246	0.09926	0.08046	0.06522	0.05287	0.04285
22	0.13807	0.11080	0.08892	0.07136	0.05727	0.04596	0.03688
23	0.12619	0.10026	0.07966	0.06329	0.05029	0.03996	0.03175
24	0.11533	0.09072	0.07136	0.05613	0.04416	0.03474	0.02732
25	0.10540	0.08208	0.06393	0.04979	0.03877	0.03020	0.02352
26	0.09633	0.07427	0.05727	0.04416	0.03405	0.02625	0.02024
27	0.08804	0.06721	0.05130	0.03916	0.02990	0.02282	0.01742
28	0.08046	0.06081	0.04596	0.03474	0.02625	0.01984	0.01500
29	0.07353	0.05502	0.04117	0.03081	0.02305	0.01725	0.01291
30	0.06721	0.04979	0.03688	0.02732	0.02024	0.01500	0.01111
35	0.04285	0.03020	0.02128	0.01500	0.01057	0.00745	0.00525
40	0.02732	0.01832	0.01228	0.00823	0.00552	0.00370	0.00248
45	0.01742	0.01111	0.00708	0.00452	0.00288	0.00184	0.00117
50	0.01111	0.00674	0.00409	0.00248	0.00150	0.00091	0.00055
60	0.00452	0.00248	0.00136	0.00075	0.00041	0.00022	0.00012
70	0.00184	0.00091	0.00045	0.00022	0.00011	0.00006	0.00003
80	0.00075	0.00034	0.00015	0.00007	0.00003	0.00001	0.00001
90	0.00030	0.00012	0.00005	0.00002	0.00001	0.00000	0.00000
100	0.00012	0.00005	0.00002	0.00001	0.00000	0.00000	0.00000

SINGLE PAYMENT PRESENT VALUE FACTORS
CONTINUOUS INTEREST RATES

YEARS	0.16	0.17	0.18	0.19	0.20	0.21	0.22
1	0.85214	0.84366	0.83527	0.82696	0.81873	0.81058	0.80252
2	0.72615	0.71177	0.69768	0.68386	0.67032	0.65705	0.64404
3	0.61878	0.60050	0.58275	0.56553	0.54881	0.53259	0.51685
4	0.52729	0.50662	0.48675	0.46767	0.44933	0.43171	0.41478
5	0.44933	0.42741	0.40657	0.38674	0.36788	0.34994	0.33287
6	0.38289	0.36059	0.33960	0.31982	0.30119	0.28365	0.26714
7	0.32628	0.30422	0.28365	0.26448	0.24660	0.22993	0.21438
8	0.27804	0.25666	0.23693	0.21871	0.20190	0.18637	0.17204
9	0.23693	0.21654	0.19790	0.18087	0.16530	0.15107	0.13807
10	0.20190	0.18268	0.16530	0.14957	0.13534	0.12246	0.11080
11	0.17204	0.15412	0.13807	0.12369	0.11080	0.09926	0.08892
12	0.14661	0.13003	0.11533	0.10228	0.09072	0.08046	0.07136
13	0.12493	0.10970	0.09633	0.08458	0.07427	0.06522	0.05727
14	0.10646	0.09255	0.08046	0.06995	0.06081	0.05287	0.04596
15	0.09072	0.07808	0.06721	0.05784	0.04979	0.04285	0.03688
16	0.07730	0.06587	0.05613	0.04783	0.04076	0.03474	0.02960
17	0.06587	0.05558	0.04689	0.03956	0.03337	0.02816	0.02375
18	0.05613	0.04689	0.03916	0.03271	0.02732	0.02282	0.01906
19	0.04783	0.03956	0.03271	0.02705	0.02237	0.01850	0.01530
20	0.04076	0.03337	0.02732	0.02237	0.01832	0.01500	0.01228
21	0.03474	0.02816	0.02282	0.01850	0.01500	0.01216	0.00985
22	0.02960	0.02375	0.01906	0.01530	0.01228	0.00985	0.00791
23	0.02522	0.02004	0.01592	0.01265	0.01005	0.00799	0.00635
24	0.02149	0.01691	0.01330	0.01046	0.00823	0.00647	0.00509
25	0.01832	0.01426	0.01111	0.00865	0.00674	0.00525	0.00409
26	0.01561	0.01203	0.00928	0.00715	0.00552	0.00425	0.00328
27	0.01330	0.01015	0.00775	0.00592	0.00452	0.00345	0.00263
28	0.01133	0.00857	0.00647	0.00489	0.00370	0.00279	0.00211
29	0.00966	0.00723	0.00541	0.00405	0.00303	0.00227	0.00170
30	0.00823	0.00610	0.00452	0.00335	0.00248	0.00184	0.00136
35	0.00370	0.00261	0.00184	0.00129	0.00091	0.00064	0.00045
40	0.00166	0.00111	0.00075	0.00050	0.00034	0.00022	0.00015
45	0.00075	0.00048	0.00030	0.00019	0.00012	0.00008	0.00005
50	0.00034	0.00020	0.00012	0.00007	0.00005	0.00003	0.00002
60	0.00007	0.00004	0.00002	0.00001	0.00001	0.00000	0.00000
70	0.00001	0.00001	0.00000	0.00000	0.00000	0.00000	0.00000

SINGLE PAYMENT PRESENT VALUE FACTORS
CONTINUOUS INTEREST RATES

YEARS	0.23	0.24	0.25	0.26	0.27	0.28	0.29
1	0.79453	0.78663	0.77880	0.77105	0.76338	0.75578	0.74826
2	0.63128	0.61878	0.60653	0.59452	0.58275	0.57121	0.55990
3	0.50158	0.48675	0.47237	0.45841	0.44486	0.43171	0.41895
4	0.39852	0.38289	0.36788	0.35345	0.33960	0.32628	0.31349
5	0.31664	0.30119	0.28650	0.27253	0.25924	0.24660	0.23457
6	0.25158	0.23693	0.22313	0.21014	0.19790	0.18637	0.17552
7	0.19989	0.18637	0.17377	0.16203	0.15107	0.14086	0.13134
8	0.15882	0.14661	0.13534	0.12493	0.11533	0.10646	0.09827
9	0.12619	0.11533	0.10540	0.09633	0.08804	0.08046	0.07353
10	0.10026	0.09072	0.08208	0.07427	0.06721	0.06081	0.05502
11	0.07966	0.07136	0.06393	0.05727	0.05130	0.04596	0.04117
12	0.06329	0.05613	0.04979	0.04416	0.03916	0.03474	0.03081
13	0.05029	0.04416	0.03877	0.03405	0.02990	0.02625	0.02305
14	0.03996	0.03474	0.03020	0.02625	0.02282	0.01984	0.01725
15	0.03175	0.02732	0.02352	0.02024	0.01742	0.01500	0.01291
16	0.02522	0.02149	0.01832	0.01561	0.01330	0.01133	0.00966
17	0.02004	0.01691	0.01426	0.01203	0.01015	0.00857	0.00723
18	0.01592	0.01330	0.01111	0.00928	0.00775	0.00647	0.00541
19	0.01265	0.01046	0.00865	0.00715	0.00592	0.00489	0.00405
20	0.01005	0.00823	0.00674	0.00552	0.00452	0.00370	0.00303
21	0.00799	0.00647	0.00525	0.00425	0.00345	0.00279	0.00227
22	0.00635	0.00509	0.00409	0.00328	0.00263	0.00211	0.00170
23	0.00504	0.00401	0.00318	0.00253	0.00201	0.00160	0.00127
24	0.00401	0.00315	0.00248	0.00195	0.00153	0.00121	0.00095
25	0.00318	0.00248	0.00193	0.00150	0.00117	0.00091	0.00071
26	0.00253	0.00195	0.00150	0.00116	0.00089	0.00069	0.00053
27	0.00201	0.00153	0.00117	0.00089	0.00068	0.00052	0.00040
28	0.00160	0.00121	0.00091	0.00069	0.00052	0.00039	0.00030
29	0.00127	0.00095	0.00071	0.00053	0.00040	0.00030	0.00022
30	0.00101	0.00075	0.00055	0.00041	0.00030	0.00022	0.00017
35	0.00032	0.00022	0.00016	0.00011	0.00008	0.00006	0.00004
40	0.00010	0.00007	0.00005	0.00003	0.00002	0.00001	0.00001
45	0.00003	0.00002	0.00001	0.00001	0.00001	0.00000	0.00000
50	0.00001	0.00001	0.00000	0.00000	0.00000	0.00000	0.00000

SINGLE PAYMENT PRESENT VALUE FACTORS
CONTINUOUS INTEREST RATES

YEARS	0.30	0.31	0.32	0.33	0.34	0.35	0.36
1	0.74082	0.73345	0.72615	0.71892	0.71177	0.70469	0.69768
2	0.54881	0.53794	0.52729	0.51685	0.50662	0.49659	0.48675
3	0.40657	0.39455	0.38289	0.37158	0.36059	0.34994	0.33960
4	0.30119	0.28938	0.27804	0.26714	0.25666	0.24660	0.23693
5	0.22313	0.21225	0.20190	0.19205	0.18268	0.17377	0.16530
6	0.16530	0.15567	0.14661	0.13807	0.13003	0.12246	0.11533
7	0.12246	0.11418	0.10646	0.09926	0.09255	0.08629	0.08046
8	0.09072	0.08374	0.07730	0.07136	0.06587	0.06081	0.05613
9	0.06721	0.06142	0.05613	0.05130	0.04689	0.04285	0.03916
10	0.04979	0.04505	0.04076	0.03688	0.03337	0.03020	0.02732
11	0.03688	0.03304	0.02960	0.02652	0.02375	0.02128	0.01906
12	0.02732	0.02423	0.02149	0.01906	0.01691	0.01500	0.01330
13	0.02024	0.01777	0.01561	0.01370	0.01203	0.01057	0.00928
14	0.01500	0.01304	0.01133	0.00985	0.00857	0.00745	0.00647
15	0.01111	0.00956	0.00823	0.00708	0.00610	0.00525	0.00452
16	0.00823	0.00701	0.00598	0.00509	0.00434	0.00370	0.00315
17	0.00610	0.00514	0.00434	0.00366	0.00309	0.00261	0.00220
18	0.00452	0.00377	0.00315	0.00263	0.00220	0.00184	0.00153
19	0.00335	0.00277	0.00229	0.00189	0.00156	0.00129	0.00107
20	0.00248	0.00203	0.00166	0.00136	0.00111	0.00091	0.00075
21	0.00184	0.00149	0.00121	0.00098	0.00079	0.00064	0.00052
22	0.00136	0.00109	0.00088	0.00070	0.00056	0.00045	0.00036
23	0.00101	0.00080	0.00064	0.00051	0.00040	0.00032	0.00025
24	0.00075	0.00059	0.00046	0.00036	0.00029	0.00022	0.00018
25	0.00055	0.00043	0.00034	0.00026	0.00020	0.00016	0.00012
26	0.00041	0.00032	0.00024	0.00019	0.00014	0.00011	0.00009
27	0.00030	0.00023	0.00018	0.00014	0.00010	0.00008	0.00006
28	0.00022	0.00017	0.00013	0.00010	0.00007	0.00006	0.00004
29	0.00017	0.00012	0.00009	0.00007	0.00005	0.00004	0.00003
30	0.00012	0.00009	0.00007	0.00005	0.00004	0.00003	0.00002
35	0.00003	0.00002	0.00001	0.00001	0.00001	0.00000	0.00000
40	0.00001	0.00000	0.00000	0.00000	0.00000	0.00000	0.00000

SINGLE PAYMENT PRESENT VALUE FACTORS
CONTINUOUS INTEREST RATES

YEARS	0.37	0.38	0.39	0.40	0.41	0.42	0.43
1	0.69073	0.68386	0.67706	0.67032	0.66365	0.65705	0.65051
2	0.47711	0.46767	0.45841	0.44933	0.44043	0.43171	0.42316
3	0.32956	0.31982	0.31037	0.30119	0.29229	0.28365	0.27527
4	0.22764	0.21871	0.21014	0.20190	0.19398	0.18637	0.17907
5	0.15724	0.14957	0.14227	0.13534	0.12873	0.12246	0.11648
6	0.10861	0.10228	0.09633	0.09072	0.08543	0.08046	0.07577
7	0.07502	0.06995	0.06522	0.06081	0.05670	0.05287	0.04929
8	0.05182	0.04783	0.04416	0.04076	0.03763	0.03474	0.03206
9	0.03579	0.03271	0.02990	0.02732	0.02497	0.02282	0.02086
10	0.02472	0.02237	0.02024	0.01832	0.01657	0.01500	0.01357
11	0.01708	0.01530	0.01370	0.01228	0.01100	0.00985	0.00883
12	0.01180	0.01046	0.00928	0.00823	0.00730	0.00647	0.00574
13	0.00815	0.00715	0.00628	0.00552	0.00484	0.00425	0.00374
14	0.00563	0.00489	0.00425	0.00370	0.00321	0.00279	0.00243
15	0.00389	0.00335	0.00288	0.00248	0.00213	0.00184	0.00158
16	0.00269	0.00229	0.00195	0.00166	0.00142	0.00121	0.00103
17	0.00185	0.00156	0.00132	0.00111	0.00094	0.00079	0.00067
18	0.00128	0.00107	0.00089	0.00075	0.00062	0.00052	0.00044
19	0.00088	0.00073	0.00061	0.00050	0.00041	0.00034	0.00028
20	0.00061	0.00050	0.00041	0.00034	0.00027	0.00022	0.00018
21	0.00042	0.00034	0.00028	0.00022	0.00018	0.00015	0.00012
22	0.00029	0.00023	0.00019	0.00015	0.00012	0.00010	0.00008
23	0.00020	0.00016	0.00013	0.00010	0.00008	0.00006	0.00005
24	0.00014	0.00011	0.00009	0.00007	0.00005	0.00004	0.00003
25	0.00010	0.00007	0.00006	0.00005	0.00004	0.00003	0.00002
26	0.00007	0.00005	0.00004	0.00003	0.00002	0.00002	0.00001
27	0.00005	0.00004	0.00003	0.00002	0.00002	0.00001	0.00001
28	0.00003	0.00002	0.00002	0.00001	0.00001	0.00001	0.00001
29	0.00002	0.00002	0.00001	0.00001	0.00001	0.00001	0.00000
30	0.00002	0.00001	0.00001	0.00001	0.00000	0.00000	0.00000

SINGLE PAYMENT PRESENT VALUE FACTORS
CONTINUOUS INTEREST RATES

YEARS	0.44	0.45	0.46	0.47	0.48	0.49	0.50
1	0.64404	0.63763	0.63128	0.62500	0.61878	0.61263	0.60653
2	0.41478	0.40657	0.39852	0.39063	0.38289	0.37531	0.36788
3	0.26714	0.25924	0.25158	0.24414	0.23693	0.22993	0.22313
4	0.17204	0.16530	0.15882	0.15259	0.14661	0.14086	0.13534
5	0.11080	0.10540	0.10026	0.09537	0.09072	0.08629	0.08208
6	0.07136	0.06721	0.06329	0.05961	0.05613	0.05287	0.04979
7	0.04596	0.04285	0.03996	0.03725	0.03474	0.03239	0.03020
8	0.02960	0.02732	0.02522	0.02328	0.02149	0.01984	0.01832
9	0.01906	0.01742	0.01592	0.01455	0.01330	0.01216	0.01111
10	0.01228	0.01111	0.01005	0.00910	0.00823	0.00745	0.00674
11	0.00791	0.00708	0.00635	0.00568	0.00509	0.00456	0.00409
12	0.00509	0.00452	0.00401	0.00355	0.00315	0.00279	0.00248
13	0.00328	0.00288	0.00253	0.00222	0.00195	0.00171	0.00150
14	0.00211	0.00184	0.00160	0.00139	0.00121	0.00105	0.00091
15	0.00136	0.00117	0.00101	0.00087	0.00075	0.00064	0.00055
16	0.00088	0.00075	0.00064	0.00054	0.00046	0.00039	0.00034
17	0.00056	0.00048	0.00040	0.00034	0.00029	0.00024	0.00020
18	0.00036	0.00030	0.00025	0.00021	0.00018	0.00015	0.00012
19	0.00023	0.00019	0.00016	0.00013	0.00011	0.00009	0.00007
20	0.00015	0.00012	0.00010	0.00008	0.00007	0.00006	0.00005
21	0.00010	0.00008	0.00006	0.00005	0.00004	0.00003	0.00003
22	0.00006	0.00005	0.00004	0.00003	0.00003	0.00002	0.00002
23	0.00004	0.00003	0.00003	0.00002	0.00002	0.00001	0.00001
24	0.00003	0.00002	0.00002	0.00001	0.00001	0.00001	0.00001
25	0.00002	0.00001	0.00001	0.00001	0.00001	0.00000	0.00000
26	0.00001	0.00001	0.00001	0.00000	0.00000	0.00000	0.00000
27	0.00001	0.00001	0.00000	0.00000	0.00000	0.00000	0.00000

DISCOUNTING SUMS WHICH OCCUR UNIFORMLY OVER A PERIOD

YEAR	CONTINUOUS INTEREST RATES					
	0.05	0.10	0.15	0.20	0.25	0.30
0 - 1	0.97541	0.95163	0.92861	0.90635	0.88480	0.86394
1 - 2	0.92784	0.86107	0.79927	0.74205	0.68908	0.64002
2 - 3	0.88259	0.77913	0.68793	0.60754	0.53666	0.47414
3 - 4	0.83954	0.70498	0.59211	0.49741	0.41795	0.35125
4 - 5	0.79860	0.63789	0.50963	0.40725	0.32550	0.26021
5 - 6	0.75965	0.57719	0.43865	0.33343	0.25350	0.19277
6 - 7	0.72260	0.52226	0.37755	0.27299	0.19742	0.14281
7 - 8	0.68736	0.47256	0.32496	0.22350	0.15375	0.10579
8 - 9	0.65384	0.42759	0.27969	0.18299	0.11974	0.07837
9 - 10	0.62195	0.38690	0.24073	0.14982	0.09326	0.05806
10 - 11	0.59162	0.35008	0.20720	0.12266	0.07263	0.04301
11 - 12	0.56276	0.31677	0.17834	0.10043	0.05656	0.03186
12 - 13	0.53532	0.28662	0.15350	0.08222	0.04405	0.02361
13 - 14	0.50921	0.25935	0.13212	0.06732	0.03431	0.01749
14 - 15	0.48438	0.23467	0.11371	0.05511	0.02672	0.01296
15 - 16	0.46075	0.21234	0.09788	0.04512	0.02081	0.00960
16 - 17	0.43828	0.19213	0.08424	0.03694	0.01621	0.00711
17 - 18	0.41691	0.17385	0.07251	0.03025	0.01262	0.00527
18 - 19	0.39657	0.15730	0.06241	0.02476	0.00983	0.00390
19 - 20	0.37723	0.14233	0.05372	0.02028	0.00765	0.00289
20 - 21	0.35883	0.12879	0.04623	0.01660	0.00596	0.00214
21 - 22	0.34133	0.11653	0.03979	0.01359	0.00464	0.00159
22 - 23	0.32469	0.10544	0.03425	0.01113	0.00362	0.00118
23 - 24	0.30885	0.09541	0.02948	0.00911	0.00282	0.00087
24 - 25	0.29379	0.08633	0.02537	0.00746	0.00219	0.00065
25 - 26	0.27946	0.07811	0.02184	0.00611	0.00171	0.00048
26 - 27	0.26583	0.07068	0.01880	0.00500	0.00133	0.00035
27 - 28	0.25287	0.06395	0.01618	0.00409	0.00104	0.00026
28 - 29	0.24053	0.05787	0.01393	0.00335	0.00081	0.00019
29 - 30	0.22880	0.05236	0.01199	0.00274	0.00063	0.00014
30 - 35	0.19742	0.03918	0.00782	0.00157	0.00032	0.00006
35 - 40	0.15375	0.02376	0.00369	0.00058	0.00009	0.00001
40 - 45	0.11974	0.01441	0.00174	0.00021	0.00003	0.00000
45 - 50	0.09326	0.00874	0.00082	0.00008	0.00001	0.00000
50 - 60	0.06460	0.00426	0.00029	0.00002	0.00000	0.00000
60 - 70	0.03918	0.00157	0.00006	0.00000	0.00000	0.00000
70 - 80	0.02376	0.00058	0.00001	0.00000	0.00000	0.00000
80 - 90	0.01441	0.00021	0.00000	0.00000	0.00000	0.00000
90 -100	0.00874	0.00008	0.00000	0.00000	0.00000	0.00000

Interpolation Table for Declining Production, Present-Value Factors

Values of U						Interpolation Fraction
1.00	2.0	4.0	8.0	16.0	32.0	0.000
1.05	2.1	4.2	8.4	16.8	33.6	0.070
1.10	2.2	4.4	8.8	17.6	35.2	0.138
1.15	2.3	4.6	9.2	18.4	36.8	0.202
1.20	2.4	4.8	9.6	19.2	38.4	0.263
1.25	2.5	5.0	10.0	20.0	40.0	0.322
1.30	2.6	5.2	10.4	20.8	41.6	0.379
1.35	2.7	5.4	10.8	21.6	43.2	0.433
1.40	2.8	5.6	11.2	22.4	44.8	0.485
1.45	2.9	5.8	11.6	23.2	46.4	0.536
1.50	3.0	6.0	12.0	24.0	48.0	0.585
1.55	3.1	6.2	12.4	24.8	49.6	0.632
1.60	3.2	6.4	12.8	25.6	51.2	0.678
1.65	3.3	6.6	13.2	26.4	52.8	0.722
1.70	3.4	6.8	13.6	27.2	54.4	0.766
1.75	3.5	7.0	14.0	28.0	56.0	0.807
1.80	3.6	7.2	14.4	28.8	57.6	0.848
1.85	3.7	7.4	14.8	29.6	59.2	0.888
1.90	3.8	7.6	15.2	30.4	60.8	0.926
1.95	3.9	7.8	15.6	31.2	62.4	0.963
2.00	4.0	8.0	16.0	32.0	64.0	1.000

EXPONENTIAL DECLINE DEFERMENT FACTORS
SUMS RECEIVED AT END OF EACH YEAR, I = 0.04

YEARS ON DECLINE	RATIO OF INITIAL TO FINAL RATE						
	1.0	2.0	4.0	8.0	16.0	32.0	64.0
1	0.96154	0.96154	0.96154	0.96154	0.96154	0.96154	0.96154
2	0.94305	0.94622	0.94921	0.95188	0.95414	0.95598	0.95743
3	0.92503	0.93057	0.93584	0.94061	0.94475	0.94821	0.95104
4	0.90747	0.91513	0.92242	0.92907	0.93489	0.93982	0.94390
5	0.89036	0.89998	0.90917	0.91757	0.92495	0.93124	0.93649
6	0.87369	0.88516	0.89614	0.90620	0.91506	0.92265	0.92901
7	0.85743	0.87067	0.88336	0.89500	0.90528	0.91411	0.92152
8	0.84159	0.85651	0.87083	0.88399	0.89564	0.90565	0.91408
9	0.82615	0.84268	0.85857	0.87318	0.88614	0.89729	0.90671
10	0.81109	0.82917	0.84656	0.86257	0.87679	0.88905	0.89941
11	0.79640	0.81596	0.83480	0.85217	0.86760	0.88092	0.89220
12	0.78209	0.80307	0.82329	0.84195	0.85856	0.87291	0.88508
13	0.76812	0.79047	0.81202	0.83194	0.84967	0.86503	0.87806
14	0.75451	0.77815	0.80099	0.82211	0.84094	0.85726	0.87112
15	0.74122	0.76612	0.79019	0.81248	0.83236	0.84961	0.86429
16	0.72827	0.75437	0.77962	0.80303	0.82393	0.84208	0.85754
17	0.71563	0.74288	0.76927	0.79375	0.81564	0.83467	0.85089
18	0.70329	0.73165	0.75914	0.78466	0.80750	0.82737	0.84433
19	0.69126	0.72067	0.74921	0.77574	0.79949	0.82019	0.83786
20	0.67951	0.70994	0.73950	0.76698	0.79163	0.81311	0.83148
21	0.66805	0.69946	0.72998	0.75839	0.78389	0.80615	0.82519
22	0.65687	0.68920	0.72066	0.74996	0.77629	0.79929	0.81899
23	0.64595	0.67917	0.71152	0.74169	0.76882	0.79254	0.81287
24	0.63529	0.66937	0.70258	0.73358	0.76148	0.78589	0.80684
25	0.62488	0.65977	0.69381	0.72561	0.75425	0.77934	0.80088
26	0.61472	0.65039	0.68522	0.71779	0.74715	0.77289	0.79502
27	0.60480	0.64122	0.67681	0.71011	0.74017	0.76654	0.78923
28	0.59511	0.63224	0.66856	0.70258	0.73330	0.76028	0.78352
29	0.58564	0.62346	0.66047	0.69518	0.72655	0.75412	0.77789
30	0.57640	0.61487	0.65255	0.68791	0.71991	0.74805	0.77233
35	0.53327	0.57457	0.61520	0.65349	0.68828	0.71902	0.74564
40	0.49482	0.53832	0.58130	0.62198	0.65910	0.69202	0.72065
45	0.46044	0.50563	0.55046	0.59307	0.63211	0.66687	0.69723
50	0.42964	0.47608	0.52235	0.56649	0.60710	0.64341	0.67523
60	0.37706	0.42497	0.47310	0.51938	0.56228	0.60092	0.63505
70	0.33421	0.38260	0.43158	0.47903	0.52334	0.56354	0.59928
80	0.29894	0.34712	0.39625	0.44419	0.48926	0.53042	0.56727
90	0.26964	0.31715	0.36594	0.41387	0.45922	0.50091	0.53846
100	0.24505	0.29160	0.33972	0.38728	0.43258	0.47447	0.51241

EXPONENTIAL DECLINE DEFERMENT FACTORS
SUMS RECEIVED AT END OF EACH YEAR, I = 0.06
RATIO OF INITIAL TO FINAL RATE

YEARS DECLINE	1.0	2.0	4.0	8.0	16.0	32.0	64.0
1	0.94340	0.94340	0.94340	0.94340	0.94340	0.94340	0.94340
2	0.91670	0.92128	0.92560	0.92945	0.93272	0.93537	0.93746
3	0.89100	0.89894	0.90648	0.91331	0.91925	0.92422	0.92828
4	0.86627	0.87713	0.88750	0.89695	0.90523	0.91226	0.91809
5	0.84247	0.85599	0.86893	0.88079	0.89122	0.90014	0.90758
6	0.81955	0.83554	0.85088	0.86496	0.87740	0.88808	0.89704
7	0.79748	0.81577	0.83335	0.84952	0.86385	0.87618	0.88657
8	0.77622	0.79666	0.81634	0.83448	0.85059	0.86449	0.87623
9	0.75574	0.77820	0.79986	0.81985	0.83765	0.85302	0.86605
10	0.73601	0.76035	0.78387	0.80563	0.82501	0.84180	0.85605
11	0.71699	0.74311	0.76838	0.79179	0.81269	0.83082	0.84623
12	0.69865	0.72644	0.75337	0.77834	0.80067	0.82007	0.83660
13	0.68097	0.71033	0.73881	0.76526	0.78895	0.80957	0.82716
14	0.66393	0.69475	0.72469	0.75254	0.77752	0.79930	0.81790
15	0.64748	0.67968	0.71100	0.74018	0.76638	0.78925	0.80883
16	0.63162	0.66510	0.69772	0.72815	0.75551	0.77943	0.79994
17	0.61631	0.65100	0.68484	0.71645	0.74491	0.76983	0.79122
18	0.60153	0.63735	0.67234	0.70506	0.73457	0.76044	0.78268
19	0.58727	0.62414	0.66021	0.69398	0.72448	0.75126	0.77430
20	0.57350	0.61135	0.64843	0.68320	0.71464	0.74228	0.76610
21	0.56019	0.59897	0.63700	0.67271	0.70504	0.73349	0.75805
22	0.54734	0.58698	0.62589	0.66249	0.69566	0.72490	0.75016
23	0.53493	0.57536	0.61511	0.65254	0.68651	0.71649	0.74242
24	0.52293	0.56410	0.60464	0.64284	0.67757	0.70825	0.73484
25	0.51133	0.55319	0.59446	0.63340	0.66885	0.70020	0.72740
26	0.50012	0.54262	0.58456	0.62420	0.66032	0.69231	0.72010
27	0.48928	0.53237	0.57494	0.61524	0.65199	0.68459	0.71294
28	0.47879	0.52242	0.56559	0.60650	0.64386	0.67703	0.70591
29	0.46865	0.51278	0.55650	0.59797	0.63591	0.66963	0.69902
30	0.45883	0.50342	0.54765	0.58967	0.62813	0.66237	0.69225
35	0.41424	0.46059	0.50686	0.55106	0.59178	0.62824	0.66023
40	0.37616	0.42354	0.47111	0.51683	0.55918	0.59731	0.63096
45	0.34346	0.39131	0.43963	0.48632	0.52982	0.56919	0.60412
50	0.31524	0.36313	0.41176	0.45902	0.50327	0.54353	0.57942
60	0.26936	0.31648	0.36483	0.41230	0.45719	0.49844	0.53555
70	0.23406	0.27973	0.32704	0.37392	0.41867	0.46015	0.49780
80	0.20636	0.25023	0.29609	0.34193	0.38605	0.42727	0.46499
90	0.18421	0.22614	0.27035	0.31489	0.35809	0.39875	0.43621
100	0.16618	0.20618	0.24866	0.29177	0.33388	0.37378	0.41078

EXPONENTIAL DECLINE DEFERMENT FACTORS
SUMS RECEIVED AT END OF EACH YEAR, I = 0.08

YEARS ON DECLINE	RATIO OF INITIAL TO FINAL RATE						
	1.0	2.0	4.0	8.0	16.0	32.0	64.0
1	0.92593	0.92593	0.92593	0.92593	0.92593	0.92593	0.92593
2	0.89163	0.89752	0.90306	0.90801	0.91221	0.91562	0.91831
3	0.85903	0.86913	0.87874	0.88746	0.89503	0.90138	0.90657
4	0.82803	0.84173	0.85484	0.86680	0.87730	0.88622	0.89362
5	0.79854	0.81546	0.83169	0.84658	0.85971	0.87095	0.88036
6	0.77048	0.79032	0.80940	0.82696	0.84251	0.85588	0.86713
7	0.74377	0.76627	0.78797	0.80799	0.82577	0.84112	0.85408
8	0.71833	0.74328	0.76739	0.78968	0.80953	0.82672	0.84128
9	0.69410	0.72129	0.74762	0.77202	0.79380	0.81270	0.82876
10	0.67101	0.70026	0.72864	0.75498	0.77856	0.79906	0.81653
11	0.64900	0.68014	0.71040	0.73856	0.76381	0.78581	0.80461
12	0.62801	0.66088	0.69289	0.72273	0.74953	0.77294	0.79298
13	0.60798	0.64245	0.67606	0.70745	0.73571	0.76044	0.78164
14	0.58887	0.62479	0.65989	0.69272	0.72233	0.74829	0.77060
15	0.57063	0.60788	0.64434	0.67851	0.70938	0.73649	0.75983
16	0.55321	0.59167	0.62938	0.66478	0.69683	0.72502	0.74934
17	0.53657	0.57613	0.61499	0.65154	0.68467	0.71388	0.73912
18	0.52066	0.56123	0.60114	0.63874	0.67290	0.70305	0.72915
19	0.50545	0.54693	0.58780	0.62638	0.66148	0.69252	0.71943
20	0.49091	0.53320	0.57496	0.61443	0.65041	0.68228	0.70995
21	0.47699	0.52003	0.56259	0.60289	0.63967	0.67232	0.70071
22	0.46367	0.50737	0.55066	0.59172	0.62926	0.66262	0.69170
23	0.45092	0.49521	0.53916	0.58092	0.61915	0.65319	0.68290
24	0.43870	0.48352	0.52807	0.57046	0.60934	0.64401	0.67431
25	0.42699	0.47228	0.51737	0.56034	0.59982	0.63507	0.66593
26	0.41577	0.46147	0.50704	0.55054	0.59057	0.62637	0.65775
27	0.40501	0.45107	0.49707	0.54105	0.58158	0.61788	0.64976
28	0.39468	0.44105	0.48743	0.53185	0.57284	0.60962	0.64196
29	0.38477	0.43141	0.47813	0.52293	0.56435	0.60156	0.63433
30	0.37526	0.42212	0.46913	0.51429	0.55609	0.59371	0.62688
35	0.33299	0.38041	0.42836	0.47475	0.51802	0.55723	0.59205
40	0.29812	0.34542	0.39360	0.44055	0.48464	0.52486	0.56081
45	0.26908	0.31581	0.36373	0.41074	0.45517	0.49597	0.53266
50	0.24467	0.29052	0.33784	0.38457	0.42900	0.47005	0.50717
60	0.20628	0.24986	0.29539	0.34088	0.38463	0.42548	0.46282
70	0.17775	0.21882	0.26218	0.30596	0.34848	0.38857	0.42557
80	0.15592	0.19447	0.23558	0.27746	0.31850	0.35754	0.39386
90	0.13875	0.17493	0.21384	0.25379	0.29326	0.33108	0.36653
100	0.12494	0.15893	0.19574	0.23383	0.27171	0.30827	0.34275

EXPONENTIAL DECLINE DEFERMENT FACTORS
SUMS RECEIVED AT END OF EACH YEAR, I = 0.10

RATIO OF INITIAL TO FINAL RATE

YEARS ON DECLINE	1.0	2.0	4.0	8.0	16.0	32.0	64.0
1	0.90909	0.90909	0.90909	0.90909	0.90909	0.90909	0.90909
2	0.86777	0.87486	0.88154	0.88750	0.89256	0.89668	0.89991
3	0.82895	0.84102	0.85251	0.86294	0.87200	0.87961	0.88584
4	0.79247	0.80870	0.82425	0.83846	0.85094	0.86156	0.87039
5	0.75816	0.77804	0.79715	0.81471	0.83022	0.84352	0.85467
6	0.72588	0.74900	0.77130	0.79186	0.81011	0.82583	0.83909
7	0.69549	0.72152	0.74668	0.76996	0.79069	0.80863	0.82382
8	0.66686	0.69550	0.72325	0.74899	0.77199	0.79195	0.80893
9	0.63989	0.67086	0.70095	0.72893	0.75401	0.77583	0.79445
10	0.61446	0.64752	0.67973	0.70975	0.73671	0.76025	0.78039
11	0.59046	0.62541	0.65952	0.69139	0.72009	0.74521	0.76676
12	0.56781	0.60444	0.64027	0.67383	0.70412	0.73069	0.75354
13	0.54641	0.58454	0.62192	0.65701	0.68875	0.71667	0.74073
14	0.52619	0.56566	0.60443	0.64091	0.67398	0.70313	0.72831
15	0.50707	0.54772	0.58774	0.62548	0.65976	0.69005	0.71627
16	0.48898	0.53068	0.57182	0.61069	0.64608	0.67741	0.70459
17	0.47186	0.51447	0.55661	0.59650	0.63291	0.66520	0.69327
18	0.45563	0.49906	0.54208	0.58289	0.62022	0.65339	0.68229
19	0.44026	0.48438	0.52819	0.56983	0.60799	0.64197	0.67163
20	0.42568	0.47040	0.51490	0.55728	0.59619	0.63092	0.66129
21	0.41184	0.45708	0.50218	0.54523	0.58482	0.62022	0.65125
22	0.39871	0.44438	0.49001	0.53363	0.57385	0.60987	0.64149
23	0.38623	0.43226	0.47834	0.52248	0.56325	0.59984	0.63202
24	0.37436	0.42070	0.46715	0.51175	0.55302	0.59012	0.62281
25	0.36308	0.40965	0.45643	0.50142	0.54313	0.58069	0.61386
26	0.35234	0.39908	0.44613	0.49147	0.53357	0.57156	0.60515
27	0.34212	0.38899	0.43625	0.48188	0.52432	0.56269	0.59669
28	0.33238	0.37932	0.42676	0.47263	0.51538	0.55409	0.58845
29	0.32309	0.37007	0.41763	0.46370	0.50672	0.54574	0.58043
30	0.31423	0.36121	0.40885	0.45509	0.49834	0.53763	0.57262
35	0.27555	0.32206	0.36964	0.41621	0.46013	0.50037	0.53647
40	0.24448	0.28998	0.33691	0.38322	0.42723	0.46785	0.50457
45	0.21917	0.26335	0.30928	0.35493	0.39863	0.43924	0.47621
50	0.19830	0.24100	0.28570	0.33045	0.37357	0.41391	0.45085
60	0.16612	0.20572	0.24771	0.29026	0.33177	0.37105	0.40742
70	0.14268	0.17928	0.21852	0.25873	0.29834	0.33621	0.37161
80	0.12494	0.15880	0.19545	0.23335	0.27102	0.30734	0.34158
90	0.11109	0.14249	0.17677	0.21249	0.24827	0.28303	0.31603
100	0.09999	0.12922	0.16135	0.19505	0.22905	0.26228	0.29404

EXPONENTIAL DECLINE DEFERMENT FACTORS
SUMS RECEIVED AT END OF EACH YEAR, I = 0.12

YEARS ON DECLINE	RATIO OF INITIAL TO FINAL RATE						
	1.0	2.0	4.0	8.0	16.0	32.0	64.0
1	0.89286	0.89286	0.89286	0.89286	0.89286	0.89286	0.89286
2	0.84502	0.85323	0.86097	0.86787	0.87372	0.87849	0.88223
3	0.80061	0.81446	0.82767	0.83966	0.85009	0.85885	0.86601
4	0.75934	0.77783	0.79555	0.81178	0.82605	0.83820	0.84831
5	0.72095	0.74342	0.76505	0.78495	0.80258	0.81771	0.83041
6	0.68523	0.71116	0.73622	0.75936	0.77996	0.79774	0.81277
7	0.65196	0.68093	0.70900	0.73503	0.75828	0.77845	0.79557
8	0.62095	0.65258	0.68332	0.71193	0.73756	0.75987	0.77890
9	0.59203	0.62599	0.65910	0.68999	0.71776	0.74202	0.76279
10	0.56502	0.60103	0.63623	0.66916	0.69886	0.72488	0.74723
11	0.53979	0.57758	0.61462	0.64938	0.68081	0.70843	0.73222
12	0.51620	0.55554	0.59421	0.63059	0.66357	0.69264	0.71774
13	0.49412	0.53481	0.57490	0.61272	0.64709	0.67747	0.70377
14	0.47344	0.51529	0.55662	0.59571	0.63134	0.66290	0.69031
15	0.45406	0.49690	0.53932	0.57953	0.61628	0.64891	0.67731
16	0.43587	0.47955	0.52291	0.56411	0.60186	0.63545	0.66477
17	0.41880	0.46318	0.50735	0.54942	0.58805	0.62252	0.65266
18	0.40276	0.44772	0.49257	0.53540	0.57481	0.61007	0.64097
19	0.38767	0.43311	0.47854	0.52202	0.56212	0.59808	0.62967
20	0.37347	0.41928	0.46519	0.50923	0.54995	0.58653	0.61875
21	0.36010	0.40619	0.45249	0.49701	0.53826	0.57541	0.60819
22	0.34748	0.39379	0.44040	0.48532	0.52704	0.56468	0.59797
23	0.33558	0.38202	0.42888	0.47413	0.51625	0.55433	0.58808
24	0.32435	0.37086	0.41789	0.46342	0.50587	0.54434	0.57850
25	0.31373	0.36025	0.40741	0.45314	0.49588	0.53470	0.56923
26	0.30368	0.35017	0.39739	0.44329	0.48627	0.52538	0.56024
27	0.29417	0.34058	0.38782	0.43383	0.47701	0.51637	0.55152
28	0.28516	0.33145	0.37867	0.42475	0.46808	0.50765	0.54307
29	0.27661	0.32276	0.36991	0.41603	0.45947	0.49922	0.53487
30	0.26851	0.31446	0.36152	0.40764	0.45116	0.49106	0.52690
35	0.23359	0.27826	0.32445	0.37014	0.41365	0.45390	0.49036
40	0.20609	0.24912	0.29401	0.33879	0.38180	0.42190	0.45852
45	0.18406	0.22528	0.26865	0.31225	0.35444	0.39408	0.43053
50	0.16609	0.20548	0.24723	0.28951	0.33072	0.36969	0.40575
60	0.13873	0.17461	0.21314	0.25265	0.29163	0.32894	0.36385
70	0.11900	0.15173	0.18726	0.22409	0.26079	0.29626	0.32979
80	0.10415	0.13413	0.16697	0.20131	0.23584	0.26949	0.30155
90	0.09259	0.12018	0.15065	0.18274	0.21524	0.24715	0.27777
100	0.08333	0.10886	0.13723	0.16731	0.19796	0.22823	0.25746

EXPONENTIAL DECLINE DEFERMENT FACTORS
SUMS RECEIVED AT END OF EACH YEAR, I = 0.14

YEARS ON DECLINE	RATIO OF INITIAL TO FINAL RATE						
	1.0	2.0	4.0	8.0	16.0	32.0	64.0
1	0.87719	0.87719	0.87719	0.87719	0.87719	0.87719	0.87719
2	0.82333	0.83257	0.84128	0.84905	0.85565	0.86101	0.86522
3	0.77388	0.78935	0.80412	0.81753	0.82920	0.83901	0.84704
4	0.72843	0.74892	0.76859	0.78662	0.80250	0.81603	0.82731
5	0.68662	0.71133	0.73516	0.75713	0.77661	0.79336	0.80745
6	0.64811	0.67642	0.70383	0.72922	0.75184	0.77142	0.78800
7	0.61261	0.64401	0.67452	0.70288	0.72827	0.75035	0.76915
8	0.57986	0.61390	0.64709	0.67805	0.70589	0.73019	0.75097
9	0.54960	0.58590	0.62141	0.65466	0.68465	0.71092	0.73349
10	0.52161	0.55985	0.59737	0.63260	0.66449	0.69253	0.71669
11	0.49570	0.53558	0.57482	0.61180	0.64536	0.67497	0.70058
12	0.47169	0.51295	0.55367	0.59215	0.62720	0.65821	0.68510
13	0.44941	0.49182	0.53381	0.57360	0.60994	0.64219	0.67025
14	0.42872	0.47208	0.51513	0.55605	0.59353	0.62689	0.65599
15	0.40948	0.45361	0.49756	0.53944	0.57791	0.61225	0.64229
16	0.39157	0.43632	0.48100	0.52371	0.56304	0.59824	0.62913
17	0.37487	0.42011	0.46539	0.50879	0.54887	0.58483	0.61647
18	0.35930	0.40489	0.45066	0.49463	0.53535	0.57198	0.60429
19	0.34476	0.39060	0.43674	0.48118	0.52245	0.55966	0.59256
20	0.33116	0.37715	0.42357	0.46840	0.51012	0.54785	0.58127
21	0.31843	0.36450	0.41111	0.45623	0.49834	0.53650	0.57039
22	0.30650	0.35257	0.39929	0.44465	0.48706	0.52560	0.55990
23	0.29531	0.34131	0.38809	0.43360	0.47626	0.51512	0.54978
24	0.28480	0.33068	0.37745	0.42306	0.46592	0.50504	0.54001
25	0.27492	0.32064	0.36734	0.41300	0.45600	0.49533	0.53058
26	0.26562	0.31113	0.35773	0.40339	0.44648	0.48599	0.52147
27	0.25686	0.30212	0.34858	0.39419	0.43734	0.47698	0.51265
28	0.24860	0.29358	0.33985	0.38539	0.42856	0.46830	0.50413
29	0.24079	0.28547	0.33153	0.37696	0.42011	0.45992	0.49588
30	0.23342	0.27777	0.32359	0.36888	0.41199	0.45183	0.48790
35	0.20200	0.24447	0.28880	0.33305	0.37558	0.41526	0.45151
40	0.17763	0.21802	0.26058	0.30346	0.34502	0.38413	0.42014
45	0.15829	0.19660	0.23730	0.27864	0.31902	0.35731	0.39283
50	0.14265	0.17894	0.21780	0.25754	0.29665	0.33399	0.36885
60	0.11900	0.15162	0.18700	0.22364	0.26014	0.29540	0.32870
70	0.10203	0.13150	0.16381	0.19762	0.23163	0.26480	0.29642
80	0.08928	0.11609	0.14573	0.17701	0.20874	0.23994	0.26992
90	0.07936	0.10391	0.13124	0.16030	0.18997	0.21935	0.24776
100	0.07143	0.09404	0.11938	0.14647	0.17430	0.20201	0.22897

EXPONENTIAL DECLINE DEFERMENT FACTORS
SUMS RECEIVED AT END OF EACH YEAR, I = 0.16

YEARS ON DECLINE	RATIO OF INITIAL TO FINAL RATE						
	1.0	2.0	4.0	8.0	16.0	32.0	64.0
1	0.86207	0.86207	0.86207	0.86207	0.86207	0.86207	0.86207
2	0.80262	0.81282	0.82243	0.83101	0.83829	0.84421	0.84886
3	0.74863	0.76558	0.78176	0.79647	0.80928	0.82005	0.82888
4	0.69954	0.72182	0.74323	0.76287	0.78019	0.79497	0.80730
5	0.65486	0.68152	0.70728	0.73106	0.75218	0.77038	0.78571
6	0.61412	0.64445	0.67388	0.70118	0.72558	0.74672	0.76467
7	0.57694	0.61034	0.64288	0.67320	0.70042	0.72414	0.74438
8	0.54295	0.57892	0.61410	0.64701	0.67668	0.70265	0.72492
9	0.51184	0.54996	0.58737	0.62250	0.65429	0.68224	0.70630
10	0.48332	0.52322	0.56251	0.59954	0.63317	0.66285	0.68851
11	0.45715	0.49851	0.53937	0.57802	0.61324	0.64443	0.67150
12	0.43309	0.47563	0.51780	0.55782	0.59442	0.62694	0.65525
13	0.41095	0.45443	0.49768	0.53885	0.57663	0.61030	0.63972
14	0.39054	0.43476	0.47888	0.52102	0.55980	0.59448	0.62487
15	0.37170	0.41647	0.46129	0.50422	0.54385	0.57941	0.61066
16	0.35428	0.39946	0.44481	0.48839	0.52874	0.56505	0.59705
17	0.33816	0.38360	0.42936	0.47345	0.51440	0.55135	0.58402
18	0.32321	0.36880	0.41484	0.45934	0.50078	0.53828	0.57152
19	0.30934	0.35498	0.40120	0.44600	0.48783	0.52579	0.55953
20	0.29644	0.34204	0.38835	0.43336	0.47551	0.51385	0.54803
21	0.28444	0.32992	0.37624	0.42138	0.46377	0.50243	0.53697
22	0.27324	0.31854	0.36481	0.41002	0.45258	0.49149	0.52634
23	0.26279	0.30786	0.35401	0.39922	0.44189	0.48101	0.51612
24	0.25303	0.29781	0.34380	0.38896	0.43169	0.47095	0.50629
25	0.24388	0.28835	0.33412	0.37919	0.42193	0.46130	0.49681
26	0.23531	0.27943	0.32495	0.36989	0.41260	0.45203	0.48768
27	0.22727	0.27101	0.31625	0.36101	0.40366	0.44312	0.47887
28	0.21972	0.26305	0.30798	0.35254	0.39509	0.43455	0.47037
29	0.21261	0.25552	0.30012	0.34445	0.38688	0.42630	0.46217
30	0.20591	0.24838	0.29263	0.33671	0.37899	0.41835	0.45424
35	0.17758	0.21775	0.26005	0.30263	0.34387	0.38266	0.41835
40	0.15584	0.19367	0.23388	0.27474	0.31467	0.35254	0.38768
45	0.13871	0.17429	0.21245	0.25152	0.29002	0.32681	0.36120
50	0.12493	0.15841	0.19459	0.23191	0.26893	0.30456	0.33809
60	0.10415	0.13395	0.16655	0.20060	0.23479	0.26807	0.29973
70	0.08928	0.11602	0.14557	0.17674	0.20833	0.23938	0.26919
80	0.07812	0.10232	0.12928	0.15794	0.18723	0.21623	0.24429
90	0.06944	0.09151	0.11627	0.14276	0.17001	0.19717	0.22361
100	0.06250	0.08277	0.10564	0.13025	0.15569	0.18119	0.20615

EXPONENTIAL DECLINE DEFERMENT FACTORS
SUMS RECEIVED AT END OF EACH YEAR, I = 0.18

YEARS ON DECLINE	RATIO OF INITIAL TO FINAL RATE						
	1.0	2.0	4.0	8.0	16.0	32.0	64.0
1	0.84746	0.84746	0.84746	0.84746	0.84746	0.84746	0.84746
2	0.78282	0.79391	0.80437	0.81369	0.82160	0.82804	0.83309
3	0.72476	0.74305	0.76052	0.77640	0.79025	0.80191	0.81146
4	0.67252	0.69638	0.71933	0.74042	0.75903	0.77494	0.78821
5	0.62543	0.65379	0.68123	0.70660	0.72917	0.74865	0.76508
6	0.58293	0.61497	0.64611	0.67507	0.70099	0.72350	0.74264
7	0.54450	0.57955	0.61377	0.64574	0.67451	0.69963	0.72112
8	0.50970	0.54720	0.58398	0.61848	0.64967	0.67705	0.70058
9	0.47811	0.51761	0.55650	0.59313	0.62638	0.65570	0.68102
10	0.44941	0.49051	0.53112	0.56953	0.60454	0.63553	0.66241
11	0.42327	0.46563	0.50765	0.54754	0.58403	0.61646	0.64470
12	0.39944	0.44277	0.48591	0.52702	0.56476	0.59842	0.62785
13	0.37765	0.42172	0.46575	0.50784	0.54663	0.58135	0.61180
14	0.35772	0.40231	0.44701	0.48990	0.52955	0.56517	0.59652
15	0.33944	0.38438	0.42958	0.47309	0.51345	0.54983	0.58195
16	0.32265	0.36778	0.41333	0.45732	0.49825	0.53526	0.56805
17	0.30720	0.35239	0.39816	0.44250	0.48389	0.52142	0.55477
18	0.29295	0.33811	0.38398	0.42855	0.47029	0.50826	0.54209
19	0.27980	0.32482	0.37070	0.41541	0.45741	0.49572	0.52995
20	0.26764	0.31245	0.35824	0.40302	0.44519	0.48377	0.51834
21	0.25637	0.30090	0.34655	0.39131	0.43359	0.47237	0.50722
22	0.24590	0.29011	0.33555	0.38024	0.42256	0.46149	0.49655
23	0.23618	0.28001	0.32520	0.36975	0.41206	0.45108	0.48632
24	0.22712	0.27055	0.31543	0.35981	0.40207	0.44113	0.47650
25	0.21868	0.26166	0.30621	0.35038	0.39253	0.43160	0.46706
26	0.21079	0.25331	0.29750	0.34141	0.38343	0.42247	0.45798
27	0.20340	0.24544	0.28924	0.33288	0.37474	0.41371	0.44924
28	0.19649	0.23803	0.28142	0.32476	0.36642	0.40531	0.44083
29	0.18999	0.23103	0.27400	0.31701	0.35846	0.39723	0.43273
30	0.18389	0.22442	0.26695	0.30962	0.35084	0.38947	0.42492
35	0.15825	0.19618	0.23643	0.27725	0.31708	0.35478	0.38971
40	0.13870	0.17414	0.21210	0.25096	0.28921	0.32575	0.35987
45	0.12338	0.15650	0.19228	0.22920	0.26583	0.30109	0.33427
50	0.11108	0.14208	0.17583	0.21090	0.24595	0.27990	0.31207
60	0.09259	0.11996	0.15013	0.18186	0.21393	0.24536	0.27546
70	0.07936	0.10380	0.13099	0.15985	0.18929	0.21841	0.24654
80	0.06944	0.09147	0.11617	0.14258	0.16974	0.19679	0.22311
90	0.06173	0.08176	0.10436	0.12869	0.15385	0.17906	0.20375
100	0.05556	0.07391	0.09473	0.11726	0.14067	0.16426	0.18748

EXPONENTIAL DECLINE DEFERMENT FACTORS
SUMS RECEIVED AT END OF EACH YEAR, I = 0.20

YEARS ON DECLINE	RATIO OF INITIAL TO FINAL RATE						
	1.0	2.0	4.0	8.0	16.0	32.0	64.0
1	0.83333	0.83333	0.83333	0.83333	0.83333	0.83333	0.83333
2	0.76389	0.77580	0.78704	0.79705	0.80556	0.81247	0.81790
3	0.70216	0.72167	0.74031	0.75728	0.77207	0.78454	0.79475
4	0.64718	0.67245	0.69679	0.71917	0.73894	0.75586	0.76999
5	0.59812	0.62794	0.65684	0.68361	0.70746	0.72807	0.74548
6	0.55425	0.58772	0.62032	0.65069	0.67793	0.70164	0.72183
7	0.51494	0.55133	0.58694	0.62028	0.65035	0.67668	0.69925
8	0.47964	0.51835	0.55640	0.59219	0.62463	0.65319	0.67780
9	0.44789	0.48841	0.52842	0.56623	0.60065	0.63109	0.65745
10	0.41925	0.46117	0.50275	0.54220	0.57827	0.61031	0.63818
11	0.39337	0.43636	0.47915	0.51993	0.55736	0.59075	0.61992
12	0.36993	0.41369	0.45742	0.49925	0.53781	0.57232	0.60261
13	0.34867	0.39295	0.43737	0.48003	0.51949	0.55496	0.58619
14	0.32933	0.37393	0.41884	0.46212	0.50231	0.53856	0.57060
15	0.31170	0.35645	0.40167	0.44542	0.48617	0.52307	0.55579
16	0.29560	0.34035	0.38575	0.42980	0.47100	0.50841	0.54170
17	0.28086	0.32550	0.37094	0.41519	0.45670	0.49453	0.52829
18	0.26734	0.31178	0.35716	0.40149	0.44322	0.48136	0.51551
19	0.25492	0.29907	0.34430	0.38863	0.43048	0.46886	0.50332
20	0.24348	0.28727	0.33228	0.37653	0.41844	0.45698	0.49169
21	0.23292	0.27630	0.32103	0.36514	0.40703	0.44567	0.48057
22	0.22316	0.26608	0.31048	0.35439	0.39622	0.43490	0.46993
23	0.21411	0.25655	0.30058	0.34424	0.38595	0.42463	0.45975
24	0.20571	0.24764	0.29126	0.33465	0.37620	0.41483	0.45000
25	0.19790	0.23930	0.28248	0.32556	0.36692	0.40547	0.44065
26	0.19063	0.23148	0.27421	0.31694	0.35807	0.39651	0.43168
27	0.18384	0.22413	0.26639	0.30875	0.34964	0.38794	0.42306
28	0.17749	0.21722	0.25899	0.30098	0.34160	0.37973	0.41477
29	0.17154	0.21070	0.25199	0.29358	0.33391	0.37186	0.40681
30	0.16596	0.20456	0.24534	0.28653	0.32655	0.36430	0.39914
35	0.14262	0.17843	0.21671	0.25577	0.29414	0.33068	0.36474
40	0.12491	0.15815	0.19401	0.23095	0.26756	0.30273	0.33579
45	0.11108	0.14198	0.17560	0.21051	0.24537	0.27913	0.31108
50	0.09999	0.12880	0.16037	0.19339	0.22658	0.25893	0.28976
60	0.08333	0.10862	0.13666	0.16632	0.19648	0.22620	0.25483
70	0.07143	0.09391	0.11906	0.14590	0.17344	0.20081	0.22740
80	0.06250	0.08270	0.10547	0.12995	0.15523	0.18055	0.20531
90	0.05556	0.07389	0.09467	0.11714	0.14049	0.16400	0.18713
100	0.05000	0.06677	0.08587	0.10663	0.12830	0.15023	0.17190

EXPONENTIAL DECLINE DEFERMENT FACTORS
SUMS RECEIVED AT END OF EACH YEAR, I = 0.25

YEARS ON DECLINE	RATIO OF INITIAL TO FINAL RATE 1.0	2.0	4.0	8.0	16.0	32.0	64.0
1	0.80000	0.80000	0.80000	0.80000	0.80000	0.80000	0.80000
2	0.72000	0.73373	0.74667	0.75821	0.76800	0.77596	0.78222
3	0.65067	0.67275	0.69388	0.71314	0.72997	0.74416	0.75581
4	0.59040	0.61852	0.64568	0.67072	0.69291	0.71194	0.72788
5	0.53786	0.57051	0.60228	0.63181	0.65822	0.68113	0.70054
6	0.49190	0.52799	0.56331	0.59637	0.62616	0.65220	0.67447
7	0.45159	0.49025	0.52830	0.56413	0.59662	0.62522	0.64986
8	0.41611	0.45666	0.49679	0.53479	0.56944	0.60012	0.62672
9	0.38479	0.42669	0.46837	0.50803	0.54440	0.57677	0.60500
10	0.35705	0.39986	0.44266	0.48359	0.52130	0.55504	0.58462
11	0.33240	0.37578	0.41934	0.46120	0.49995	0.53479	0.56547
12	0.31043	0.35409	0.39813	0.44065	0.48018	0.51589	0.54748
13	0.29078	0.33449	0.37879	0.42174	0.46185	0.49823	0.53055
14	0.27315	0.31674	0.36110	0.40430	0.44480	0.48169	0.51460
15	0.25728	0.30061	0.34489	0.38817	0.42892	0.46619	0.49955
16	0.24296	0.28591	0.32998	0.37323	0.41411	0.45162	0.48534
17	0.23000	0.27248	0.31624	0.35935	0.40025	0.43792	0.47189
18	0.21822	0.26017	0.30355	0.34644	0.38727	0.42501	0.45916
19	0.20749	0.24887	0.29180	0.33439	0.37508	0.41282	0.44709
20	0.19769	0.23845	0.28089	0.32314	0.36363	0.40131	0.43562
21	0.18872	0.22883	0.27074	0.31260	0.35285	0.39041	0.42473
22	0.18048	0.21993	0.26128	0.30271	0.34267	0.38008	0.41436
23	0.17289	0.21166	0.25244	0.29342	0.33307	0.37028	0.40448
24	0.16588	0.20398	0.24417	0.28468	0.32398	0.36097	0.39505
25	0.15940	0.19682	0.23641	0.27643	0.31537	0.35211	0.38605
26	0.15338	0.19013	0.22912	0.26865	0.30720	0.34368	0.37745
27	0.14779	0.18388	0.22227	0.26128	0.29944	0.33563	0.36923
28	0.14258	0.17801	0.21580	0.25431	0.29206	0.32796	0.36135
29	0.13772	0.17250	0.20970	0.24770	0.28503	0.32062	0.35380
30	0.13317	0.16732	0.20393	0.24142	0.27834	0.31361	0.34656
35	0.11424	0.14544	0.17925	0.21424	0.24907	0.28267	0.31437
40	0.09999	0.12859	0.15988	0.19256	0.22536	0.25728	0.28765
45	0.08889	0.11523	0.14429	0.17485	0.20577	0.23607	0.26510
50	0.08000	0.10439	0.13146	0.16013	0.18931	0.21809	0.24583
60	0.06667	0.08785	0.11162	0.13705	0.16320	0.18926	0.21463
70	0.05714	0.07584	0.09698	0.11978	0.14342	0.16715	0.19045
80	0.05000	0.06671	0.08573	0.10638	0.12791	0.14967	0.17116
90	0.04444	0.05955	0.07682	0.09567	0.11543	0.13550	0.15543
100	0.04000	0.05377	0.06959	0.08692	0.10517	0.12378	0.14234

EXPONENTIAL DECLINE DEFERMENT FACTORS
SUMS RECEIVED AT END OF EACH YEAR, I = 0.30

YEARS ON DECLINE	RATIO OF INITIAL TO FINAL RATE						
	1.0	2.0	4.0	8.0	16.0	32.0	64.0
1	0.76923	0.76923	0.76923	0.76923	0.76923	0.76923	0.76923
2	0.68047	0.69570	0.71006	0.72286	0.73373	0.74256	0.74951
3	0.60537	0.62947	0.65256	0.67365	0.69209	0.70766	0.72046
4	0.54156	0.57177	0.60102	0.62806	0.65208	0.67273	0.69007
5	0.48711	0.52168	0.55542	0.58690	0.61515	0.63973	0.66064
6	0.44046	0.47812	0.51515	0.54996	0.58146	0.60911	0.63287
7	0.40030	0.44012	0.47952	0.51681	0.55080	0.58086	0.60690
8	0.36559	0.40684	0.44791	0.48703	0.52290	0.55485	0.58271
9	0.33544	0.37758	0.41977	0.46019	0.49748	0.53089	0.56020
10	0.30915	0.35174	0.39462	0.43593	0.47426	0.50879	0.53925
11	0.28612	0.32883	0.37207	0.41394	0.45299	0.48836	0.51973
12	0.26586	0.30844	0.35176	0.39394	0.43346	0.46944	0.50151
13	0.24794	0.29021	0.33342	0.37569	0.41549	0.45189	0.48448
14	0.23205	0.27385	0.31679	0.35898	0.39890	0.43557	0.46854
15	0.21788	0.25911	0.30166	0.34365	0.38355	0.42035	0.45359
16	0.20520	0.24579	0.28785	0.32954	0.36931	0.40615	0.43955
17	0.19381	0.23369	0.27520	0.31651	0.35607	0.39285	0.42633
18	0.18354	0.22268	0.26359	0.30444	0.34373	0.38039	0.41388
19	0.17424	0.21262	0.25288	0.29325	0.33221	0.36869	0.40212
20	0.16579	0.20340	0.24300	0.28284	0.32142	0.35767	0.39101
21	0.15809	0.19492	0.23384	0.27313	0.31131	0.34729	0.38048
22	0.15104	0.18711	0.22534	0.26406	0.30181	0.33749	0.37051
23	0.14458	0.17988	0.21742	0.25557	0.29286	0.32823	0.36104
24	0.13863	0.17318	0.21004	0.24760	0.28443	0.31946	0.35204
25	0.13314	0.16695	0.20313	0.24011	0.27647	0.31114	0.34348
26	0.12807	0.16115	0.19667	0.23306	0.26894	0.30324	0.33532
27	0.12335	0.15574	0.19059	0.22641	0.26181	0.29574	0.32754
28	0.11897	0.15068	0.18488	0.22013	0.25504	0.28859	0.32011
29	0.11489	0.14593	0.17950	0.21418	0.24862	0.28178	0.31301
30	0.11107	0.14147	0.17443	0.20855	0.24251	0.27528	0.30621
35	0.09523	0.12270	0.15281	0.18430	0.21596	0.24683	0.27623
40	0.08333	0.10833	0.13596	0.16510	0.19465	0.22369	0.25158
45	0.07407	0.09696	0.12245	0.14952	0.17717	0.20452	0.23097
50	0.06667	0.08776	0.11138	0.13663	0.16256	0.18837	0.21347
60	0.05556	0.07375	0.09433	0.11653	0.13956	0.16268	0.18538
70	0.04762	0.06360	0.08181	0.10159	0.12225	0.14316	0.16383
80	0.04167	0.05591	0.07222	0.09004	0.10877	0.12782	0.14676
90	0.03704	0.04987	0.06464	0.08085	0.09796	0.11545	0.13291
100	0.03333	0.04501	0.05850	0.07337	0.08910	0.10526	0.12145

EXPONENTIAL DECLINE DEFERMENT FACTORS
SUMS RECEIVED AT END OF EACH YEAR, I = 0.35

YEARS ON DECLINE	RATIO OF INITIAL TO FINAL RATE						
	1.0	2.0	4.0	8.0	16.0	32.0	64.0
1	0.74074	0.74074	0.74074	0.74074	0.74074	0.74074	0.74074
2	0.64472	0.66119	0.67673	0.69058	0.70233	0.71189	0.71940
3	0.56529	0.59096	0.61560	0.63811	0.65784	0.67451	0.68824
4	0.49924	0.53095	0.56173	0.59025	0.61564	0.63753	0.65595
5	0.44399	0.47978	0.51484	0.54765	0.57719	0.60298	0.62499
6	0.39753	0.43602	0.47404	0.50992	0.54252	0.57125	0.59603
7	0.35822	0.39844	0.43844	0.47648	0.51131	0.54227	0.56920
8	0.32477	0.36598	0.40725	0.44677	0.48321	0.51583	0.54441
9	0.29615	0.33781	0.37980	0.42026	0.45782	0.49167	0.52153
10	0.27150	0.31322	0.35552	0.39653	0.43483	0.46955	0.50037
11	0.25017	0.29165	0.33395	0.37520	0.41394	0.44926	0.48078
12	0.23160	0.27262	0.31469	0.35594	0.39489	0.43059	0.46262
13	0.21534	0.25576	0.29743	0.33850	0.37747	0.41337	0.44574
14	0.20103	0.24074	0.28188	0.32263	0.36148	0.39744	0.43002
15	0.18836	0.22729	0.26781	0.30814	0.34676	0.38267	0.41534
16	0.17710	0.21520	0.25505	0.29487	0.33318	0.36895	0.40162
17	0.16704	0.20429	0.24341	0.28268	0.32061	0.35616	0.38877
18	0.15801	0.19440	0.23277	0.27144	0.30894	0.34422	0.37670
19	0.14987	0.18539	0.22300	0.26105	0.29808	0.33305	0.36535
20	0.14250	0.17717	0.21401	0.25141	0.28795	0.32257	0.35467
21	0.13581	0.16963	0.20570	0.24246	0.27849	0.31273	0.34458
22	0.12969	0.16269	0.19802	0.23412	0.26962	0.30347	0.33505
23	0.12410	0.15630	0.19088	0.22633	0.26130	0.29474	0.32603
24	0.11896	0.15038	0.18423	0.21904	0.25347	0.28650	0.31747
25	0.11422	0.14489	0.17803	0.21220	0.24610	0.27870	0.30936
26	0.10985	0.13979	0.17223	0.20578	0.23914	0.27132	0.30165
27	0.10579	0.13503	0.16680	0.19973	0.23257	0.26431	0.29431
28	0.10202	0.13058	0.16169	0.19403	0.22635	0.25766	0.28732
29	0.09851	0.12642	0.15689	0.18864	0.22045	0.25133	0.28065
30	0.09523	0.12251	0.15237	0.18355	0.21485	0.24531	0.27428
35	0.08163	0.10611	0.13316	0.16170	0.19063	0.21905	0.24633
40	0.07143	0.09358	0.11826	0.14450	0.17131	0.19786	0.22355
45	0.06349	0.08369	0.10635	0.13061	0.15555	0.18041	0.20462
50	0.05714	0.07569	0.09662	0.11915	0.14244	0.16579	0.18864
60	0.04762	0.06355	0.08168	0.10136	0.12190	0.14266	0.16315
70	0.04082	0.05476	0.07074	0.08820	0.10653	0.12519	0.14373
80	0.03571	0.04811	0.06238	0.07806	0.09461	0.11153	0.12844
90	0.03175	0.04290	0.05579	0.07001	0.08508	0.10056	0.11609
100	0.02857	0.03871	0.05046	0.06347	0.07730	0.09155	0.10591

EXPONENTIAL DECLINE DEFERMENT FACTORS
SUMS RECEIVED AT END OF EACH YEAR, I = 0.40

YEARS ON DECLINE	RATIO OF INITIAL TO FINAL RATE						
	1.0	2.0	4.0	8.0	16.0	32.0	64.0
1	0.71429	0.71429	0.71429	0.71429	0.71429	0.71429	0.71429
2	0.61224	0.62975	0.64626	0.66098	0.67347	0.68363	0.69161
3	0.52964	0.55652	0.58235	0.60600	0.62673	0.64429	0.65875
4	0.46231	0.49506	0.52694	0.55654	0.58295	0.60577	0.62501
5	0.40703	0.44354	0.47941	0.51309	0.54351	0.57014	0.59294
6	0.36133	0.40014	0.43862	0.47508	0.50833	0.53774	0.56320
7	0.32326	0.36338	0.40346	0.44176	0.47698	0.50841	0.53587
8	0.29132	0.33203	0.37300	0.41243	0.44898	0.48186	0.51080
9	0.26433	0.30512	0.34645	0.38652	0.42390	0.45777	0.48780
10	0.24136	0.28186	0.32318	0.36349	0.40135	0.43587	0.46668
11	0.22166	0.26163	0.30266	0.34294	0.38100	0.41590	0.44724
12	0.20466	0.24393	0.28448	0.32450	0.36254	0.39763	0.42930
13	0.18988	0.22834	0.26826	0.30789	0.34575	0.38086	0.41271
14	0.17696	0.21453	0.25374	0.29286	0.33042	0.36542	0.39733
15	0.16560	0.20224	0.24067	0.27920	0.31637	0.35117	0.38303
16	0.15553	0.19123	0.22885	0.26674	0.30345	0.33797	0.36971
17	0.14658	0.18133	0.21812	0.25533	0.29154	0.32572	0.35728
18	0.13856	0.17238	0.20833	0.24484	0.28051	0.31432	0.34565
19	0.13136	0.16426	0.19938	0.23518	0.27029	0.30368	0.33474
20	0.12485	0.15686	0.19115	0.22625	0.26078	0.29374	0.32450
21	0.11895	0.15009	0.18358	0.21796	0.25191	0.28442	0.31486
22	0.11357	0.14387	0.17657	0.21026	0.24363	0.27568	0.30578
23	0.10865	0.13815	0.17008	0.20308	0.23587	0.26745	0.29720
24	0.10413	0.13286	0.16405	0.19638	0.22858	0.25970	0.28908
25	0.09998	0.12796	0.15843	0.19010	0.22174	0.25238	0.28140
26	0.09614	0.12341	0.15318	0.18421	0.21529	0.24547	0.27411
27	0.09258	0.11917	0.14827	0.17867	0.20920	0.23892	0.26719
28	0.08928	0.11521	0.14366	0.17346	0.20345	0.23271	0.26061
29	0.08620	0.11150	0.13933	0.16854	0.19801	0.22682	0.25435
30	0.08333	0.10803	0.13526	0.16389	0.19285	0.22121	0.24838
35	0.07143	0.09346	0.11799	0.14404	0.17061	0.19689	0.22228
40	0.06250	0.08236	0.10463	0.12847	0.15297	0.17738	0.20114
45	0.05556	0.07361	0.09399	0.11594	0.13863	0.16138	0.18366
50	0.05000	0.06655	0.08532	0.10563	0.12675	0.14803	0.16898
60	0.04167	0.05583	0.07202	0.08969	0.10821	0.12702	0.14568
70	0.03571	0.04808	0.06231	0.07792	0.09439	0.11123	0.12803
80	0.03125	0.04222	0.05491	0.06889	0.08371	0.09893	0.11419
90	0.02778	0.03764	0.04908	0.06173	0.07520	0.08907	0.10305
100	0.02500	0.03395	0.04436	0.05592	0.06825	0.08101	0.09389

EXPONENTIAL DECLINE DEFERMENT FACTORS
SUMS RECEIVED AT END OF EACH YEAR, I = 0.45

YEARS ON DECLINE	RATIO OF INITIAL TO FINAL RATE						
	1.0	2.0	4.0	8.0	16.0	32.0	64.0
1	0.68966	0.68966	0.68966	0.68966	0.68966	0.68966	0.68966
2	0.58264	0.60100	0.61831	0.63375	0.64685	0.65750	0.66587
3	0.49777	0.52557	0.55232	0.57684	0.59837	0.61661	0.63167
4	0.42988	0.46333	0.49595	0.52632	0.55347	0.57697	0.59682
5	0.37511	0.41195	0.44827	0.48247	0.51345	0.54065	0.56399
6	0.33052	0.36928	0.40784	0.44452	0.47809	0.50789	0.53377
7	0.29390	0.33358	0.37339	0.41159	0.44686	0.47848	0.50619
8	0.26356	0.30346	0.34382	0.38286	0.41920	0.45203	0.48107
9	0.23820	0.27786	0.31828	0.35766	0.39459	0.42821	0.45816
10	0.21681	0.25592	0.29606	0.33542	0.37260	0.40667	0.43722
11	0.19863	0.23698	0.27659	0.31570	0.35286	0.38713	0.41805
12	0.18304	0.22050	0.25943	0.29809	0.33505	0.36933	0.40045
13	0.16958	0.20607	0.24421	0.28231	0.31892	0.35307	0.38423
14	0.15786	0.19334	0.23063	0.26808	0.30425	0.33816	0.36925
15	0.14759	0.18205	0.21846	0.25519	0.29086	0.32444	0.35538
16	0.13853	0.17198	0.20749	0.24348	0.27858	0.31178	0.34250
17	0.13048	0.16294	0.19755	0.23278	0.26729	0.30006	0.33051
18	0.12330	0.15480	0.18851	0.22298	0.25687	0.28919	0.31932
19	0.11686	0.14741	0.18026	0.21396	0.24723	0.27907	0.30886
20	0.11105	0.14070	0.17269	0.20565	0.23829	0.26963	0.29906
21	0.10578	0.13457	0.16573	0.19795	0.22996	0.26081	0.28986
22	0.10098	0.12894	0.15931	0.19081	0.22220	0.25254	0.28121
23	0.09660	0.12377	0.15337	0.18416	0.21494	0.24478	0.27305
24	0.09258	0.11899	0.14785	0.17796	0.20814	0.23748	0.26536
25	0.08888	0.11457	0.14272	0.17216	0.20176	0.23061	0.25808
26	0.08546	0.11046	0.13793	0.16673	0.19576	0.22412	0.25119
27	0.08230	0.10664	0.13345	0.16163	0.19010	0.21798	0.24466
28	0.07936	0.10307	0.12925	0.15683	0.18476	0.21217	0.23845
29	0.07663	0.09973	0.12531	0.15231	0.17971	0.20666	0.23256
30	0.07407	0.09661	0.12160	0.14804	0.17493	0.20143	0.22695
35	0.06349	0.08351	0.10592	0.12985	0.15440	0.17880	0.20251
40	0.05556	0.07355	0.09383	0.11564	0.13817	0.16074	0.18281
45	0.04938	0.06570	0.08421	0.10423	0.12503	0.14599	0.16660
50	0.04444	0.05937	0.07638	0.09487	0.11418	0.13372	0.15303
60	0.03704	0.04978	0.06441	0.08042	0.09728	0.11447	0.13159
70	0.03175	0.04285	0.05568	0.06980	0.08474	0.10007	0.11542
80	0.02778	0.03762	0.04903	0.06165	0.07506	0.08888	0.10279
90	0.02469	0.03353	0.04380	0.05520	0.06737	0.07994	0.09265
100	0.02222	0.03024	0.03958	0.04998	0.06110	0.07264	0.08433

EXPONENTIAL DECLINE DEFERMENT FACTORS
SUMS RECEIVED AT END OF EACH YEAR, I = 0.50

RATIO OF INITIAL TO FINAL RATE

YEARS ON DECLINE	1.0	2.0	4.0	8.0	16.0	32.0	64.0
1	0.66667	0.66667	0.66667	0.66667	0.66667	0.66667	0.66667
2	0.55556	0.57462	0.59259	0.60862	0.62222	0.63328	0.64198
3	0.46914	0.49763	0.52508	0.55026	0.57240	0.59119	0.60670
4	0.40123	0.43511	0.46821	0.49909	0.52675	0.55074	0.57105
5	0.34733	0.38423	0.42071	0.45517	0.48646	0.51400	0.53771
6	0.30407	0.34251	0.38090	0.41754	0.45118	0.48114	0.50724
7	0.26899	0.30800	0.34730	0.38516	0.42026	0.45183	0.47961
8	0.24025	0.27917	0.31872	0.35714	0.39306	0.42565	0.45459
9	0.21644	0.25486	0.29421	0.33273	0.36902	0.40220	0.43189
10	0.19653	0.23418	0.27302	0.31131	0.34765	0.38111	0.41125
11	0.17972	0.21643	0.25456	0.29241	0.32855	0.36206	0.39243
12	0.16538	0.20107	0.23836	0.27561	0.31141	0.34478	0.37522
13	0.15306	0.18767	0.22406	0.26061	0.29593	0.32905	0.35942
14	0.14237	0.17590	0.21134	0.24713	0.28191	0.31467	0.34487
15	0.13303	0.16548	0.19997	0.23497	0.26913	0.30149	0.33144
16	0.12481	0.15621	0.18974	0.22393	0.25746	0.28935	0.31901
17	0.11753	0.14791	0.18050	0.21388	0.24675	0.27815	0.30746
18	0.11104	0.14044	0.17212	0.20469	0.23690	0.26777	0.29672
19	0.10522	0.13369	0.16447	0.19625	0.22779	0.25814	0.28669
20	0.09997	0.12755	0.15747	0.18848	0.21936	0.24918	0.27732
21	0.09522	0.12194	0.15105	0.18130	0.21153	0.24081	0.26854
22	0.09090	0.11681	0.14512	0.17464	0.20424	0.23299	0.26029
23	0.08695	0.11209	0.13964	0.16846	0.19743	0.22566	0.25254
24	0.08333	0.10774	0.13456	0.16270	0.19106	0.21877	0.24523
25	0.08000	0.10371	0.12984	0.15732	0.18509	0.21229	0.23833
26	0.07692	0.09997	0.12543	0.15228	0.17948	0.20618	0.23180
27	0.07407	0.09649	0.12132	0.14755	0.17420	0.20041	0.22563
28	0.07143	0.09324	0.11746	0.14311	0.16922	0.19496	0.21977
29	0.06896	0.09021	0.11385	0.13893	0.16451	0.18980	0.21421
30	0.06667	0.08737	0.11045	0.13499	0.16007	0.18490	0.20892
35	0.05714	0.07548	0.09609	0.11821	0.14100	0.16376	0.18596
40	0.05000	0.06644	0.08504	0.10514	0.12599	0.14695	0.16754
45	0.04444	0.05933	0.07627	0.09467	0.11387	0.13327	0.15244
50	0.04000	0.05359	0.06914	0.08610	0.10387	0.12192	0.13984
60	0.03333	0.04491	0.05825	0.07289	0.08836	0.10418	0.11999
70	0.02857	0.03865	0.05032	0.06320	0.07688	0.09094	0.10507
80	0.02500	0.03392	0.04429	0.05578	0.06803	0.08069	0.09346
90	0.02222	0.03022	0.03955	0.04993	0.06102	0.07251	0.08415
100	0.02000	0.02725	0.03573	0.04518	0.05531	0.06584	0.07653

EXPONENTIAL DECLINE DEFERMENT FACTORS
CONTINUOUS INTEREST, J = 0.04

YEARS ON DECLINE	RATIO OF INITIAL TO FINAL RATE						
	1.0	2.0	4.0	8.0	16.0	32.0	64.0
1	0.98027	0.98251	0.98466	0.98662	0.98835	0.98984	0.99109
2	0.96105	0.96546	0.96967	0.97353	0.97694	0.97986	0.98232
3	0.94233	0.94882	0.95502	0.96071	0.96574	0.97006	0.97370
4	0.92410	0.93259	0.94071	0.94817	0.95476	0.96043	0.96521
5	0.90635	0.91675	0.92672	0.93588	0.94399	0.95097	0.95686
6	0.88905	0.90130	0.91305	0.92385	0.93343	0.94167	0.94864
7	0.87220	0.88622	0.89969	0.91208	0.92306	0.93254	0.94055
8	0.85578	0.87151	0.88662	0.90054	0.91289	0.92356	0.93259
9	0.83979	0.85715	0.87385	0.88924	0.90292	0.91474	0.92475
10	0.82420	0.84313	0.86136	0.87817	0.89313	0.90606	0.91703
11	0.80901	0.82944	0.84914	0.86733	0.88352	0.89753	0.90943
12	0.79420	0.81608	0.83720	0.85671	0.87409	0.88915	0.90195
13	0.77977	0.80304	0.82551	0.84630	0.86483	0.88091	0.89458
14	0.76570	0.79030	0.81408	0.83609	0.85574	0.87280	0.88733
15	0.75198	0.77786	0.80290	0.82609	0.84682	0.86483	0.88018
16	0.73861	0.76571	0.79195	0.81629	0.83806	0.85699	0.87314
17	0.72556	0.75384	0.78125	0.80669	0.82946	0.84928	0.86620
18	0.71284	0.74225	0.77077	0.79727	0.82101	0.84170	0.85937
19	0.70044	0.73092	0.76051	0.78803	0.81271	0.83423	0.85264
20	0.68834	0.71985	0.75047	0.77898	0.80456	0.82689	0.84600
21	0.67654	0.70904	0.74065	0.77010	0.79656	0.81967	0.83947
22	0.66502	0.69847	0.73103	0.76139	0.78869	0.81256	0.83303
23	0.65378	0.68813	0.72161	0.75285	0.78096	0.80556	0.82668
24	0.64282	0.67804	0.71238	0.74447	0.77337	0.79868	0.82042
25	0.63212	0.66816	0.70335	0.73624	0.76590	0.79190	0.81425
26	0.62168	0.65851	0.69450	0.72818	0.75856	0.78523	0.80817
27	0.61149	0.64907	0.68583	0.72026	0.75135	0.77866	0.80217
28	0.60154	0.63985	0.67734	0.71249	0.74427	0.77219	0.79626
29	0.59182	0.63082	0.66902	0.70487	0.73730	0.76582	0.79043
30	0.58234	0.62199	0.66087	0.69738	0.73044	0.75955	0.78468
35	0.53815	0.58064	0.62249	0.66196	0.69785	0.72958	0.75710
40	0.49881	0.54351	0.58772	0.62959	0.66782	0.70177	0.73132
45	0.46372	0.51009	0.55613	0.59993	0.64009	0.67589	0.70718
50	0.43233	0.47992	0.52738	0.57270	0.61443	0.65177	0.68453
60	0.37887	0.42785	0.47711	0.52452	0.56851	0.60817	0.64324
70	0.33543	0.38479	0.43482	0.48334	0.52869	0.56988	0.60655
80	0.29976	0.34883	0.39892	0.44785	0.49391	0.53602	0.57376
90	0.27019	0.31850	0.36816	0.41701	0.46329	0.50589	0.54429
100	0.24542	0.29268	0.34159	0.39001	0.43618	0.47892	0.51768

EXPONENTIAL DECLINE DEFERMENT FACTORS
CONTINUOUS INTEREST, J = 0.06

YEARS ON DECLINE	RATIO OF INITIAL TO FINAL RATE						
	1.0	2.0	4.0	8.0	16.0	32.0	64.0
1	0.97059	0.97393	0.97712	0.98004	0.98262	0.98483	0.98669
2	0.94233	0.94882	0.95502	0.96071	0.96574	0.97006	0.97370
3	0.91517	0.92462	0.93368	0.94199	0.94935	0.95568	0.96102
4	0.88905	0.90130	0.91305	0.92385	0.93343	0.94167	0.94864
5	0.86394	0.87882	0.89312	0.90628	0.91795	0.92803	0.93656
6	0.83979	0.85715	0.87385	0.88924	0.90292	0.91474	0.92475
7	0.81656	0.83625	0.85522	0.87272	0.88830	0.90178	0.91322
8	0.79420	0.81608	0.83720	0.85671	0.87409	0.88915	0.90195
9	0.77269	0.79663	0.81976	0.84117	0.86027	0.87684	0.89094
10	0.75198	0.77786	0.80290	0.82609	0.84682	0.86483	0.88018
11	0.73204	0.75974	0.78657	0.81147	0.83374	0.85312	0.86966
12	0.71284	0.74225	0.77077	0.79727	0.82101	0.84170	0.85937
13	0.69435	0.72535	0.75547	0.78348	0.80862	0.83055	0.84931
14	0.67654	0.70904	0.74065	0.77010	0.79656	0.81967	0.83947
15	0.65937	0.69327	0.72629	0.75710	0.78481	0.80905	0.82984
16	0.64282	0.67804	0.71238	0.74447	0.77337	0.79868	0.82042
17	0.62687	0.66331	0.69890	0.73219	0.76222	0.78855	0.81120
18	0.61149	0.64907	0.68583	0.72026	0.75135	0.77866	0.80217
19	0.59665	0.63531	0.67316	0.70866	0.74077	0.76899	0.79333
20	0.58234	0.62199	0.66087	0.69738	0.73044	0.75955	0.78468
21	0.56853	0.60911	0.64895	0.68641	0.72038	0.75032	0.77621
22	0.55520	0.59664	0.63738	0.67574	0.71057	0.74130	0.76791
23	0.54233	0.58458	0.62616	0.66536	0.70099	0.73248	0.75977
24	0.52991	0.57289	0.61526	0.65525	0.69165	0.72386	0.75180
25	0.51791	0.56158	0.60468	0.64541	0.68253	0.71542	0.74399
26	0.50632	0.55062	0.59440	0.63583	0.67364	0.70717	0.73634
27	0.49512	0.54001	0.58442	0.62650	0.66495	0.69909	0.72883
28	0.48430	0.52972	0.57472	0.61742	0.65647	0.69119	0.72147
29	0.47384	0.51975	0.56530	0.60856	0.64818	0.68346	0.71426
30	0.46372	0.51009	0.55613	0.59993	0.64009	0.67589	0.70718
35	0.41788	0.46593	0.51396	0.55991	0.60231	0.64032	0.67373
40	0.37887	0.42785	0.47711	0.52452	0.56851	0.60817	0.64324
45	0.34548	0.39483	0.44474	0.49306	0.53814	0.57901	0.61533
50	0.31674	0.36603	0.41616	0.46496	0.51073	0.55245	0.58970
60	0.27019	0.31850	0.36816	0.41701	0.46329	0.50589	0.54429
70	0.23452	0.28119	0.32964	0.37775	0.42375	0.46647	0.50532
80	0.20662	0.25133	0.29817	0.34509	0.39034	0.43270	0.47153
90	0.18435	0.22700	0.27206	0.31756	0.36177	0.40347	0.44196
100	0.16625	0.20686	0.25009	0.29405	0.33707	0.37792	0.41587

EXPONENTIAL DECLINE DEFERMENT FACTORS
CONTINUOUS INTEREST, J = 0.08

YEARS ON DECLINE	RATIO OF INITIAL TO FINAL RATE						
	1.0	2.0	4.0	8.0	16.0	32.0	64.0
1	0.96105	0.96546	0.96967	0.97353	0.97694	0.97986	0.98232
2	0.92410	0.93259	0.94071	0.94817	0.95476	0.96043	0.96521
3	0.88905	0.90130	0.91305	0.92385	0.93343	0.94167	0.94864
4	0.85578	0.87151	0.88662	0.90054	0.91289	0.92356	0.93259
5	0.82420	0.84313	0.86136	0.87817	0.89313	0.90606	0.91703
6	0.79420	0.81608	0.83720	0.85671	0.87409	0.88915	0.90195
7	0.76570	0.79030	0.81408	0.83609	0.85574	0.87280	0.88733
8	0.73861	0.76571	0.79195	0.81629	0.83806	0.85699	0.87314
9	0.71284	0.74225	0.77077	0.79727	0.82101	0.84170	0.85937
10	0.68834	0.71985	0.75047	0.77898	0.80456	0.82689	0.84600
11	0.66502	0.69847	0.73103	0.76139	0.78869	0.81256	0.83303
12	0.64282	0.67804	0.71238	0.74447	0.77337	0.79868	0.82042
13	0.62168	0.65851	0.69450	0.72818	0.75856	0.78523	0.80817
14	0.60154	0.63985	0.67734	0.71249	0.74427	0.77219	0.79626
15	0.58234	0.62199	0.66087	0.69738	0.73044	0.75955	0.78468
16	0.56403	0.60491	0.64506	0.68282	0.71708	0.74729	0.77342
17	0.54657	0.58856	0.62986	0.66879	0.70416	0.73540	0.76246
18	0.52991	0.57289	0.61526	0.65525	0.69165	0.72386	0.75180
19	0.51401	0.55789	0.60122	0.64219	0.67954	0.71265	0.74142
20	0.49881	0.54351	0.58772	0.62959	0.66782	0.70177	0.73132
21	0.48430	0.52972	0.57472	0.61742	0.65647	0.69119	0.72147
22	0.47043	0.51650	0.56221	0.60566	0.64547	0.68092	0.71188
23	0.45716	0.50380	0.55017	0.59430	0.63480	0.67093	0.70253
24	0.44448	0.49162	0.53856	C.58332	0.62446	0.66122	0.69342
25	0.43233	0.47992	0.52738	0.57270	0.61443	0.65177	0.68453
26	0.42071	0.46867	0.51659	0.56243	0.60469	0.64258	0.67587
27	0.40957	0.45787	0.50619	0.55249	0.59525	0.63363	0.66741
28	0.39890	0.44747	0.49615	0.54287	0.58607	0.62492	0.65916
29	0.38868	0.43747	0.48646	0.53355	0.57716	0.61644	0.65110
30	0.37887	0.42785	0.47711	0.52452	0.56851	0.60817	0.64324
35	0.33543	0.38479	0.43482	0.48334	0.52869	0.56988	0.60655
40	0.29976	0.34883	0.39892	0.44785	0.49391	0.53602	0.57376
45	0.27019	0.31850	0.36816	0.41701	0.46329	0.50589	0.54429
50	0.24542	0.29268	0.34159	0.39001	0.43618	0.47892	0.51768
60	0.20662	0.25133	0.29817	0.34509	0.39034	0.43270	0.47153
70	0.17791	0.21988	0.26433	0.30932	0.35315	0.39458	0.43290
80	0.15599	0.19528	0.23729	0.28021	0.32239	0.36260	0.40012
90	0.13879	0.17557	0.21523	0.25608	0.29654	0.33542	0.37194
100	0.12496	0.15944	0.19691	0.23577	0.27453	0.31202	0.34747

EXPONENTIAL DECLINE DEFERMENT FACTORS
CONTINUOUS INTEREST, J = 0.10

YEARS ON DECLINE	RATIO OF INITIAL TO FINAL RATE						
	1.0	2.0	4.0	8.0	16.0	32.0	64.0
1	0.95163	0.95708	0.96231	0.96709	0.97131	0.97494	0.97799
2	0.90635	0.91675	0.92672	0.93588	0.94399	0.95097	0.95686
3	0.86394	0.87882	0.89312	0.90628	0.91795	0.92803	0.93656
4	0.82420	0.84313	0.86136	0.87817	0.89313	0.90606	0.91703
5	0.78694	0.80952	0.83132	0.85147	0.86944	0.88501	0.89825
6	0.75198	0.77786	0.80290	0.82609	0.84682	0.86483	0.88018
7	0.71916	0.74801	0.77598	0.80196	0.82522	0.84547	0.86277
8	0.68834	0.71985	0.75047	0.77898	0.80456	0.82689	0.84600
9	0.65937	0.69327	0.72629	0.75710	0.78481	0.80905	0.82984
10	0.63212	0.66816	0.70335	0.73624	0.76590	0.79190	0.81425
11	0.60648	0.64443	0.68157	0.71636	0.74779	0.77541	0.79920
12	0.58234	0.62199	0.66087	0.69738	0.73044	0.75955	0.78468
13	0.55959	0.60075	0.64120	0.67927	0.71381	0.74428	0.77065
14	0.53815	0.58064	0.62249	0.66196	0.69785	0.72958	0.75710
15	0.51791	0.56158	0.60468	0.64541	0.68253	0.71542	0.74399
16	0.49881	0.54351	0.58772	0.62959	0.66782	0.70177	0.73132
17	0.48077	0.52636	0.57155	0.61444	0.65368	0.68860	0.71905
18	0.46372	0.51009	0.55613	0.59993	0.64009	0.67589	0.70718
19	0.44760	0.49462	0.54142	0.58603	0.62701	0.66362	0.69568
20	0.43233	0.47992	0.52738	0.57270	0.61443	0.65177	0.68453
21	0.41788	0.46593	0.51396	0.55991	0.60231	0.64032	0.67373
22	0.40418	0.45262	0.50113	0.54764	0.59063	0.62925	0.66326
23	0.39119	0.43994	0.48885	0.53585	0.57937	0.61854	0.65310
24	0.37887	0.42785	0.47711	0.52452	0.56851	0.60817	0.64324
25	0.36717	0.41633	0.46586	0.51363	0.55803	0.59814	0.63366
26	0.35605	0.40533	0.45508	0.50315	0.54791	0.58842	0.62436
27	0.34548	0.39483	0.44474	0.49306	0.53814	0.57901	0.61533
28	0.33543	0.38479	0.43482	0.48334	0.52869	0.56988	0.60655
29	0.32585	0.37520	0.42530	0.47398	0.51956	0.56103	0.59801
30	0.31674	0.36603	0.41616	0.46496	0.51073	0.55245	0.58970
35	0.27709	0.32562	0.37543	0.42433	0.47059	0.51311	0.55137
40	0.24542	0.29268	0.34159	0.39001	0.43618	0.47892	0.51768
45	0.21975	0.26546	0.31314	0.36070	0.40637	0.44896	0.48784
50	0.19865	0.24268	0.28894	0.33541	0.38033	0.42250	0.46124
60	0.16625	0.20686	0.25009	0.29405	0.33707	0.37792	0.41587
70	0.14273	0.18012	0.22036	0.26172	0.30261	0.34182	0.37861
80	0.12496	0.15944	0.19691	0.23577	0.27453	0.31202	0.34747
90	0.11110	0.14301	0.17796	0.21449	0.25121	0.28699	0.32107
100	0.10000	0.12964	0.16233	0.19674	0.23154	0.26568	0.29839

EXPONENTIAL DECLINE DEFERMENT FACTORS
CONTINUOUS INTEREST, J = 0.12

RATIO OF INITIAL TO FINAL RATE

YEARS ON DECLINE	1.0	2.0	4.0	8.0	16.0	32.0	64.0
1	0.94233	0.94882	0.95502	0.96071	0.96574	0.97006	0.97370
2	0.88905	0.90130	0.91305	0.92385	0.93343	0.94167	0.94864
3	0.83979	0.85715	0.87385	0.88924	0.90292	0.91474	0.92475
4	0.79420	0.81608	0.83720	0.85671	0.87409	0.88915	0.90195
5	0.75198	0.77786	0.80290	0.82609	0.84682	0.86483	0.88018
6	0.71284	0.74225	0.77077	0.79727	0.82101	0.84170	0.85937
7	0.67654	0.70904	0.74065	0.77010	0.79656	0.81967	0.83947
8	0.64282	0.67804	0.71238	0.74447	0.77337	0.79868	0.82042
9	0.61149	0.64907	0.68583	0.72026	0.75135	0.77866	0.80217
10	0.58234	0.62199	0.66087	0.69738	0.73044	0.75955	0.78468
11	0.55520	0.59664	0.63738	0.67574	0.71057	0.74130	0.76791
12	0.52991	0.57289	0.61526	0.65525	0.69165	0.72386	0.75180
13	0.50632	0.55062	0.59440	0.63583	0.67364	0.70717	0.73634
14	0.48430	0.52972	0.57472	0.61742	0.65647	0.69119	0.72147
15	0.46372	0.51009	0.55613	0.59993	0.64009	0.67589	0.70718
16	0.44448	0.49162	0.53856	0.58332	0.62446	0.66122	0.69342
17	0.42646	0.47424	0.52194	0.56752	0.60953	0.64714	0.68017
18	0.40957	0.45787	0.50619	0.55249	0.59525	0.63363	0.66741
19	0.39374	0.44243	0.49127	0.53817	0.58159	0.62065	0.65511
20	0.37887	0.42785	0.47711	0.52452	0.56851	0.60817	0.64324
21	0.36490	0.41409	0.46367	0.51150	0.55598	0.59617	0.63178
22	0.35176	0.40107	0.45089	0.49906	0.54396	0.58462	0.62072
23	0.33939	0.38875	0.43874	0.48718	0.53243	0.57350	0.61003
24	0.32773	0.37709	0.42717	0.47583	0.52136	0.56278	0.59970
25	0.31674	0.36603	0.41616	0.46496	0.51073	0.55245	0.58970
26	0.30636	0.35553	0.40565	0.45455	0.50050	0.54247	0.58003
27	0.29655	0.34556	0.39563	0.44457	0.49067	0.53285	0.57067
28	0.28728	0.33609	0.38606	0.43501	0.48120	0.52356	0.56160
29	0.27850	0.32708	0.37691	0.42583	0.47208	0.51457	0.55281
30	0.27019	0.31850	0.36816	0.41701	0.46329	0.50589	0.54429
35	0.23452	0.28119	0.32964	0.37775	0.42375	0.46647	0.50532
40	0.20662	0.25133	0.29817	0.34509	0.39034	0.43270	0.47153
45	0.18435	0.22700	0.27206	0.31756	0.36177	0.40347	0.44196
50	0.16625	0.20686	0.25009	0.29405	0.33707	0.37792	0.41587
60	0.13879	0.17557	0.21523	0.25608	0.29654	0.33542	0.37194
70	0.11902	0.15244	0.18887	0.22677	0.26470	0.30150	0.33641
80	0.10416	0.13468	0.16824	0.20348	0.23903	0.27381	0.30707
90	0.09259	0.12062	0.15168	0.18452	0.21790	0.25078	0.28243
100	0.08333	0.10922	0.13808	0.16879	0.20020	0.23132	0.26146

EXPONENTIAL DECLINE DEFERMENT FACTORS
CONTINUOUS INTEREST, J = 0.14

YEARS ON DECLINE	RATIO OF INITIAL TO FINAL RATE						
	1.0	2.0	4.0	8.0	16.0	32.0	64.0
1	0.93316	0.94065	0.94783	0.95441	0.96022	0.96522	0.96944
2	0.87220	0.88622	0.89969	0.91208	0.92306	0.93254	0.94055
3	0.81656	0.83625	0.85522	0.87272	0.88830	0.90178	0.91322
4	0.76570	0.79030	0.81408	0.83609	0.85574	0.87280	0.88733
5	0.71916	0.74801	0.77598	0.80196	0.82522	0.84547	0.86277
6	0.67654	0.70904	0.74065	0.77010	0.79656	0.81967	0.83947
7	0.63744	0.67307	0.70784	0.74034	0.76962	0.79527	0.81732
8	0.60154	0.63985	0.67734	0.71249	0.74427	0.77219	0.79626
9	0.56853	0.60911	0.64895	0.68641	0.72038	0.75032	0.77621
10	0.53815	0.58064	0.62249	0.66196	0.69785	0.72958	0.75710
11	0.51014	0.55424	0.59780	0.63900	0.67658	0.70990	0.73887
12	0.48430	0.52972	0.57472	0.61742	0.65647	0.69119	0.72147
13	0.46043	0.50693	0.55314	0.59710	0.63744	0.67340	0.70485
14	0.43834	0.48571	0.53292	0.57797	0.61941	0.65646	0.68895
15	0.41788	0.46593	0.51396	0.55991	0.60231	0.64032	0.67373
16	0.39890	0.44747	0.49615	0.54287	0.58607	0.62492	0.65916
17	0.38128	0.43022	0.47942	0.52675	0.57065	0.61022	0.64518
18	0.36490	0.41409	0.46367	0.51150	0.55598	0.59617	0.63178
19	0.34964	0.39897	0.44882	0.49705	0.54201	0.58274	0.61891
20	0.33543	0.38479	0.43482	0.48334	0.52869	0.56988	0.60655
21	0.32215	0.37148	0.42160	0.47033	0.51599	0.55757	0.59466
22	0.30975	0.35897	0.40910	0.45797	0.50387	0.54576	0.58322
23	0.29815	0.34719	0.39727	0.44620	0.49228	0.53443	0.57221
24	0.28728	0.33609	0.38606	0.43501	0.48120	0.52356	0.56160
25	0.27709	0.32562	0.37543	0.42433	0.47059	0.51311	0.55137
26	0.26751	0.31573	0.36533	0.41415	0.46043	0.50306	0.54151
27	0.25851	0.30638	0.35574	0.40443	0.45069	0.49339	0.53199
28	0.25004	0.29753	0.34661	0.39514	0.44135	0.48408	0.52279
29	0.24206	0.28914	0.33792	0.38625	0.43237	0.47512	0.51391
30	0.23452	0.28119	0.32964	0.37775	0.42375	0.46647	0.50532
35	0.20256	0.24693	0.29349	0.34018	0.38527	0.42754	0.46633
40	0.17791	0.21988	0.26433	0.30932	0.35315	0.39458	0.43290
45	0.15844	0.19805	0.24037	0.28355	0.32594	0.36631	0.40394
50	0.14273	0.18012	0.22036	0.26172	0.30261	0.34182	0.37861
60	0.11902	0.15244	0.18887	0.22677	0.26470	0.30150	0.33641
70	0.10204	0.13211	0.16523	0.20005	0.23523	0.26968	0.30267
80	0.08928	0.11656	0.14686	0.17896	0.21166	0.24394	0.27508
90	0.07936	0.10429	0.13216	0.16189	0.19238	0.22268	0.25210
100	0.07143	0.09435	0.12013	0.14780	0.17633	0.20483	0.23266

EXPONENTIAL DECLINE DEFERMENT FACTORS
CONTINUOUS INTEREST, J = 0.16

YEARS ON DECLINE	RATIO OF INITIAL TO FINAL RATE						
	1.0	2.0	4.0	8.0	16.0	32.0	64.0
1	0.92410	0.93259	0.94071	0.94817	0.95476	0.96043	0.96521
2	0.85578	0.87151	0.88662	0.90054	0.91289	0.92356	0.93259
3	0.79420	0.81608	0.83720	0.85671	0.87409	0.88915	0.90195
4	0.73861	0.76571	0.79195	0.81629	0.83806	0.85699	0.87314
5	0.68834	0.71985	0.75047	0.77898	0.80456	0.82689	0.84600
6	0.64282	0.67804	0.71238	0.74447	0.77337	0.79868	0.82042
7	0.60154	0.63985	0.67734	0.71249	0.74427	0.77219	0.79626
8	0.56403	0.60491	0.64506	0.68282	0.71708	0.74729	0.77342
9	0.52991	0.57289	0.61526	0.65525	0.69165	0.72386	0.75180
10	0.49881	0.54351	0.58772	0.62959	0.66782	0.70177	0.73132
11	0.47043	0.51650	0.56221	0.60566	0.64547	0.68092	0.71188
12	0.44448	0.49162	0.53856	0.58332	0.62446	0.66122	0.69342
13	0.42071	0.46867	0.51659	0.56243	0.60469	0.64258	0.67587
14	0.39890	0.44747	0.49615	0.54287	0.58607	0.62492	0.65916
15	0.37887	0.42785	0.47711	0.52452	0.56851	0.60817	0.64324
16	0.36043	0.40967	0.45933	0.50729	0.55192	0.59227	0.62805
17	0.34343	0.39279	0.44272	0.49109	0.53622	0.57716	0.61355
18	0.32773	0.37709	0.42717	0.47583	0.52136	0.56278	0.59970
19	0.31321	0.36247	0.41260	0.46144	0.50728	0.54908	0.58644
20	0.29976	0.34883	0.39892	0.44785	0.49391	0.53602	0.57376
21	0.28728	0.33609	0.38606	0.43501	0.48120	0.52356	0.56160
22	0.27568	0.32417	0.37395	0.42285	0.46912	0.51165	0.54994
23	0.26489	0.31300	0.36254	0.41133	0.45761	0.50026	0.53875
24	0.25482	0.30253	0.35177	0.40040	0.44664	0.48936	0.52801
25	0.24542	0.29268	0.34159	0.39001	0.43618	0.47892	0.51768
26	0.23663	0.28342	0.33197	0.38014	0.42618	0.46891	0.50774
27	0.22840	0.27469	0.32284	0.37074	0.41663	0.45931	0.49818
28	0.22068	0.26646	0.31419	0.36179	0.40749	0.45009	0.48897
29	0.21344	0.25868	0.30598	0.35325	0.39873	0.44123	0.48009
30	0.20662	0.25133	0.29817	0.34509	0.39034	0.43270	0.47153
35	0.17791	0.21988	0.26433	0.30932	0.35315	0.39458	0.43290
40	0.15599	0.19528	0.23729	0.28021	0.32239	0.36260	0.40012
45	0.13879	0.17557	0.21523	0.25608	0.29654	0.33542	0.37194
50	0.12496	0.15944	0.19691	0.23577	0.27453	0.31202	0.34747
60	0.10416	0.13468	0.16824	0.20348	0.23903	0.27381	0.30707
70	0.08928	0.11656	0.14686	0.17896	0.21166	0.24394	0.27508
80	0.07812	0.10274	0.13029	0.15972	0.18991	0.21994	0.24913
90	0.06944	0.09185	0.11709	0.14421	0.17222	0.20025	0.22765
100	0.06250	0.08305	0.10631	0.13145	0.15754	0.18379	0.20958

EXPONENTIAL DECLINE DEFERMENT FACTORS
CONTINUOUS INTEREST, J = 0.18

RATIO OF INITIAL TO FINAL RATE

YEARS ON DECLINE	1.0	2.0	4.0	8.0	16.0	32.0	64.0
1	0.91517	0.92462	0.93368	0.94199	0.94935	0.95568	0.96102
2	0.83979	0.85715	0.87385	0.88924	0.90292	0.91474	0.92475
3	0.77269	0.79663	0.81976	0.84117	0.86027	0.87684	0.89094
4	0.71284	0.74225	0.77077	0.79727	0.82101	0.84170	0.85937
5	0.65937	0.69327	0.72629	0.75710	0.78481	0.80905	0.82984
6	0.61149	0.64907	0.68583	0.72026	0.75135	0.77866	0.80217
7	0.56853	0.60911	0.64895	0.68641	0.72038	0.75032	0.77621
8	0.52991	0.57289	0.61526	0.65525	0.69165	0.72386	0.75180
9	0.49512	0.54001	0.58442	0.62650	0.66495	0.69909	0.72883
10	0.46372	0.51009	0.55613	0.59993	0.64009	0.67589	0.70718
11	0.43532	0.48280	0.53014	0.57532	0.61691	0.65411	0.68673
12	0.40957	0.45787	0.50619	0.55249	0.59525	0.63363	0.66741
13	0.38618	0.43503	0.48409	0.53126	0.57498	0.61435	0.64912
14	0.36490	0.41409	0.46367	0.51150	0.55598	0.59617	0.63178
15	0.34548	0.39483	0.44474	0.49306	0.53814	0.57901	0.61533
16	0.32773	0.37709	0.42717	0.47583	0.52136	0.56278	0.59970
17	0.31147	0.36071	0.41084	0.45970	0.50557	0.54742	0.58483
18	0.29655	0.34556	0.39563	0.44457	0.49067	0.53285	0.57067
19	0.28283	0.33153	0.38143	0.43037	0.47660	0.51903	0.55717
20	0.27019	0.31850	0.36816	0.41701	0.46329	0.50589	0.54429
21	0.25851	0.30638	0.35574	0.40443	0.45069	0.49339	0.53199
22	0.24771	0.29509	0.34409	0.39256	0.43875	0.48149	0.52022
23	0.23770	0.28455	0.33314	0.38135	0.42741	0.47014	0.50897
24	0.22840	0.27469	0.32284	0.37074	0.41663	0.45931	0.49818
25	0.21975	0.26546	0.31314	0.36070	0.40637	0.44896	0.48784
26	0.21169	0.25681	0.30399	0.35118	0.39660	0.43906	0.47792
27	0.20417	0.24867	0.29534	0.34213	0.38729	0.42959	0.46839
28	0.19713	0.24102	0.28716	0.33353	0.37839	0.42052	0.45924
29	0.19054	0.23381	0.27941	0.32535	0.36990	0.41182	0.45043
30	0.18435	0.22700	0.27206	0.31756	0.36177	0.40347	0.44196
35	0.15844	0.19805	0.24037	0.28355	0.32594	0.36631	0.40394
40	0.13879	0.17557	0.21523	0.25608	0.29654	0.33542	0.37194
45	0.12342	0.15763	0.19483	0.23345	0.27200	0.30932	0.34464
50	0.11110	0.14301	0.17796	0.21449	0.25121	0.28699	0.32107
60	0.09259	0.12062	0.15168	0.18452	0.21790	0.25078	0.28243
70	0.07936	0.10429	0.13216	0.16189	0.19238	0.22268	0.25210
80	0.06944	0.09185	0.11709	0.14421	0.17222	0.20025	0.22765
90	0.06173	0.08206	0.10510	0.13001	0.15588	0.18192	0.20752
100	0.05556	0.07416	0.09535	0.11836	0.14237	0.16666	0.19066

EXPONENTIAL DECLINE DEFERMENT FACTORS
CONTINUOUS INTEREST, J = 0.20

RATIO OF INITIAL TO FINAL RATE

YEARS ON DECLINE	1.0	2.0	4.0	8.0	16.0	32.0	64.0
1	0.90635	0.91675	0.92672	0.93588	0.94399	0.95097	0.95686
2	0.82420	0.84313	0.86136	0.87817	0.89313	0.90606	0.91703
3	0.75198	0.77786	0.80290	0.82609	0.84682	0.86483	0.88018
4	0.68834	0.71985	0.75047	0.77898	0.80456	0.82689	0.84600
5	0.63212	0.66816	0.70335	0.73624	0.76590	0.79190	0.81425
6	0.58234	0.62199	0.66087	0.69738	0.73044	0.75955	0.78468
7	0.53815	0.58064	0.62249	0.66196	0.69785	0.72958	0.75710
8	0.49881	0.54351	0.58772	0.62959	0.66782	0.70177	0.73132
9	0.46372	0.51009	0.55613	0.59993	0.64009	0.67589	0.70718
10	0.43233	0.47992	0.52738	0.57270	0.61443	0.65177	0.68453
11	0.40418	0.45262	0.50113	0.54764	0.59063	0.62925	0.66326
12	0.37887	0.42785	0.47711	0.52452	0.56851	0.60817	0.64324
13	0.35605	0.40533	0.45508	0.50315	0.54791	0.58842	0.62436
14	0.33543	0.38479	0.43482	0.48334	0.52869	0.56988	0.60655
15	0.31674	0.36603	0.41616	0.46496	0.51073	0.55245	0.58970
16	0.29976	0.34883	0.39892	0.44785	0.49391	0.53602	0.57376
17	0.28430	0.33304	0.38296	0.43190	0.47812	0.52053	0.55864
18	0.27019	0.31850	0.36816	0.41701	0.46329	0.50589	0.54429
19	0.25727	0.30508	0.35441	C.40308	0.44933	0.49204	0.53065
20	0.24542	0.29268	0.34159	0.39001	0.43618	0.47892	0.51768
21	0.23452	0.28119	0.32964	0.37775	0.42375	0.46647	0.50532
22	0.22448	0.27052	0.31846	0.36621	0.41201	0.45465	0.49353
23	0.21521	0.26059	0.30799	0.35535	0.40089	0.44341	0.48228
24	0.20662	0.25133	0.29817	0.34509	0.39034	0.43270	0.47153
25	0.19865	0.24268	0.28894	0.33541	0.38033	0.42250	0.46124
26	0.19125	0.23459	0.28026	0.32624	0.37082	0.41277	0.45139
27	0.18435	0.22700	0.27206	0.31756	0.36177	0.40347	0.44196
28	0.17791	0.21988	0.26433	0.30932	0.35315	0.39458	0.43290
29	0.17189	0.21318	0.25702	0.30149	0.34492	0.38607	0.42421
30	0.16625	0.20686	0.25009	0.29405	0.33707	0.37792	0.41587
35	0.14273	0.18012	0.22036	0.26172	0.30261	0.34182	0.37861
40	0.12496	0.15944	0.19691	0.23577	0.27453	0.31202	0.34747
45	0.11110	0.14301	0.17796	0.21449	0.25121	0.28699	0.32107
50	0.10000	0.12964	0.16233	0.19674	0.23154	0.26568	0.29839
60	0.08333	0.10922	0.13808	0.16879	0.20020	0.23132	0.26146
70	0.07143	0.09435	0.12013	0.14780	0.17633	0.20483	0.23266
80	0.06250	0.08305	0.10631	0.13145	0.15754	0.18379	0.20958
90	0.05556	0.07416	0.09535	0.11836	0.14237	0.16666	0.19066
100	0.05000	0.06699	0.08643	0.10763	0.12987	0.15246	0.17488

EXPONENTIAL DECLINE DEFERMENT FACTORS
CONTINUOUS INTEREST, J = 0.25

RATIO OF INITIAL TO FINAL RATE

YEARS ON DECLINE	1.0	2.0	4.0	8.0	16.0	32.0	64.0
1	0.88480	0.89750	0.90968	0.92089	0.93082	0.93937	0.94661
2	0.78694	0.80952	0.83132	0.85147	0.86944	0.88501	0.89825
3	0.70351	0.73373	0.76306	0.79033	0.81477	0.83609	0.85431
4	0.63212	0.66816	0.70335	0.73624	0.76590	0.79190	0.81425
5	0.57080	0.61123	0.65091	0.68822	0.72204	0.75185	0.77761
6	0.51791	0.56158	0.60468	0.64541	0.68253	0.71542	0.74399
7	0.47213	0.51812	0.56375	0.60711	0.64682	0.68219	0.71307
8	0.43233	0.47992	0.52738	0.57270	0.61443	0.65177	0.68453
9	0.39760	0.44620	0.49492	0.54169	0.58495	0.62385	0.65814
10	0.36717	0.41633	0.46586	0.51363	0.55803	0.59814	0.63366
11	0.34039	0.38975	0.43973	0.48815	0.53338	0.57441	0.61091
12	0.31674	0.36603	0.41616	0.46496	0.51073	0.55245	0.58970
13	0.29576	0.34475	0.39481	0.44376	0.48987	0.53206	0.56990
14	0.27709	0.32562	0.37543	0.42433	0.47059	0.51311	0.55137
15	0.26040	0.30834	0.35775	0.40647	0.45275	0.49543	0.53400
16	0.24542	0.29268	0.34159	0.39001	0.43618	0.47892	0.51768
17	0.23194	0.27845	0.32678	0.37480	0.42076	0.46346	0.50232
18	0.21975	0.26546	0.31314	0.36070	0.40637	0.44896	0.48784
19	0.20870	0.25358	0.30057	0.34760	0.39293	0.43533	0.47417
20	0.19865	0.24268	0.28894	0.33541	0.38033	0.42250	0.46124
21	0.18948	0.23265	0.27816	0.32403	0.36852	0.41040	0.44900
22	0.18108	0.22339	0.26814	0.31339	0.35741	0.39897	0.43738
23	0.17336	0.21482	0.25881	0.30341	0.34694	0.38816	0.42635
24	0.16625	0.20686	0.25009	0.29405	0.33707	0.37792	0.41587
25	0.15969	0.19947	0.24194	0.28524	0.32774	0.36820	0.40588
26	0.15361	0.19258	0.23429	0.27695	0.31891	0.35897	0.39636
27	0.14797	0.18614	0.22711	0.26912	0.31055	0.35019	0.38728
28	0.14273	0.18012	0.22036	0.26172	0.30261	0.34182	0.37861
29	0.13783	0.17447	0.21399	0.25471	0.29506	0.33385	0.37031
30	0.13326	0.16915	0.20798	0.24807	0.28789	0.32624	0.36237
35	0.11427	0.14679	0.18235	0.21944	0.25666	0.29286	0.32729
40	0.10000	0.12964	0.16233	0.19674	0.23154	0.26568	0.29839
45	0.08889	0.11607	0.14628	0.17829	0.21090	0.24311	0.27419
50	0.08000	0.10508	0.13311	0.16300	0.19364	0.22408	0.25361
60	0.06667	0.08834	0.11280	0.13914	0.16640	0.19374	0.22052
70	0.05714	0.07620	0.09787	0.12138	0.14588	0.17064	0.19507
80	0.05000	0.06699	0.08643	0.10763	0.12987	0.15246	0.17488
90	0.04444	0.05977	0.07738	0.09669	0.11702	0.13778	0.15848
100	0.04000	0.05396	0.07005	0.08776	0.10649	0.12568	0.14489

EXPONENTIAL DECLINE DEFERMENT FACTORS
CONTINUOUS INTEREST, J = 0.30

YEARS ON DECLINE	RATIO OF INITIAL TO FINAL RATE						
	1.0	2.0	4.0	8.0	16.0	32.0	64.0
1	0.86394	0.87882	0.89312	0.90628	0.91795	0.92803	0.93656
2	0.75198	0.77786	0.80290	0.82609	0.84682	0.86483	0.88018
3	0.65937	0.69327	0.72629	0.75710	0.78481	0.80905	0.82984
4	0.58234	0.62199	0.66087	0.69738	0.73044	0.75955	0.78468
5	0.51791	0.56158	0.60468	0.64541	0.68253	0.71542	0.74399
6	0.46372	0.51009	0.55613	0.59993	0.64009	0.67589	0.70718
7	0.41788	0.46593	0.51396	0.55991	0.60231	0.64032	0.67373
8	0.37887	0.42785	0.47711	0.52452	0.56851	0.60817	0.64324
9	0.34548	0.39483	0.44474	0.49306	0.53814	0.57901	0.61533
10	0.31674	0.36603	0.41616	0.46496	0.51073	0.55245	0.58970
11	0.29185	0.34077	0.39079	0.43974	0.48589	0.52816	0.56610
12	0.27019	0.31850	0.36816	0.41701	0.46329	0.50589	0.54429
13	0.25122	0.29876	0.34789	0.39644	0.44266	0.48539	0.52409
14	0.23452	0.28119	0.32964	0.37775	0.42375	0.46647	0.50532
15	0.21975	0.26546	0.31314	0.36070	0.40637	0.44896	0.48784
16	0.20662	0.25133	0.29817	0.34509	0.39034	0.43270	0.47153
17	0.19488	0.23857	0.28453	0.33076	0.37552	0.41758	0.45626
18	0.18435	0.22700	0.27206	0.31756	0.36177	0.40347	0.44196
19	0.17485	0.21648	0.26062	0.30536	0.34899	0.39028	0.42851
20	0.16625	0.20686	0.25009	0.29405	0.33707	0.37792	0.41587
21	0.15844	0.19805	0.24037	0.28355	0.32594	0.36631	0.40394
22	0.15131	0.18995	0.23137	0.27376	0.31551	0.35540	0.39268
23	0.14478	0.18248	0.22301	0.26463	0.30573	0.34512	0.38203
24	0.13879	0.17557	0.21523	0.25608	0.29654	0.33542	0.37194
25	0.13326	0.16915	0.20798	0.24807	0.28789	0.32624	0.36237
26	0.12815	0.16319	0.20119	0.24054	0.27972	0.31755	0.35328
27	0.12342	0.15763	0.19483	0.23345	0.27200	0.30932	0.34464
28	0.11902	0.15244	0.18887	0.22677	0.26470	0.30150	0.33641
29	0.11492	0.14757	0.18325	0.22046	0.25778	0.29406	0.32856
30	0.11110	0.14301	0.17796	0.21449	0.25121	0.28699	0.32107
35	0.09524	0.12385	0.15551	0.18892	0.22282	0.25616	0.28821
40	0.08333	0.10922	0.13808	0.16879	0.20020	0.23132	0.26146
45	0.07407	0.09767	0.12417	0.15254	0.18174	0.21087	0.23925
50	0.06667	0.08834	0.11280	0.13914	0.16640	0.19374	0.22052
60	0.05556	0.07416	0.09535	0.11836	0.14237	0.16666	0.19066
70	0.04762	0.06390	0.08257	0.10297	0.12440	0.14623	0.16793
80	0.04167	0.05614	0.07281	0.09113	0.11046	0.13025	0.15004
90	0.03704	0.05006	0.06512	0.08172	0.09933	0.11743	0.13559
100	0.03333	0.04517	0.05889	0.07408	0.09024	0.10690	0.12368

EXPONENTIAL DECLINE DEFERMENT FACTORS
CONTINUOUS INTEREST, J = 0.35

YEARS ON DECLINE	RATIO OF INITIAL TO FINAL RATE						
	1.0	2.0	4.0	8.0	16.0	32.0	64.0
1	0.84375	0.86070	0.87702	0.89204	0.90539	0.91693	0.92670
2	0.71916	0.74801	0.77598	0.80196	0.82522	0.84547	0.86277
3	0.61911	0.65613	0.69232	0.72618	0.75675	0.78357	0.80666
4	0.53815	0.58064	0.62249	0.66196	0.69785	0.72958	0.75710
5	0.47213	0.51812	0.56375	0.60711	0.64682	0.68219	0.71307
6	0.41788	0.46593	0.51396	0.55991	0.60231	0.64032	0.67373
7	0.37294	0.42202	0.47142	0.51902	0.56322	0.60312	0.63841
8	0.33543	0.38479	0.43482	0.48334	0.52869	0.56988	0.60655
9	0.30386	0.35299	0.40310	0.45201	0.49801	0.54004	0.57766
10	0.27709	0.32562	0.37543	0.42433	0.47059	0.51311	0.55137
11	0.25421	0.30189	0.35112	0.39973	0.44597	0.48869	0.52735
12	0.23452	0.28119	0.32964	0.37775	0.42375	0.46647	0.50532
13	0.21746	0.26300	0.31055	0.35800	0.40361	0.44617	0.48504
14	0.20256	0.24693	0.29349	0.34018	0.38527	0.42754	0.46633
15	0.18948	0.23265	0.27816	0.32403	0.36852	0.41040	0.44900
16	0.17791	0.21988	0.26433	0.30932	0.35315	0.39458	0.43290
17	0.16763	0.20841	0.25179	0.29588	0.33900	0.37992	0.41792
18	0.15844	0.19805	0.24037	0.28355	0.32594	0.36631	0.40394
19	0.15018	0.18867	0.22993	0.27220	0.31384	0.35365	0.39086
20	0.14273	0.18012	0.22036	0.26172	0.30261	0.34182	0.37861
21	0.13597	0.17230	0.21154	0.25201	0.29215	0.33076	0.36710
22	0.12981	0.16513	0.20340	0.24300	0.28239	0.32040	0.35626
23	0.12418	0.15853	0.19587	0.23460	0.27326	0.31066	0.34605
24	0.11902	0.15244	0.18887	0.22677	0.26470	0.30150	0.33641
25	0.11427	0.14679	0.18235	0.21944	0.25666	0.29286	0.32729
26	0.10988	0.14155	0.17626	0.21258	0.24910	0.28470	0.31865
27	0.10581	0.13667	0.17057	0.20612	0.24196	0.27699	0.31045
28	0.10204	0.13211	0.16523	0.20005	0.23523	0.26968	0.30267
29	0.09852	0.12785	0.16022	0.19433	0.22886	0.26275	0.29526
30	0.09524	0.12385	0.15551	0.18892	0.22282	0.25616	0.28821
35	0.08163	0.10711	0.13555	0.16585	0.19687	0.22764	0.25748
40	0.07143	0.09435	0.12013	0.14780	0.17633	0.20483	0.23266
45	0.06349	0.08431	0.10786	0.13329	0.15967	0.18618	0.21221
50	0.05714	0.07620	0.09787	0.12138	0.14588	0.17064	0.19507
60	0.04762	0.06390	0.08257	0.10297	0.12440	0.14623	0.16793
70	0.04082	0.05503	0.07140	0.08941	0.10844	0.12793	0.14742
80	0.03571	0.04831	0.06290	0.07901	0.09611	0.11370	0.13138
90	0.03175	0.04306	0.05621	0.07077	0.08629	0.10232	0.11848
100	0.02857	0.03884	0.05080	0.06409	0.07830	0.09301	0.10789

EXPONENTIAL DECLINE DEFERMENT FACTORS
CONTINUOUS INTEREST, J = 0.40

YEARS ON DECLINE	RATIO OF INITIAL TO FINAL RATE						
	1.0	2.0	4.0	8.0	16.0	32.0	64.0
1	0.82420	0.84313	0.86136	0.87817	0.89313	0.90606	0.91703
2	0.68834	0.71985	0.75047	0.77898	0.80456	0.82689	0.84600
3	0.58234	0.62199	0.66087	0.69738	0.73044	0.75955	0.78468
4	0.49881	0.54351	0.58772	0.62959	0.66782	0.70177	0.73132
5	0.43233	0.47992	0.52738	0.57270	0.61443	0.65177	0.68453
6	0.37887	0.42785	0.47711	0.52452	0.56851	0.60817	0.64324
7	0.33543	0.38479	0.43482	0.48334	0.52869	0.56988	0.60655
8	0.29976	0.34883	0.39892	0.44785	0.49391	0.53602	0.57376
9	0.27019	0.31850	0.36816	0.41701	0.46329	0.50589	0.54429
10	0.24542	0.29268	0.34159	0.39001	0.43618	0.47892	0.51768
11	0.22448	0.27052	0.31846	0.36621	0.41201	0.45465	0.49353
12	0.20662	0.25133	0.29817	0.34509	0.39034	0.43270	0.47153
13	0.19125	0.23459	0.28026	0.32624	0.37082	0.41277	0.45139
14	0.17791	0.21988	0.26433	0.30932	0.35315	0.39458	0.43290
15	0.16625	0.20686	0.25009	0.29405	0.33707	0.37792	0.41587
16	0.15599	0.19528	0.23729	0.28021	0.32239	0.36260	0.40012
17	0.14690	0.18491	0.22573	0.26760	0.30893	0.34848	0.38552
18	0.13879	0.17557	0.21523	0.25608	0.29654	0.33542	0.37194
19	0.13151	0.16712	0.20566	0.24551	0.28511	0.32329	0.35929
20	0.12496	0.15944	0.19691	0.23577	0.27453	0.31202	0.34747
21	0.11902	0.15244	0.18887	0.22677	0.26470	0.30150	0.33641
22	0.11362	0.14602	0.18145	0.21844	0.25555	0.29167	0.32602
23	0.10868	0.14012	0.17460	0.21069	0.24702	0.28246	0.31626
24	0.10416	0.13468	0.16824	0.20348	0.23903	0.27381	0.30707
25	0.10000	0.12964	0.16233	0.19674	0.23154	0.26568	0.29839
26	0.09615	0.12497	0.15682	0.19043	0.22451	0.25801	0.29019
27	0.09259	0.12062	0.15168	0.18452	0.21790	0.25078	0.28243
28	0.08928	0.11656	0.14686	0.17896	0.21166	0.24394	0.27508
29	0.08621	0.11277	0.14233	0.17373	0.20577	0.23746	0.26810
30	0.08333	0.10922	0.13808	0.16879	0.20020	0.23132	0.26146
35	0.07143	0.09435	0.12013	0.14780	0.17633	0.20483	0.23266
40	0.06250	0.08305	0.10631	0.13145	0.15754	0.18379	0.20958
45	0.05556	0.07416	0.09535	0.11836	0.14237	0.16666	0.19066
50	0.05000	0.06699	0.08643	0.10763	0.12987	0.15246	0.17488
60	0.04167	0.05614	0.07281	0.09113	0.11046	0.13025	0.15004
70	0.03571	0.04831	0.06290	0.07901	0.09611	0.11370	0.13138
80	0.03125	0.04240	0.05536	0.06973	0.08505	0.10087	0.11684
90	0.02778	0.03778	0.04944	0.06241	0.07628	0.09065	0.10520
100	0.02500	0.03407	0.04466	0.05648	0.06914	0.08231	0.09567

EXPONENTIAL DECLINE DEFERMENT FACTORS
CONTINUOUS INTEREST, J = 0.45

YEARS ON DECLINE	RATIO OF INITIAL TO FINAL RATE						
	1.0	2.0	4.0	8.0	16.0	32.0	64.0
1	0.80527	0.82607	0.84613	0.86465	0.88115	0.89542	0.90755
2	0.65937	0.69327	0.72629	0.75710	0.78481	0.80905	0.82984
3	0.54871	0.59056	0.63173	0.67052	0.70575	0.73687	0.76382
4	0.46372	0.51009	0.55613	0.59993	0.64009	0.67589	0.70718
5	0.39760	0.44620	0.49492	0.54169	0.58495	0.62385	0.65814
6	0.34548	0.39483	0.44474	0.49306	0.53814	0.57901	0.61533
7	0.30386	0.35299	0.40310	0.45201	0.49801	0.54004	0.57766
8	0.27019	0.31850	0.36816	0.41701	0.46329	0.50589	0.54429
9	0.24261	0.28973	0.33853	0.38688	0.43300	0.47575	0.51453
10	0.21975	0.26546	0.31314	0.36070	0.40637	0.44896	0.48784
11	0.20059	0.24479	0.29120	0.33778	0.38279	0.42501	0.46377
12	0.18435	0.22700	0.27206	0.31756	0.36177	0.40347	0.44196
13	0.17045	0.21156	0.25525	0.29960	0.34292	0.38400	0.42210
14	0.15844	0.19805	0.24037	0.28355	0.32594	0.36631	0.40394
15	0.14797	0.18614	0.22711	0.26912	0.31055	0.35019	0.38728
16	0.13879	0.17557	0.21523	0.25608	0.29654	0.33542	0.37194
17	0.13066	0.16612	0.20453	0.24424	0.28374	0.32184	0.35777
18	0.12342	0.15763	0.19483	0.23345	0.27200	0.30932	0.34464
19	0.11694	0.14997	0.18602	0.22357	0.26119	0.29774	0.33244
20	0.11110	0.14301	0.17796	0.21449	0.25121	0.28699	0.32107
21	0.10581	0.13667	0.17057	0.20612	0.24196	0.27699	0.31045
22	0.10101	0.13086	0.16377	0.19838	0.23337	0.26766	0.30051
23	0.09662	0.12553	0.15749	0.19120	0.22537	0.25895	0.29119
24	0.09259	0.12062	0.15168	0.18452	0.21790	0.25078	0.28243
25	0.08889	0.11607	0.14628	0.17829	0.21090	0.24311	0.27419
26	0.08547	0.11186	0.14125	0.17247	0.20435	0.23590	0.26641
27	0.08230	0.10794	0.13655	0.16701	0.19818	0.22910	0.25905
28	0.07936	0.10429	0.13216	0.16189	0.19238	0.22268	0.25210
29	0.07663	0.10087	0.12804	0.15708	0.18691	0.21661	0.24551
30	0.07407	0.09767	0.12417	0.15254	0.18174	0.21087	0.23925
35	0.06349	0.08431	0.10786	0.13329	0.15967	0.18618	0.21221
40	0.05556	0.07416	0.09535	0.11836	0.14237	0.16666	0.19066
45	0.04938	0.06619	0.08543	0.10643	0.12846	0.15085	0.17309
50	0.04444	0.05977	0.07738	0.09669	0.11702	0.13778	0.15848
60	0.03704	0.05006	0.06512	0.08172	0.09933	0.11743	0.13559
70	0.03175	0.04306	0.05621	0.07077	0.08629	0.10232	0.11848
80	0.02778	0.03778	0.04944	0.06241	0.07628	0.09065	0.10520
90	0.02469	0.03365	0.04413	0.05581	0.06834	0.08137	0.09460
100	0.02222	0.03034	0.03985	0.05048	0.06191	0.07382	0.08594

EXPONENTIAL DECLINE DEFERMENT FACTORS
CONTINUOUS INTEREST, J = 0.50

YEARS ON DECLINE	RATIO OF INITIAL TO FINAL RATE						
	1.0	2.0	4.0	8.0	16.0	32.0	64.0
1	0.78694	0.80952	0.83132	0.85147	0.86944	0.88501	0.89825
2	0.63212	0.66816	0.70335	0.73624	0.76590	0.79190	0.81425
3	0.51791	0.56158	0.60468	0.64541	0.68253	0.71542	0.74399
4	0.43233	0.47992	0.52738	0.57270	0.61443	0.65177	0.68453
5	0.36717	0.41633	0.46586	0.51363	0.55803	0.59814	0.63366
6	0.31674	0.36603	0.41616	0.46496	0.51073	0.55245	0.58970
7	0.27709	0.32562	0.37543	0.42433	0.47059	0.51311	0.55137
8	0.24542	0.29268	0.34159	0.39001	0.43618	0.47892	0.51768
9	0.21975	0.26546	0.31314	0.36070	0.40637	0.44896	0.48784
10	0.19865	0.24268	0.28894	0.33541	0.38033	0.42250	0.46124
11	0.18108	0.22339	0.26814	0.31339	0.35741	0.39897	0.43738
12	0.16625	0.20686	0.25009	0.29405	0.33707	0.37792	0.41587
13	0.15361	0.19258	0.23429	0.27695	0.31891	0.35897	0.39636
14	0.14273	0.18012	0.22036	C.26172	0.30261	0.34182	0.37861
15	0.13326	0.16915	0.20798	0.24807	0.28789	0.32624	0.36237
16	0.12496	0.15944	0.19691	0.23577	0.27453	0.31202	0.34747
17	0.11762	0.15078	0.18696	0.22463	0.26235	0.29898	0.33375
18	0.11110	0.14301	0.17796	0.21449	0.25121	0.28699	0.32107
19	0.10526	0.13600	0.16979	0.20523	0.24098	0.27592	0.30931
20	0.10000	0.12964	0.16233	0.19674	0.23154	0.26568	0.29839
21	0.09524	0.12385	0.15551	0.18892	0.22282	0.25616	0.28821
22	0.09091	0.11856	0.14923	0.18170	0.21473	0.24731	0.27871
23	0.08696	0.11369	0.14344	0.17501	0.20721	0.23905	0.26981
24	0.08333	0.10922	0.13808	0.16879	0.20020	0.23132	0.26146
25	0.08000	0.10508	0.13311	0.16300	0.19364	0.22408	0.25361
26	0.07692	0.10124	0.12848	0.15760	0.18750	0.21727	0.24622
27	0.07407	0.09767	0.12417	0.15254	0.18174	0.21087	0.23925
28	0.07143	0.09435	0.12013	0.14780	0.17633	0.20483	0.23266
29	0.06897	0.09124	0.11635	0.14334	0.17122	0.19913	0.22643
30	0.06667	0.08834	0.11280	0.13914	0.16640	0.19374	0.22052
35	0.05714	0.07620	0.09787	0.12138	0.14588	0.17064	0.19507
40	0.05000	0.06699	0.08643	0.10763	0.12987	0.15246	0.17488
45	0.04444	0.05977	0.07738	0.09669	0.11702	0.13778	0.15848
50	0.04000	0.05396	0.07005	0.08776	0.10649	0.12568	0.14489
60	0.03333	0.04517	0.05889	0.07408	0.09024	0.10690	0.12368
70	0.02857	0.03884	0.05080	0.06409	0.07830	0.09301	0.10789
80	0.02500	0.03407	0.04466	0.05648	0.06914	0.08231	0.09567
90	0.02222	0.03034	0.03985	0.05048	0.06191	0.07382	0.08594
100	0.02000	0.02735	0.03597	0.04563	0.05604	0.06691	0.07801

HYPERBOLIC (B=0.5) DECLINE PRESENT VALUE FACTORS
CONTINUOUS INTEREST, J = 0.04

YEARS ON DECLINE	RATIO OF INITIAL TO FINAL RATE 1.0	2.0	4.0	8.0	16.0	32.0	64.0
1	0.98026	0.98251	0.98466	0.98662	0.98836	0.98984	0.99109
2	0.96105	0.96547	0.96980	0.97391	0.97771	0.98110	0.98405
3	0.94233	0.94885	0.95523	0.96134	0.96703	0.97220	0.97676
4	0.92410	0.93263	0.94100	0.94902	0.95652	0.96337	0.96946
5	0.90635	0.91680	0.92708	0.93695	0.94620	0.95467	0.96224
6	0.88905	0.90137	0.91349	0.92514	0.93608	0.94612	0.95512
7	0.87220	0.88630	0.90020	0.91358	0.92616	0.93772	0.94811
8	0.85578	0.87160	0.88721	0.90226	0.91643	0.92947	0.94121
9	0.83979	0.85725	0.87451	0.89118	0.90689	0.92136	0.93442
10	0.82420	0.84325	0.86210	0.88032	0.89753	0.91340	0.92773
11	0.80901	0.82958	0.84997	0.86970	0.88835	0.90558	0.92116
12	0.79420	0.81623	0.83810	0.85929	0.87934	0.89790	0.91469
13	0.77977	0.80320	0.82649	0.84909	0.87051	0.89035	0.90832
14	0.76570	0.79048	0.81514	0.83910	0.86184	0.88293	0.90206
15	0.75198	0.77805	0.80403	0.82931	0.85333	0.87564	0.89589
16	0.73861	0.76592	0.79317	0.81972	0.84499	0.86847	0.88982
17	0.72556	0.75406	0.78254	0.81033	0.83679	0.86142	0.88385
18	0.71284	0.74249	0.77215	0.80112	0.82875	0.85450	0.87797
19	0.70044	0.73117	0.76197	0.79209	0.82085	0.84769	0.87218
20	0.68834	0.72012	0.75201	0.78324	0.81310	0.84100	0.86648
21	0.67654	0.70932	0.74226	0.77456	0.80549	0.83441	0.86086
22	0.66502	0.69877	0.73272	0.76606	0.79801	0.82794	0.85533
23	0.65378	0.68845	0.72338	0.75772	0.79067	0.82157	0.84989
24	0.64282	0.67837	0.71423	0.74954	0.78346	0.81530	0.84452
25	0.63212	0.66852	0.70528	0.74151	0.77638	0.80914	0.83923
26	0.62168	0.65888	0.69651	0.73364	0.76942	0.80308	0.83403
27	0.61149	0.64946	0.68792	0.72592	0.76258	0.79711	0.82889
28	0.60154	0.64025	0.67950	0.71835	0.75586	0.79123	0.82384
29	0.59182	0.63124	0.67126	0.71091	0.74926	0.78545	0.81885
30	0.58234	0.62242	0.66318	0.70362	0.74276	0.77976	0.81394
35	0.53815	0.58115	0.62517	0.66911	0.71191	0.75260	0.79039
40	0.49881	0.54410	0.59075	0.63761	0.68353	0.72743	0.76841
45	0.46372	0.51075	0.55950	0.60877	0.65734	0.70404	0.74785
50	0.43233	0.48065	0.53105	0.58231	0.63313	0.68225	0.72857
60	0.37887	0.42871	0.48135	0.53552	0.58982	0.64287	0.69339
70	0.33542	0.38577	0.43955	0.49553	0.55226	0.60825	0.66207
80	0.29976	0.34989	0.40406	0.46106	0.51941	0.57756	0.63398
90	0.27019	0.31964	0.37365	0.43107	0.49045	0.55017	0.60862
100	0.24542	0.29389	0.34737	0.40479	0.46472	0.52554	0.58557

HYPERBOLIC (B=0.5) DECLINE PRESENT VALUE FACTORS
CONTINUOUS INTEREST, J = 0.06

YEARS ON DECLINE	RATIO OF INITIAL TO FINAL RATE						
	1.0	2.0	4.0	8.0	16.0	32.0	64.0
1	0.97059	0.97393	0.97712	0.98004	0.98262	0.98483	0.98669
2	0.94233	0.94885	0.95522	0.96129	0.96689	0.97191	0.97628
3	0.91517	0.92467	0.93400	0.94293	0.95128	0.95886	0.96557
4	0.88905	0.90137	0.91349	0.92513	0.93605	0.94605	0.95496
5	0.86394	0.87891	0.89367	0.90788	0.92125	0.93353	0.94454
6	0.83979	0.85725	0.87451	0.89117	0.90688	0.92134	0.93435
7	0.81656	0.83637	0.85600	0.87498	0.89291	0.90945	0.92438
8	0.79420	0.81623	0.83810	0.85928	0.87934	0.89788	0.91465
9	0.77269	0.79680	0.82078	0.84407	0.86615	0.88661	0.90515
10	0.75198	0.77805	0.80403	0.82931	0.85333	0.87563	0.89587
11	0.73204	0.75996	0.78783	0.81500	0.84087	0.86493	0.88681
12	0.71284	0.74249	0.77215	0.80112	0.82875	0.85450	0.87796
13	0.69435	0.72562	0.75696	0.78764	0.81696	0.84433	0.86931
14	0.67654	0.70932	0.74226	0.77456	0.80549	0.83441	0.86085
15	0.65937	0.69358	0.72803	0.76187	0.79433	0.82474	0.85259
16	0.64282	0.67837	0.71423	0.74954	0.78346	0.81530	0.84451
17	0.62687	0.66367	0.70087	0.73756	0.77288	0.80609	0.83662
18	0.61149	0.64946	0.68792	0.72592	0.76258	0.79711	0.82889
19	0.59665	0.63572	0.67536	0.71461	0.75254	0.78833	0.82133
20	0.58234	0.62242	0.66318	0.70362	0.74276	0.77976	0.81393
21	0.56853	0.60957	0.65137	0.69293	0.73323	0.77139	0.80669
22	0.55520	0.59712	0.63992	0.68253	0.72394	0.76322	0.79961
23	0.54233	0.58508	0.62880	0.67242	0.71488	0.75523	0.79267
24	0.52991	0.57342	0.61801	0.66258	0.70605	0.74742	0.78587
25	0.51791	0.56213	0.60754	0.65301	0.69743	0.73978	0.77921
26	0.50632	0.55120	0.59736	0.64369	0.68902	0.73232	0.77268
27	0.49512	0.54061	0.58749	0.63461	0.68082	0.72501	0.76629
28	0.48430	0.53034	0.57789	0.62577	0.67280	0.71787	0.76002
29	0.47384	0.52039	0.56856	0.61716	0.66498	0.71088	0.75387
30	0.46372	0.51075	0.55950	0.60878	0.65734	0.70404	0.74785
35	0.41738	0.46670	0.51778	0.56989	0.62169	0.67190	0.71937
40	0.37837	0.42871	0.48135	0.53552	0.58982	0.64287	0.69339
45	0.34548	0.39577	0.44936	0.50497	0.56117	0.61650	0.66958
50	0.31674	0.36705	0.42111	0.47768	0.53531	0.59246	0.64766
60	0.27019	0.31964	0.37365	0.43107	0.49045	0.55017	0.60862
70	0.23452	0.28242	0.33554	0.39283	0.45291	0.51414	0.57480
80	0.20662	0.25263	0.30436	0.36093	0.42102	0.48302	0.54516
90	0.18435	0.22834	0.27845	0.33394	0.39359	0.45584	0.51891
100	0.16625	0.20823	0.25661	0.31080	0.36971	0.43184	0.49543

HYPERBOLIC (B=0.5) DECLINE PRESENT VALUE FACTORS
CONTINUOUS INTEREST, J = 0.08

YEARS ON DECLINE	RATIO OF INITIAL TO FINAL RATE						
	1.0	2.0	4.0	8.0	16.0	32.0	64.0
1	0.96105	0.96546	0.96967	0.97353	0.97694	0.97986	0.98232
2	0.92410	0.93262	0.94097	0.94893	0.95629	0.96289	0.96863
3	0.88905	0.90136	0.91348	0.92511	0.93599	0.94590	0.95467
4	0.85578	0.87160	0.88721	0.90225	0.91638	0.92935	0.94094
5	0.82420	0.84325	0.86210	0.88032	0.89750	0.91333	0.92756
6	0.79420	0.81623	0.83810	0.85928	0.87933	0.89785	0.91457
7	0.76570	0.79048	0.81514	0.83910	0.86183	0.88290	0.90198
8	0.73861	0.76592	0.79317	0.81972	0.84498	0.86845	0.88977
9	0.71284	0.74249	0.77215	0.80112	0.82874	0.85448	0.87793
10	0.68834	0.72012	0.75201	0.78324	0.81310	0.84098	0.86644
11	0.66502	0.69877	0.73272	0.76606	0.79801	0.82793	0.85531
12	0.64282	0.67837	0.71423	0.74954	0.78346	0.81530	0.84450
13	0.62168	0.65888	0.69651	0.73364	0.76942	0.80307	0.83401
14	0.60154	0.64025	0.67950	0.71835	0.75586	0.79123	0.82382
15	0.58234	0.62242	0.66318	0.70362	0.74276	0.77976	0.81393
16	0.56403	0.60537	0.64752	0.68943	0.73011	0.76865	0.80431
17	0.54657	0.58905	0.63247	0.67576	0.71788	0.75787	0.79496
18	0.52991	0.57342	0.61801	0.66258	0.70605	0.74742	0.78586
19	0.51401	0.55845	0.60411	0.64987	0.69461	0.73727	0.77702
20	0.49881	0.54410	0.59075	0.63761	0.68353	0.72743	0.76840
21	0.48430	0.53034	0.57789	0.62577	0.67280	0.71787	0.76002
22	0.47043	0.51714	0.56551	0.61434	0.66242	0.70858	0.75185
23	0.45716	0.50448	0.55359	0.60330	0.65235	0.69955	0.74389
24	0.44448	0.49232	0.54212	0.59263	0.64259	0.69078	0.73613
25	0.43233	0.48065	0.53105	0.58231	0.63313	0.68224	0.72856
26	0.42071	0.46943	0.52039	0.57233	0.62395	0.67394	0.72118
27	0.40957	0.45865	0.51010	0.56268	0.61504	0.66586	0.71398
28	0.39890	0.44828	0.50018	0.55333	0.60639	0.65799	0.70696
29	0.38868	0.43831	0.49060	0.54428	0.59798	0.65033	0.70009
30	0.37887	0.42871	0.48135	0.53552	0.58982	0.64287	0.69339
35	0.33542	0.38577	0.43955	0.49553	0.55226	0.60825	0.66207
40	0.29976	0.34989	0.40406	0.46106	0.51941	0.57756	0.63398
45	0.27019	0.31964	0.37365	0.43107	0.49045	0.55017	0.60862
50	0.24542	0.29389	0.34737	0.40479	0.46472	0.52554	0.58557
60	0.20662	0.25263	0.30436	0.36093	0.42102	0.48302	0.54516
70	0.17791	0.22123	0.27077	0.32584	0.38527	0.44752	0.51080
80	0.15599	0.19665	0.24386	0.29712	0.35543	0.41734	0.48110
90	0.13879	0.17695	0.22185	0.27318	0.33011	0.39132	0.45511
100	0.12496	0.16082	0.20351	0.25290	0.30832	0.36860	0.43212

HYPERBOLIC (B=0.5) DECLINE PRESENT VALUE FACTORS
CONTINUOUS INTEREST, J = 0.10

YEARS ON DECLINE	RATIO OF INITIAL TO FINAL RATE						
	1.0	2.0	4.0	8.0	16.0	32.0	64.0
1	0.95163	0.95708	0.96231	0.96709	0.97131	0.97494	0.97799
2	0.90635	0.91680	0.92706	0.93684	0.94590	0.95403	0.96112
3	0.86394	0.87890	0.89366	0.90785	0.92115	0.93328	0.94404
4	0.82420	0.84325	0.86210	0.88031	0.89747	0.91324	0.92738
5	0.78694	0.80968	0.83226	0.85415	0.87487	0.89401	0.91127
6	0.75198	0.77805	0.80403	0.82931	0.85331	0.87558	0.89574
7	0.71916	0.74824	0.77732	0.80570	0.83274	0.85791	0.88079
8	0.68834	0.72012	0.75201	0.78324	0.81309	0.84097	0.86640
9	0.65937	0.69358	0.72802	0.76187	0.79432	0.82472	0.85254
10	0.63212	0.66852	0.70528	0.74151	0.77637	0.80912	0.83919
11	0.60648	0.64483	0.68369	0.72211	0.75920	0.79415	0.82632
12	0.58234	0.62242	0.66318	0.70362	0.74276	0.77976	0.81391
13	0.55959	0.60123	0.64370	0.68597	0.72701	0.76592	0.80193
14	0.53815	0.58115	0.62517	0.66911	0.71191	0.75260	0.79037
15	0.51791	0.56213	0.60754	0.65301	0.69743	0.73978	0.77920
16	0.49881	0.54410	0.59075	0.63761	0.68353	0.72742	0.76840
17	0.48077	0.52699	0.57475	0.62288	0.67018	0.71552	0.75795
18	0.46372	0.51075	0.55950	0.60877	0.65734	0.70403	0.74784
19	0.44760	0.49532	0.54495	0.59526	0.64500	0.69295	0.73805
20	0.43233	0.48065	0.53105	0.58231	0.63313	0.68224	0.72856
21	0.41788	0.46670	0.51778	0.56989	0.62169	0.67190	0.71936
22	0.40418	0.45342	0.50510	0.55797	0.61068	0.66190	0.71045
23	0.39119	0.44077	0.49296	0.54652	0.60006	0.65223	0.70179
24	0.37887	0.42871	0.48135	0.53552	0.58982	0.64287	0.69339
25	0.36717	0.41722	0.47023	0.52494	0.57994	0.63380	0.68522
26	0.35605	0.40625	0.45957	0.51476	0.57039	0.62502	0.67729
27	0.34548	0.39577	0.44936	0.50497	0.56117	0.61650	0.66957
28	0.33542	0.38577	0.43955	0.49553	0.55226	0.60825	0.66207
29	0.32585	0.37620	0.43014	0.48644	0.54364	0.60023	0.65477
30	0.31674	0.36705	0.42111	0.47768	0.53531	0.59246	0.64766
35	0.27709	0.32674	0.38084	0.43819	0.49736	0.55674	0.61472
40	0.24542	0.29389	0.34737	0.40479	0.46472	0.52554	0.58556
45	0.21975	0.26673	0.31920	0.37619	0.43634	0.49803	0.55952
50	0.19865	0.24400	0.29521	0.35145	0.41143	0.47357	0.53607
60	0.16625	0.20823	0.25661	0.31079	0.36971	0.43184	0.49543
70	0.14273	0.18150	0.22697	0.27878	0.33607	0.39748	0.46130
80	0.12496	0.16082	0.20351	0.25290	0.30832	0.36860	0.43212
90	0.11110	0.14436	0.18448	0.23151	0.28500	0.34392	0.40679
100	0.10000	0.13096	0.16873	0.21353	0.26508	0.32255	0.38456

HYPERBOLIC (B=0.5) DECLINE PRESENT VALUE FACTORS
CONTINUOUS INTEREST, J = 0.12

RATIO OF INITIAL TO FINAL RATE

YEARS ON DECLINE	1.0	2.0	4.0	8.0	16.0	32.0	64.0
1	0.94233	0.94882	0.95502	0.96071	0.96574	0.97006	0.97370
2	0.88905	0.90136	0.91345	0.92500	0.93571	0.94534	0.95372
3	0.83979	0.85725	0.87450	0.89113	0.90674	0.92100	0.93367
4	0.79420	0.81623	0.83809	0.85926	0.87927	0.89770	0.91425
5	0.75198	0.77805	0.80403	0.82930	0.85329	0.87552	0.89562
6	0.71284	0.74249	0.77214	0.80111	0.82872	0.85443	0.87778
7	0.67654	0.70932	0.74226	0.77456	0.80547	0.83436	0.86073
8	0.64282	0.67837	0.71423	0.74953	0.78345	0.81527	0.84443
9	0.61149	0.64946	0.68791	0.72592	0.76257	0.79708	0.82882
10	0.58234	0.62242	0.66318	0.70362	0.74276	0.77975	0.81388
11	0.55520	0.59712	0.63992	0.68253	0.72394	0.76320	0.79957
12	0.52991	0.57342	0.61801	0.66258	0.70605	0.74741	0.78584
13	0.50632	0.55120	0.59736	0.64369	0.68902	0.73231	0.77266
14	0.48430	0.53034	0.57789	0.62577	0.67280	0.71786	0.76000
15	0.46372	0.51075	0.55950	0.60877	0.65734	0.70403	0.74783
16	0.44448	0.49232	0.54212	0.59263	0.64259	0.69077	0.73612
17	0.42646	0.47499	0.52567	0.57728	0.62850	0.67806	0.72484
18	0.40957	0.45865	0.51010	0.56268	0.61504	0.66586	0.71398
19	0.39373	0.44325	0.49535	0.54877	0.60215	0.65413	0.70350
20	0.37887	0.42871	0.48135	0.53552	0.58982	0.64286	0.69338
21	0.36490	0.41498	0.46806	0.52287	0.57800	0.63202	0.68361
22	0.35176	0.40200	0.45544	0.51080	0.56667	0.62158	0.67417
23	0.33939	0.38971	0.44343	0.49927	0.55579	0.61152	0.66504
24	0.32773	0.37808	0.43200	0.48823	0.54535	0.60182	0.65621
25	0.31674	0.36705	0.42111	0.47768	0.53531	0.59246	0.64765
26	0.30636	0.35658	0.41072	0.46756	0.52565	0.58342	0.63937
27	0.29655	0.34664	0.40081	0.45787	0.51635	0.57468	0.63133
28	0.28728	0.33719	0.39135	0.44857	0.50740	0.56624	0.62353
29	0.27850	0.32820	0.38231	0.43965	0.49877	0.55807	0.61596
30	0.27019	0.31964	0.37365	0.43107	0.49045	0.55017	0.60861
35	0.23452	0.28242	0.33554	0.39283	0.45291	0.51414	0.57480
40	0.20662	0.25263	0.30436	0.36093	0.42102	0.48302	0.54516
45	0.18435	0.22834	0.27845	0.33394	0.39359	0.45584	0.51890
50	0.16625	0.20823	0.25661	0.31079	0.36971	0.43184	0.49543
60	0.13879	0.17695	0.22185	0.27318	0.33010	0.39132	0.45511
70	0.11902	0.15380	0.19544	0.24387	0.29852	0.35828	0.42157
80	0.10416	0.13601	0.17469	0.22037	0.27269	0.33075	0.39312
90	0.09259	0.12191	0.15796	0.20108	0.25112	0.30739	0.36861
100	0.08333	0.11046	0.14417	0.18495	0.23282	0.28728	0.34722

HYPERBOLIC (B=0.5) DECLINE PRESENT VALUE FACTORS
CONTINUOUS INTEREST, J = 0.14

RATIO OF INITIAL TO FINAL RATE

YEARS ON DECLINE	1.0	2.0	4.0	8.0	16.0	32.0	64.0
1	0.93316	0.94065	0.94783	0.95441	0.96022	0.96522	0.96944
2	0.87220	0.88630	0.90016	0.91342	0.92572	0.93679	0.94645
3	0.81656	0.83637	0.85599	0.87492	0.89274	0.90904	0.92355
4	0.76570	0.79048	0.81513	0.83907	0.86175	0.88270	0.90154
5	0.71916	0.74824	0.77731	0.80568	0.83270	0.85781	0.88057
6	0.67654	0.70932	0.74226	0.77456	0.80546	0.83433	0.86064
7	0.63744	0.67342	0.70973	0.74550	0.77988	0.81215	0.84172
8	0.60154	0.64025	0.67950	0.71834	0.75585	0.79119	0.82373
9	0.56853	0.60957	0.65137	0.69293	0.73322	0.77136	0.80662
10	0.53815	0.58115	0.62517	0.66911	0.71191	0.75258	0.79032
11	0.51014	0.55480	0.60072	0.64677	0.69180	0.73477	0.77480
12	0.48430	0.53034	0.57789	0.62577	0.67280	0.71785	0.75998
13	0.46043	0.50760	0.55653	0.60603	0.65483	0.70178	0.74583
14	0.43834	0.48643	0.53653	0.58743	0.63782	0.68647	0.73230
15	0.41738	0.46670	0.51778	0.56989	0.62169	0.67189	0.71935
16	0.39890	0.44828	0.50018	0.55333	0.60638	0.65799	0.70694
17	0.38128	0.43108	0.48363	0.53768	0.59184	0.64471	0.69504
18	0.36490	0.41498	0.46806	0.52287	0.57800	0.63202	0.68361
19	0.34964	0.39991	0.45339	0.50884	0.56482	0.61987	0.67263
20	0.33542	0.38577	0.43955	0.49553	0.55226	0.60824	0.66206
21	0.32215	0.37249	0.42648	0.48290	0.54028	0.59709	0.65189
22	0.30975	0.36001	0.41413	0.47089	0.52883	0.58639	0.64210
23	0.29815	0.34826	0.40243	0.45946	0.51788	0.57612	0.63265
24	0.28728	0.33719	0.39135	0.44857	0.50740	0.56624	0.62353
25	0.27709	0.32674	0.38084	0.43819	0.49736	0.55674	0.61472
26	0.26751	0.31688	0.37085	0.42829	0.48774	0.54759	0.60621
27	0.25851	0.30755	0.36136	0.41883	0.47850	0.53877	0.59797
28	0.25004	0.29872	0.35234	0.40978	0.46964	0.53027	0.59000
29	0.24206	0.29036	0.34374	0.40112	0.46111	0.52206	0.58228
30	0.23452	0.28242	0.33554	0.39283	0.45291	0.51414	0.57480
35	0.20256	0.24824	0.29972	0.35612	0.41617	0.47824	0.54057
40	0.17791	0.22123	0.27077	0.32584	0.38527	0.44752	0.51080
45	0.15844	0.19943	0.24693	0.30042	0.35888	0.42086	0.48459
50	0.14273	0.18150	0.22697	0.27878	0.33607	0.39748	0.46130
60	0.11902	0.15380	0.19544	0.24387	0.29852	0.35828	0.42157
70	0.10204	0.13343	0.17166	0.21690	0.26883	0.32659	0.38879
80	0.08928	0.11783	0.15308	0.19539	0.24470	0.30036	0.36116
90	0.07936	0.10550	0.13815	0.17783	0.22467	0.27823	0.33750
100	0.07143	0.09551	0.12589	0.16321	0.20775	0.25928	0.31696

HYPERBOLIC (B=0.5) DECLINE PRESENT VALUE FACTORS
CONTINUOUS INTEREST, J = 0.16

YEARS ON DECLINE	RATIO OF INITIAL TO FINAL RATE						
	1.0	2.0	4.0	8.0	16.0	32.0	64.0
1	0.92410	0.93259	0.94071	0.94817	0.95476	0.96043	0.96521
2	0.85578	0.87159	0.88716	0.90207	0.91593	0.92840	0.93929
3	0.79420	0.81623	0.83808	0.85922	0.87914	0.89740	0.91368
4	0.73861	0.76592	0.79316	0.81969	0.84488	0.86820	0.88923
5	0.68834	0.72012	0.75201	0.78322	0.81304	0.84084	0.86610
6	0.64282	0.67837	0.71423	0.74953	0.78343	0.81520	0.84427
7	0.60154	0.64025	0.67950	0.71834	0.75584	0.79117	0.82366
8	0.56403	0.60537	0.64752	0.68943	0.73009	0.76860	0.80419
9	0.52991	0.57342	0.61801	0.66258	0.70604	0.74738	0.78578
10	0.49881	0.54410	0.59075	0.63761	0.68352	0.72740	0.76834
11	0.47043	0.51714	0.56551	0.61434	0.66241	0.70856	0.75180
12	0.44448	0.49232	0.54212	0.59263	0.64259	0.69076	0.73609
13	0.42071	0.46943	0.52039	0.57233	0.62394	0.67393	0.72115
14	0.39890	0.44828	0.50018	0.55333	0.60638	0.65798	0.70693
15	0.37887	0.42871	0.48135	0.53552	0.58982	0.64286	0.69337
16	0.36043	0.41058	0.46378	0.51879	0.57417	0.62849	0.68042
17	0.34343	0.39374	0.44736	0.50305	0.55937	0.61483	0.66804
18	0.32773	0.37808	0.43200	0.48823	0.54534	0.60181	0.65620
19	0.31321	0.36350	0.41759	0.47426	0.53204	0.58940	0.64485
20	0.29976	0.34989	0.40406	0.46106	0.51941	0.57756	0.63397
21	0.28728	0.33719	0.39135	0.44857	0.50740	0.56624	0.62352
22	0.27568	0.32530	0.37938	0.43675	0.49596	0.55541	0.61348
23	0.26489	0.31416	0.36809	0.42554	0.48506	0.54503	0.60382
24	0.25482	0.30371	0.35744	0.41490	0.47466	0.53509	0.59452
25	0.24542	0.29389	0.34737	0.40479	0.46472	0.52554	0.58556
26	0.23663	0.28465	0.33784	0.39516	0.45522	0.51637	0.57691
27	0.22840	0.27594	0.32881	0.38600	0.44612	0.50756	0.56857
28	0.22068	0.26773	0.32024	0.37725	0.43741	0.49907	0.56050
29	0.21344	0.25997	0.31210	0.36891	0.42905	0.49090	0.55270
30	0.20662	0.25263	0.30436	0.36093	0.42102	0.48302	0.54516
35	0.17791	0.22123	0.27077	0.32584	0.38527	0.44752	0.51079
40	0.15599	0.19665	0.24386	0.29712	0.35543	0.41734	0.48110
45	0.13879	0.17695	0.22185	0.27318	0.33010	0.39132	0.45511
50	0.12496	0.16082	0.20351	0.25290	0.30832	0.36860	0.43212
60	0.10416	0.13601	0.17469	0.22037	0.27269	0.33075	0.39312
70	0.08928	0.11783	0.15308	0.19539	0.24470	0.30036	0.36116
80	0.07812	0.10395	0.13625	0.17558	0.22208	0.27534	0.33439
90	0.06944	0.09300	0.12278	0.15947	0.20339	0.25435	0.31158
100	0.06250	0.08413	0.11174	0.14610	0.18767	0.23645	0.29186

HYPERBOLIC (B=0.5) DECLINE PRESENT VALUE FACTORS
CONTINUOUS INTEREST, J = 0.18

YEARS ON DECLINE	RATIO OF INITIAL TO FINAL RATE						
	1.0	2.0	4.0	8.0	16.0	32.0	64.0
1	0.91517	0.92462	0.93368	0.94199	0.94935	0.95568	0.96102
2	0.83979	0.85724	0.87446	0.89096	0.90632	0.92016	0.93225
3	0.77269	0.79680	0.82076	0.84399	0.86592	0.88606	0.90404
4	0.71284	0.74248	0.77214	0.80108	0.82864	0.85420	0.87730
5	0.65937	0.69358	0.72802	0.76185	0.79426	0.82456	0.85218
6	0.61149	0.64946	0.68791	0.72591	0.76254	0.79699	0.82861
7	0.56853	0.60957	0.65137	0.69292	0.73321	0.77132	0.80650
8	0.52991	0.57342	0.61801	0.66258	0.70603	0.74736	0.78573
9	0.49512	0.54061	0.58748	0.63461	0.68080	0.72497	0.76618
10	0.46372	0.51075	0.55950	0.60877	0.65733	0.70401	0.74776
11	0.43532	0.48352	0.53378	0.58486	0.63546	0.68433	0.73038
12	0.40957	0.45865	0.51010	0.56268	0.61503	0.66584	0.71393
13	0.38618	0.43588	0.48826	0.54207	0.59592	0.64843	0.69836
14	0.36490	0.41498	0.46806	0.52287	0.57800	0.63201	0.68359
15	0.34548	0.39577	0.44936	0.50497	0.56117	0.61649	0.66955
16	0.32773	0.37808	0.43200	0.48823	0.54534	0.60181	0.65619
17	0.31147	0.36174	0.41585	0.47257	0.53043	0.58789	0.64346
18	0.29655	0.34664	0.40081	0.45787	0.51635	0.57468	0.63131
19	0.28283	0.33264	0.38678	0.44407	0.50304	0.56212	0.61971
20	0.27019	0.31964	0.37365	0.43107	0.49045	0.55016	0.60860
21	0.25851	0.30755	0.36136	0.41883	0.47850	0.53877	0.59797
22	0.24771	0.29629	0.34984	0.40727	0.46716	0.52789	0.58777
23	0.23770	0.28577	0.33900	0.39634	0.45639	0.51750	0.57797
24	0.22840	0.27594	0.32881	0.38600	0.44612	0.50756	0.56856
25	0.21975	0.26673	0.31920	0.37619	0.43634	0.49803	0.55951
26	0.21169	0.25809	0.31013	0.36688	0.42701	0.48890	0.55079
27	0.20417	0.24998	0.30156	0.35803	0.41810	0.48014	0.54239
28	0.19713	0.24234	0.29344	0.34961	0.40957	0.47172	0.53429
29	0.19053	0.23514	0.28575	0.34159	0.40141	0.46363	0.52646
30	0.18435	0.22834	0.27845	0.33394	0.39359	0.45584	0.51890
35	0.15844	0.19943	0.24693	0.30042	0.35888	0.42086	0.48459
40	0.13879	0.17695	0.22185	0.27318	0.33010	0.39132	0.45511
45	0.12342	0.15900	0.20143	0.25058	0.30581	0.36596	0.42942
50	0.11110	0.14436	0.18448	0.23151	0.28500	0.34392	0.40679
60	0.09259	0.12191	0.15796	0.20108	0.25112	0.30739	0.36861
70	0.07936	0.10550	0.13815	0.17783	0.22467	0.27823	0.33750
80	0.06944	0.09300	0.12278	0.15947	0.20339	0.25435	0.31158
90	0.06173	0.08314	0.11050	0.14458	0.18587	0.23439	0.28958
100	0.05556	0.07518	0.10046	0.13226	0.17119	0.21744	0.27065

HYPERBOLIC (B=0.5) DECLINE PRESENT VALUE FACTORS

CONTINUOUS INTEREST, J = 0.20

YEARS ON DECLINE	RATIO OF INITIAL TO FINAL RATE						
	1.0	2.0	4.0	8.0	16.0	32.0	64.0
1	0.90635	0.91675	0.92672	0.93588	0.94399	0.95097	0.95686
2	0.82420	0.84324	0.86204	0.88009	0.89689	0.91206	0.92533
3	0.75198	0.77805	0.80401	0.82923	0.85308	0.87501	0.89462
4	0.68834	0.72012	0.75200	0.78320	0.81297	0.84066	0.86573
5	0.63212	0.66851	0.70527	0.74149	0.77631	0.80894	0.83876
6	0.58234	0.62242	0.66318	0.70360	0.74272	0.77964	0.81362
7	0.53815	0.58115	0.62517	0.66910	0.71189	0.75252	0.79017
8	0.49881	0.54410	0.59075	0.63760	0.68351	0.72737	0.76825
9	0.46372	0.51075	0.55950	0.60877	0.65733	0.70399	0.74773
10	0.43233	0.48065	0.53105	0.58231	0.63312	0.68221	0.72848
11	0.40418	0.45342	0.50510	0.55797	0.61067	0.66188	0.71038
12	0.37887	0.42871	0.48135	0.53552	0.58981	0.64285	0.69333
13	0.35605	0.40625	0.45957	0.51476	0.57039	0.62500	0.67725
14	0.33542	0.38577	0.43955	0.49553	0.55226	0.60823	0.66204
15	0.31674	0.36705	0.42110	0.47768	0.53530	0.59245	0.64763
16	0.29976	0.34989	0.40406	0.46106	0.51941	0.57755	0.63396
17	0.28430	0.33414	0.38829	0.44556	0.50449	0.56348	0.62096
18	0.27019	0.31964	0.37365	0.43107	0.49045	0.55016	0.60860
19	0.25727	0.30626	0.36005	0.41751	0.47721	0.53753	0.59681
20	0.24542	0.29389	0.34737	0.40479	0.46472	0.52554	0.58555
21	0.23452	0.28242	0.33554	0.39283	0.45291	0.51413	0.57479
22	0.22448	0.27178	0.32447	0.38157	0.44172	0.50327	0.56449
23	0.21521	0.26187	0.31410	0.37096	0.43111	0.49291	0.55462
24	0.20662	0.25263	0.30436	0.36093	0.42102	0.48302	0.54515
25	0.19865	0.24400	0.29521	0.35145	0.41143	0.47356	0.53606
26	0.19125	0.23592	0.28659	0.34246	0.40230	0.46451	0.52732
27	0.18435	0.22834	0.27845	0.33394	0.39359	0.45584	0.51890
28	0.17791	0.22123	0.27077	0.32584	0.38527	0.44751	0.51079
29	0.17189	0.21454	0.26350	0.31813	0.37732	0.43952	0.50297
30	0.16625	0.20823	0.25661	0.31079	0.36971	0.43184	0.49543
35	0.14273	0.18150	0.22697	0.27878	0.33607	0.39748	0.46130
40	0.12496	0.16082	0.20351	0.25290	0.30832	0.36860	0.43211
45	0.11110	0.14436	0.18448	0.23151	0.28500	0.34392	0.40679
50	0.10000	0.13095	0.16873	0.21353	0.26508	0.32255	0.38456
60	0.08333	0.11046	0.14417	0.18495	0.23282	0.28728	0.34722
70	0.07143	0.09551	0.12589	0.16321	0.20775	0.25928	0.31696
80	0.06250	0.08413	0.11174	0.14610	0.18767	0.23645	0.29186
90	0.05556	0.07518	0.10046	0.13226	0.17119	0.21744	0.27065
100	0.05000	0.06795	0.09126	0.12084	0.15742	0.20135	0.25245

HYPERBOLIC (B=0.5) DECLINE PRESENT VALUE FACTORS
CONTINUOUS INTEREST, J = 0.25

RATIO OF INITIAL TO FINAL RATE

YEARS ON DECLINE	1.0	2.0	4.0	8.0	16.0	32.0	64.0
1	0.88480	0.89750	0.90968	0.92089	0.93082	0.93937	0.94661
2	0.78694	0.80967	0.83218	0.85386	0.87411	0.89242	0.90848
3	0.70351	0.73397	0.76447	0.79422	0.82249	0.84860	0.87203
4	0.63212	0.66851	0.70526	0.74146	0.77622	0.80872	0.83830
5	0.57080	0.61168	0.65331	0.69466	0.73471	0.77253	0.80730
6	0.51791	0.56213	0.60753	0.65299	0.69738	0.73962	0.77881
7	0.47213	0.51876	0.56703	0.61574	0.66366	0.70962	0.75258
8	0.43233	0.48065	0.53105	0.58231	0.63310	0.68217	0.72836
9	0.39760	0.44702	0.49896	0.55218	0.60530	0.65697	0.70594
10	0.36717	0.41722	0.47023	0.52494	0.57992	0.63376	0.68511
11	0.34039	0.39071	0.44440	0.50020	0.55667	0.61231	0.66571
12	0.31674	0.36705	0.42110	0.47767	0.53530	0.59243	0.64759
13	0.29576	0.34583	0.40001	0.45708	0.51559	0.57395	0.63061
14	0.27709	0.32674	0.38084	0.43819	0.49736	0.55672	0.61468
15	0.26040	0.30951	0.36336	0.42082	0.48045	0.54062	0.59968
16	0.24542	0.29389	0.34737	0.40479	0.46472	0.52553	0.58553
17	0.23194	0.27969	0.33270	0.38995	0.45005	0.51136	0.57216
18	0.21975	0.26673	0.31920	0.37619	0.43634	0.49803	0.55949
19	0.20870	0.25488	0.30674	0.36339	0.42350	0.48545	0.54747
20	0.19865	0.24400	0.29521	0.35145	0.41143	0.47356	0.53605
21	0.18948	0.23398	0.28451	0.34029	0.40008	0.46230	0.52517
22	0.18108	0.22473	0.27456	0.32983	0.38938	0.45163	0.51480
23	0.17336	0.21617	0.26528	0.32002	0.37927	0.44149	0.50489
24	0.16625	0.20823	0.25661	0.31079	0.36971	0.43184	0.49542
25	0.15969	0.20084	0.24849	0.30210	0.36064	0.42265	0.48635
26	0.15361	0.19396	0.24087	0.29389	0.35204	0.41388	0.47766
27	0.14797	0.18752	0.23371	0.28613	0.34386	0.40550	0.46931
28	0.14273	0.18150	0.22697	0.27878	0.33607	0.39748	0.46130
29	0.13783	0.17585	0.22060	0.27181	0.32865	0.38981	0.45359
30	0.13326	0.17053	0.21459	0.26519	0.32156	0.38245	0.44616
35	0.11427	0.14815	0.18889	0.23650	0.29047	0.34975	0.41280
40	0.10000	0.13096	0.16873	0.21353	0.26508	0.32255	0.38456
45	0.08889	0.11734	0.15249	0.19471	0.24392	0.29950	0.36026
50	0.08000	0.10630	0.13912	0.17898	0.22599	0.27969	0.33908
60	0.06667	0.08946	0.11839	0.15417	0.19719	0.24731	0.30386
70	0.05714	0.07724	0.10306	0.13547	0.17503	0.22189	0.27564
80	0.05000	0.06795	0.09126	0.12084	0.15742	0.20135	0.25245
90	0.04444	0.06066	0.08189	0.10909	0.14308	0.18438	0.23302
100	0.04000	0.05478	0.07427	0.09943	0.13117	0.17011	0.21647

HYPERBOLIC (B=0.5) DECLINE PRESENT VALUE FACTORS

CONTINUOUS INTEREST, J = 0.30

YEARS ON DECLINE	RATIO OF INITIAL TO FINAL RATE						
	1.0	2.0	4.0	8.0	16.0	32.0	64.0
1	0.86394	0.87882	0.89312	0.90628	0.91795	0.92803	0.93656
2	0.75198	0.77804	0.80394	0.82895	0.85238	0.87362	0.89229
3	0.65937	0.69358	0.72799	0.76173	0.79394	0.82380	0.85069
4	0.58234	0.62242	0.66317	0.70356	0.74257	0.77926	0.81282
5	0.51791	0.56213	0.60753	0.65297	0.69732	0.73948	0.77850
6	0.46372	0.51075	0.55949	0.60875	0.65728	0.70384	0.74737
7	0.41788	0.46670	0.51778	0.56988	0.62165	0.67177	0.71903
8	0.37887	0.42871	0.48135	0.53551	0.58979	0.64277	0.69315
9	0.34548	0.39577	0.44936	0.50496	0.56115	0.61643	0.66939
10	0.31674	0.36705	0.42110	0.47767	0.53529	0.59240	0.64752
11	0.29185	0.34185	0.39603	0.45317	0.51182	0.57039	0.62729
12	0.27019	0.31964	0.37365	0.43107	0.49044	0.55014	0.60853
13	0.25122	0.29996	0.35360	0.41105	0.47087	0.53144	0.59106
14	0.23452	0.28242	0.33554	0.39283	0.45290	0.51412	0.57475
15	0.21975	0.26673	0.31920	0.37619	0.43634	0.49802	0.55947
16	0.20662	0.25263	0.30436	0.36093	0.42102	0.48301	0.54512
17	0.19488	0.23989	0.29083	0.34689	0.40681	0.46898	0.53162
18	0.18435	0.22834	0.27845	0.33393	0.39358	0.45583	0.51888
19	0.17485	0.21783	0.26708	0.32194	0.38124	0.44347	0.50683
20	0.16625	0.20823	0.25661	0.31079	0.36970	0.43184	0.49541
21	0.15844	0.19943	0.24693	0.30042	0.35888	0.42086	0.48457
22	0.15131	0.19133	0.23796	0.29074	0.34871	0.41048	0.47427
23	0.14478	0.18386	0.22962	0.28167	0.33914	0.40064	0.46446
24	0.13879	0.17695	0.22185	0.27318	0.33010	0.39131	0.45510
25	0.13326	0.17053	0.21459	0.26519	0.32156	0.38245	0.44616
26	0.12815	0.16457	0.20780	0.25767	0.31348	0.37400	0.43761
27	0.12342	0.15900	0.20143	0.25058	0.30581	0.36596	0.42942
28	0.11902	0.15380	0.19544	0.24387	0.29852	0.35828	0.42156
29	0.11492	0.14893	0.18980	0.23753	0.29159	0.35094	0.41403
30	0.11110	0.14436	0.18448	0.23151	0.28499	0.34392	0.40679
35	0.09524	0.12515	0.16183	0.20557	0.25618	0.31289	0.37441
40	0.08333	0.11046	0.14417	0.18495	0.23282	0.28728	0.34722
45	0.07407	0.09886	0.13001	0.16814	0.21348	0.26573	0.32398
50	0.06667	0.08946	0.11839	0.15417	0.19719	0.24731	0.30386
60	0.05556	0.07518	0.10046	0.13226	0.17119	0.21744	0.27065
70	0.04762	0.06483	0.08726	0.11585	0.15135	0.19419	0.24428
80	0.04167	0.05699	0.07714	0.10308	0.13568	0.17554	0.22279
90	0.03704	0.05084	0.06912	0.09286	0.12299	0.16022	0.20489
100	0.03333	0.04589	0.06262	0.08449	0.11249	0.14741	0.18973

HYPERBOLIC (B=0.5) DECLINE PRESENT VALUE FACTORS
CONTINUOUS INTEREST, J = 0.35

YEARS ON DECLINE	RATIO OF INITIAL TO FINAL RATE						
	1.0	2.0	4.0	8.0	16.0	32.0	64.0
1	0.84375	0.86070	0.87702	0.89204	0.90539	0.91693	0.92670
2	0.71916	0.74822	0.77720	0.80527	0.83164	0.85560	0.87670
3	0.61911	0.65650	0.69430	0.73154	0.76724	0.80048	0.83052
4	0.53815	0.58115	0.62515	0.66904	0.71169	0.75201	0.78908
5	0.47213	0.51876	0.56702	0.61571	0.66357	0.70937	0.75203
6	0.41788	0.46670	0.51778	0.56987	0.62161	0.67167	0.71881
7	0.37294	0.42290	0.47573	0.53016	0.58478	0.63814	0.68888
8	0.33542	0.38577	0.43955	0.49552	0.55223	0.60814	0.66179
9	0.30386	0.35404	0.40820	0.46509	0.52326	0.58113	0.63712
10	0.27709	0.32674	0.38084	0.43819	0.49734	0.55668	0.61456
11	0.25421	0.30308	0.35679	0.41425	0.47401	0.53443	0.59383
12	0.23452	0.28242	0.33554	0.39283	0.45290	0.51410	0.57470
13	0.21746	0.26428	0.31663	0.37355	0.43370	0.49543	0.55698
14	0.20256	0.24824	0.29972	0.35612	0.41616	0.47822	0.54050
15	0.18948	0.23398	0.28451	0.34029	0.40008	0.46229	0.52513
16	0.17791	0.22123	0.27077	0.32584	0.38526	0.44750	0.51075
17	0.16763	0.20977	0.25830	0.31260	0.37157	0.43372	0.49725
18	0.15844	0.19943	0.24693	0.30042	0.35888	0.42085	0.48456
19	0.15018	0.19004	0.23652	0.28918	0.34708	0.40880	0.47259
20	0.14273	0.18150	0.22697	0.27878	0.33607	0.39748	0.46128
21	0.13597	0.17368	0.21816	0.26912	0.32577	0.38682	0.45057
22	0.12981	0.16651	0.21001	0.26013	0.31612	0.37677	0.44040
23	0.12418	0.15990	0.20246	0.25173	0.30706	0.36727	0.43075
24	0.11902	0.15380	0.19544	0.24387	0.29852	0.35828	0.42156
25	0.11427	0.14815	0.18889	0.23650	0.29047	0.34975	0.41279
26	0.10988	0.14290	0.18277	0.22957	0.28286	0.34165	0.40443
27	0.10581	0.13800	0.17704	0.22305	0.27566	0.33394	0.39643
28	0.10204	0.13343	0.17166	0.21690	0.26883	0.32659	0.38878
29	0.09852	0.12916	0.16660	0.21108	0.26234	0.31958	0.38144
30	0.09524	0.12515	0.16183	0.20557	0.25617	0.31289	0.37441
35	0.08163	0.10834	0.14160	0.18192	0.22935	0.28343	0.34309
40	0.07143	0.09551	0.12589	0.16321	0.20775	0.25928	0.31696
45	0.06349	0.08541	0.11333	0.14803	0.18996	0.23907	0.29476
50	0.05714	0.07724	0.10306	0.13547	0.17503	0.22189	0.27564
60	0.04762	0.06483	0.08726	0.11585	0.15135	0.19419	0.24428
70	0.04082	0.05586	0.07567	0.10122	0.13339	0.17278	0.21958
80	0.03571	0.04908	0.06681	0.08989	0.11928	0.15571	0.19956
90	0.03175	0.04376	0.05980	0.08085	0.10789	0.14175	0.18299
100	0.02857	0.03948	0.05413	0.07347	0.09850	0.13013	0.16902

HYPERBOLIC (B=0.5) DECLINE PRESENT VALUE FACTORS
CONTINUOUS INTEREST, J = 0.40

RATIO OF INITIAL TO FINAL RATE

YEARS ON DECLINE	1.0	2.0	4.0	8.0	16.0	32.0	64.0
1	0.82420	0.84313	0.86136	0.87817	0.89313	0.90606	0.91703
2	0.68834	0.72010	0.75187	0.78275	0.81182	0.83832	0.86170
3	0.58234	0.62242	0.66314	0.70344	0.74224	0.77851	0.81141
4	0.49881	0.54410	0.59073	0.63753	0.68327	0.72675	0.76691
5	0.43233	0.48065	0.53104	0.58227	0.63298	0.68184	0.72762
6	0.37887	0.42871	0.48135	0.53549	0.58973	0.64260	0.69275
7	0.33542	0.38577	0.43955	0.49552	0.55220	0.60807	0.66162
8	0.29976	0.34989	0.40406	0.46104	0.51937	0.57744	0.63365
9	0.27019	0.31964	0.37365	0.43106	0.49042	0.55007	0.60837
10	0.24542	0.29389	0.34737	0.40478	0.46470	0.52547	0.58537
11	0.22448	0.27178	0.32447	0.38157	0.44170	0.50322	0.56435
12	0.20662	0.25263	0.30436	0.36093	0.42101	0.48298	0.54504
13	0.19125	0.23592	0.28659	0.34246	0.40229	0.46448	0.52722
14	0.17791	0.22123	0.27077	0.32583	0.38526	0.44749	0.51072
15	0.16625	0.20823	0.25661	0.31079	0.36970	0.43182	0.49537
16	0.15599	0.19665	0.24386	0.29712	0.35542	0.41732	0.48105
17	0.14690	0.18629	0.23233	0.28463	0.34226	0.40385	0.46764
18	0.13879	0.17695	0.22185	0.27317	0.33010	0.39130	0.45507
19	0.13151	0.16850	0.21228	0.26263	0.31882	0.37958	0.44324
20	0.12496	0.16082	0.20351	0.25289	0.30832	0.36859	0.43209
21	0.11902	0.15380	0.19544	0.24387	0.29852	0.35827	0.42155
22	0.11362	0.14737	0.18799	0.23549	0.28936	0.34856	0.41157
23	0.10868	0.14146	0.18110	0.22767	0.28076	0.33940	0.40210
24	0.10416	0.13601	0.17469	0.22037	0.27269	0.33074	0.39310
25	0.10000	0.13096	0.16873	0.21353	0.26508	0.32254	0.38454
26	0.09615	0.12627	0.16317	0.20712	0.25791	0.31477	0.37638
27	0.09259	0.12191	0.15796	0.20108	0.25112	0.30738	0.36860
28	0.08928	0.11783	0.15308	0.19539	0.24470	0.30035	0.36115
29	0.08621	0.11403	0.14849	0.19003	0.23861	0.29366	0.35403
30	0.08333	0.11046	0.14417	0.18495	0.23282	0.28727	0.34721
35	0.07143	0.09551	0.12589	0.16321	0.20775	0.25928	0.31696
40	0.06250	0.08413	0.11174	0.14610	0.18767	0.23645	0.29186
45	0.05556	0.07518	0.10046	0.13226	0.17119	0.21744	0.27065
50	0.05000	0.06795	0.09126	0.12084	0.15742	0.20135	0.25245
60	0.04167	0.05699	0.07714	0.10308	0.13568	0.17554	0.22278
70	0.03571	0.04908	0.06681	0.08989	0.11928	0.15571	0.19956
80	0.03125	0.04309	0.05892	0.07971	0.10644	0.13997	0.18085
90	0.02778	0.03841	0.05270	0.07160	0.09612	0.12716	0.16542
100	0.02500	0.03464	0.04767	0.06500	0.08763	0.11652	0.15247

HYPERBOLIC (B=0.5) DECLINE PRESENT VALUE FACTORS
CONTINUOUS INTEREST, J = 0.45

RATIO OF INITIAL TO FINAL RATE

YEARS ON DECLINE	1.0	2.0	4.0	8.0	16.0	32.0	64.0
1	0.80527	0.82607	0.84613	0.86465	0.88115	0.89543	0.90755
2	0.65937	0.69356	0.72787	0.76131	0.79289	0.82174	0.84725
3	0.54871	0.59105	0.63426	0.67724	0.71879	0.75779	0.79328
4	0.46372	0.51074	0.55947	0.60868	0.65705	0.70327	0.74617
5	0.39760	0.44701	0.49895	0.55213	0.60515	0.65657	0.70502
6	0.34548	0.39577	0.44935	0.50494	0.56107	0.61621	0.66885
7	0.30386	0.35404	0.40820	0.46508	0.52322	0.58100	0.63683
8	0.27019	0.31964	0.37365	0.43106	0.49040	0.55002	0.60824
9	0.24261	0.29094	0.34433	0.40172	0.46167	0.52253	0.58255
10	0.21975	0.26673	0.31920	0.37618	0.43632	0.49795	0.55930
11	0.20059	0.24610	0.29745	0.35376	0.41377	0.47583	0.53814
12	0.18435	0.22834	0.27845	0.33393	0.39357	0.45579	0.51877
13	0.17045	0.21293	0.26174	0.31626	0.37537	0.43754	0.50096
14	0.15844	0.19943	0.24693	0.30042	0.35887	0.42083	0.48450
15	0.14797	0.18752	0.23371	0.28613	0.34385	0.40547	0.46925
16	0.13879	0.17695	0.22185	0.27317	0.33010	0.39129	0.45505
17	0.13066	0.16750	0.21114	0.26137	0.31746	0.37816	0.44179
18	0.12342	0.15900	0.20143	0.25057	0.30580	0.36595	0.42938
19	0.11694	0.15133	0.19258	0.24066	0.29501	0.35456	0.41773
20	0.11110	0.14436	0.18448	0.23151	0.28499	0.34391	0.40676
21	0.10581	0.13800	0.17704	0.22305	0.27566	0.33393	0.39641
22	0.10101	0.13218	0.17018	0.21520	0.26694	0.32455	0.38663
23	0.09662	0.12684	0.16384	0.20790	0.25878	0.31571	0.37737
24	0.09259	0.12191	0.15796	0.20108	0.25112	0.30738	0.36859
25	0.08889	0.11734	0.15249	0.19471	0.24392	0.29950	0.36024
26	0.08547	0.11311	0.14739	0.18873	0.23713	0.29203	0.35229
27	0.08230	0.10918	0.14262	0.18312	0.23073	0.28495	0.34472
28	0.07936	0.10550	0.13815	0.17783	0.22467	0.27822	0.33749
29	0.07663	0.10207	0.13395	0.17285	0.21893	0.27182	0.33058
30	0.07407	0.09886	0.13001	0.16814	0.21348	0.26572	0.32397
35	0.06349	0.08541	0.11333	0.14803	0.18996	0.23907	0.29476
40	0.05556	0.07518	0.10046	0.13226	0.17119	0.21744	0.27064
45	0.04938	0.06714	0.09023	0.11955	0.15586	0.19951	0.25035
50	0.04444	0.06066	0.08189	0.10909	0.14308	0.18438	0.23302
60	0.03704	0.05084	0.06912	0.09286	0.12299	0.16022	0.20488
70	0.03175	0.04376	0.05980	0.08085	0.10789	0.14175	0.18299
80	0.02778	0.03841	0.05270	0.07160	0.09612	0.12716	0.16542
90	0.02469	0.03423	0.04711	0.06426	0.08667	0.11532	0.15099
100	0.02222	0.03086	0.04259	0.05828	0.07893	0.10552	0.13892

HYPERBOLIC (B=0.5) DECLINE PRESENT VALUE FACTORS
CONTINUOUS INTEREST, J = 0.50

YEARS ON DECLINE	RATIO OF INITIAL TO FINAL RATE						
	1.0	2.0	4.0	8.0	16.0	32.0	64.0
1	0.78694	0.80952	0.83132	0.85147	0.86944	0.88501	0.89825
2	0.63212	0.66849	0.70510	0.74089	0.77478	0.80581	0.83331
3	0.51791	0.56212	0.60748	0.65278	0.69677	0.73822	0.77606
4	0.43233	0.48065	0.53103	0.58220	0.63279	0.68139	0.72670
5	0.36717	0.41722	0.47022	0.52488	0.57975	0.63329	0.68403
6	0.31674	0.36704	0.42110	0.47764	0.53519	0.59213	0.64685
7	0.27709	0.32674	0.38083	0.43817	0.49728	0.55651	0.61416
8	0.24542	0.29389	0.34737	0.40477	0.46467	0.52538	0.58515
9	0.21975	0.26673	0.31920	0.37618	0.43630	0.49792	0.55920
10	0.19865	0.24400	0.29521	0.35144	0.41140	0.47348	0.53583
11	0.18108	0.22473	0.27456	0.32983	0.38936	0.45156	0.51462
12	0.16625	0.20823	0.25661	0.31079	0.36969	0.43179	0.49528
13	0.15361	0.19396	0.24087	0.29389	0.35202	0.41384	0.47754
14	0.14273	0.18150	0.22697	0.27878	0.33606	0.39745	0.46120
15	0.13326	0.17053	0.21459	0.26519	0.32155	0.38242	0.44608
16	0.12496	0.16082	0.20351	0.25289	0.30831	0.36858	0.43204
17	0.11762	0.15214	0.19352	0.24172	0.29617	0.35578	0.41896
18	0.11110	0.14436	0.18448	0.23151	0.28499	0.34390	0.40674
19	0.10526	0.13733	0.17625	0.22215	0.27466	0.33285	0.39528
20	0.10000	0.13096	0.16873	0.21353	0.26508	0.32254	0.38452
21	0.09524	0.12515	0.16183	0.20557	0.25617	0.31288	0.37438
22	0.09091	0.11984	0.15548	0.19819	0.24787	0.30382	0.36481
23	0.08696	0.11496	0.14961	0.19134	0.24010	0.29530	0.35577
24	0.08333	0.11046	0.14417	0.18495	0.23282	0.28727	0.34719
25	0.08000	0.10630	0.13912	0.17898	0.22598	0.27969	0.33906
26	0.07692	0.10244	0.13441	0.17339	0.21955	0.27252	0.33133
27	0.07407	0.09886	0.13001	0.16814	0.21348	0.26572	0.32397
28	0.07143	0.09551	0.12589	0.16321	0.20775	0.25927	0.31695
29	0.06897	0.09239	0.12202	0.15856	0.20233	0.25314	0.31025
30	0.06667	0.08946	0.11839	0.15417	0.19718	0.24731	0.30385
35	0.05714	0.07724	0.10306	0.13547	0.17503	0.22189	0.27563
40	0.05000	0.06795	0.09126	0.12084	0.15742	0.20135	0.25245
45	0.04444	0.06066	0.08189	0.10909	0.14308	0.18438	0.23301
50	0.04000	0.05478	0.07427	0.09943	0.13117	0.17011	0.21647
60	0.03333	0.04589	0.06262	0.08449	0.11249	0.14741	0.18973
70	0.02857	0.03948	0.05413	0.07347	0.09850	0.13013	0.16901
80	0.02500	0.03464	0.04767	0.06500	0.08763	0.11652	0.15246
90	0.02222	0.03086	0.04259	0.05828	0.07893	0.10552	0.13892
100	0.02000	0.02783	0.03849	0.05283	0.07181	0.09643	0.12762

HARMONIC (B=1.0) DECLINE PRESENT VALUE FACTORS
CONTINUOUS INTEREST, J = 0.04

YEARS ON DECLINE	RATIO OF INITIAL TO FINAL RATE						
	1.0	2.0	4.0	8.0	16.0	32.0	64.0
1	0.98026	0.98251	0.98466	0.98662	0.98836	0.98984	0.99109
2	0.96105	0.96546	0.96968	0.97356	0.97698	0.97991	0.98239
3	0.94233	0.94883	0.95506	0.96078	0.96584	0.97019	0.97386
4	0.92410	0.93260	0.94077	0.94828	0.95494	0.96067	0.96550
5	0.90635	0.91677	0.92681	0.93606	0.94427	0.95134	0.95731
6	0.88905	0.90133	0.91318	0.92411	0.93382	0.94220	0.94928
7	0.87220	0.88627	0.89986	0.91242	0.92360	0.93325	0.94141
8	0.85578	0.87157	0.88684	0.90098	0.91358	0.92448	0.93370
9	0.83979	0.85722	0.87412	0.88979	0.90378	0.91588	0.92614
10	0.82420	0.84322	0.86168	0.87884	0.89417	0.90746	0.91872
11	0.80901	0.82955	0.84953	0.86812	0.88476	0.89920	0.91145
12	0.79420	0.81620	0.83765	0.85764	0.87554	0.89110	0.90432
13	0.77977	0.80318	0.82603	0.84737	0.86651	0.88316	0.89732
14	0.76570	0.79046	0.81468	0.83732	0.85767	0.87538	0.89046
15	0.75198	0.77804	0.80357	0.82748	0.84899	0.86775	0.88372
16	0.73861	0.76591	0.79271	0.81784	0.84049	0.86026	0.87711
17	0.72556	0.75406	0.78208	0.80841	0.83216	0.85292	0.87063
18	0.71284	0.74249	0.77169	0.79917	0.82400	0.84571	0.86426
19	0.70044	0.73118	0.76152	0.79012	0.81599	0.83864	0.85801
20	0.68834	0.72014	0.75158	0.78125	0.80814	0.83171	0.85188
21	0.67654	0.70935	0.74184	0.77256	0.80044	0.82490	0.84585
22	0.66502	0.69880	0.73231	0.76405	0.79288	0.81822	0.83994
23	0.65378	0.68850	0.72299	0.75571	0.78547	0.81166	0.83413
24	0.64282	0.67842	0.71386	0.74753	0.77821	0.80522	0.82842
25	0.63212	0.66858	0.70493	0.73952	0.77108	0.79890	0.82281
26	0.62168	0.65895	0.69618	0.73166	0.76408	0.79269	0.81730
27	0.61149	0.64954	0.68762	0.72396	0.75721	0.78659	0.81189
28	0.60154	0.64034	0.67923	0.71641	0.75047	0.78060	0.80657
29	0.59182	0.63134	0.67102	0.70901	0.74386	0.77472	0.80134
30	0.58234	0.62254	0.66297	0.70174	0.73736	0.76894	0.79620
35	0.53815	0.58132	0.62512	0.66745	0.70660	0.74151	0.77177
40	0.49881	0.54433	0.59089	0.63624	0.67846	0.71631	0.74927
45	0.46372	0.51104	0.55985	0.60775	0.65264	0.69310	0.72849
50	0.43233	0.48100	0.53162	0.58167	0.62889	0.67167	0.70924
60	0.37887	0.42918	0.48236	0.53572	0.58670	0.63338	0.67471
70	0.33542	0.38634	0.44100	0.49663	0.55043	0.60019	0.64461
80	0.29976	0.35057	0.40593	0.46304	0.51894	0.57116	0.61813
90	0.27019	0.32040	0.37590	0.43392	0.49136	0.54552	0.59461
100	0.24542	0.29473	0.34997	0.40845	0.46699	0.52269	0.57356

HARMONIC (B=1.0) DECLINE PRESENT VALUE FACTORS
CONTINUOUS INTEREST, J = 0.06

YEARS ON DECLINE	RATIO OF INITIAL TO FINAL RATE						
	1.0	2.0	4.0	8.0	16.0	32.0	64.0
1	0.97059	0.97393	0.97712	0.98004	0.98262	0.98483	0.98669
2	0.94233	0.94883	0.95506	0.96077	0.96583	0.97018	0.97384
3	0.91517	0.92464	0.93375	0.94214	0.94957	0.95597	0.96137
4	0.88905	0.90133	0.91318	0.92411	0.93382	0.94220	0.94927
5	0.86394	0.87887	0.89331	0.90667	0.91856	0.92884	0.93753
6	0.83979	0.85722	0.87412	0.88979	0.90377	0.91588	0.92613
7	0.81656	0.83634	0.85557	0.87345	0.88944	0.90330	0.91506
8	0.79420	0.81620	0.83765	0.85764	0.87554	0.89110	0.90431
9	0.77269	0.79678	0.82032	0.84232	0.86207	0.87925	0.89387
10	0.75198	0.77804	0.80357	0.82748	0.84899	0.86774	0.88372
11	0.73204	0.75995	0.78737	0.81310	0.83631	0.85657	0.87385
12	0.71284	0.74249	0.77169	0.79917	0.82399	0.84571	0.86426
13	0.69435	0.72563	0.75652	0.78566	0.81204	0.83516	0.85493
14	0.67654	0.70935	0.74184	0.77256	0.80043	0.82490	0.84585
15	0.65937	0.69362	0.72763	0.75986	0.78916	0.81492	0.83701
16	0.64282	0.67842	0.71386	0.74753	0.77821	0.80522	0.82841
17	0.62687	0.66374	0.70053	0.73557	0.76756	0.79578	0.82004
18	0.61149	0.64954	0.68762	0.72396	0.75721	0.78659	0.81189
19	0.59665	0.63581	0.67510	0.71269	0.74715	0.77765	0.80394
20	0.58234	0.62254	0.66297	0.70174	0.73736	0.76894	0.79620
21	0.56853	0.60969	0.65121	0.69111	0.72784	0.76046	0.78865
22	0.55520	0.59727	0.63980	0.68078	0.71858	0.75220	0.78130
23	0.54233	0.58524	0.62874	0.67074	0.70956	0.74414	0.77412
24	0.52991	0.57360	0.61800	0.66098	0.70078	0.73630	0.76712
25	0.51791	0.56233	0.60758	0.65149	0.69223	0.72864	0.76029
26	0.50632	0.55141	0.59747	0.64225	0.68390	0.72118	0.75362
27	0.49512	0.54084	0.58765	0.63327	0.67578	0.71390	0.74712
28	0.48430	0.53059	0.57811	0.62454	0.66787	0.70680	0.74076
29	0.47384	0.52066	0.56885	0.61603	0.66016	0.69987	0.73455
30	0.46372	0.51104	0.55985	0.60775	0.65264	0.69310	0.72849
35	0.41788	0.46708	0.51846	0.56945	0.61771	0.66155	0.70013
40	0.37887	0.42918	0.48236	0.53572	0.58670	0.63338	0.67471
45	0.34548	0.39632	0.45070	0.50584	0.55901	0.60807	0.65177
50	0.31674	0.36767	0.42276	0.47922	0.53415	0.58521	0.63096
60	0.27019	0.32040	0.37590	0.43392	0.49136	0.54552	0.59461
70	0.23452	0.28329	0.33830	0.39688	0.45583	0.51219	0.56383
80	0.20662	0.25358	0.30755	0.36605	0.42583	0.48373	0.53733
90	0.18435	0.22937	0.28200	0.33999	0.40012	0.45907	0.51419
100	0.16625	0.20930	0.26044	0.31766	0.37780	0.43745	0.49374

HARMONIC (B=1.0) DECLINE PRESENT VALUE FACTORS
CONTINUOUS INTEREST, J = 0.08

YEARS ON DECLINE	RATIO OF INITIAL TO FINAL RATE						
	1.0	2.0	4.0	8.0	16.0	32.0	64.0
1	0.96105	0.96546	0.96967	0.97353	0.97694	0.97986	0.98232
2	0.92410	0.93260	0.94077	0.94827	0.95492	0.96064	0.96546
3	0.88905	0.90133	0.91318	0.92410	0.93381	0.94218	0.94925
4	0.85578	0.87157	0.88684	0.90098	0.91357	0.92446	0.93367
5	0.82420	0.84322	0.86168	0.87884	0.89416	0.90744	0.91870
6	0.79420	0.81620	0.83765	0.85763	0.87554	0.89109	0.90430
7	0.76570	0.79046	0.81468	0.83732	0.85766	0.87537	0.89044
8	0.73861	0.76591	0.79271	0.81784	0.84049	0.86025	0.87710
9	0.71284	0.74249	0.77169	0.79917	0.82399	0.84570	0.86425
10	0.68834	0.72014	0.75158	0.78125	0.80813	0.83170	0.85186
11	0.66502	0.69880	0.73231	0.76405	0.79288	0.81821	0.83992
12	0.64282	0.67842	0.71386	0.74753	0.77820	0.80522	0.82841
13	0.62168	0.65895	0.69618	0.73166	0.76408	0.79269	0.81729
14	0.60154	0.64034	0.67923	0.71641	0.75047	0.78060	0.80656
15	0.58234	0.62254	0.66297	0.70174	0.73736	0.76894	0.79620
16	0.56403	0.60551	0.64737	0.68763	0.72472	0.75768	0.78618
17	0.54657	0.58921	0.63239	0.67405	0.71254	0.74680	0.77649
18	0.52991	0.57360	0.61800	0.66098	0.70078	0.73629	0.76712
19	0.51401	0.55865	0.60418	0.64838	0.68942	0.72614	0.75805
20	0.49881	0.54433	0.59089	0.63624	0.67846	0.71631	0.74926
21	0.48430	0.53059	0.57811	0.62454	0.66787	0.70680	0.74076
22	0.47043	0.51742	0.56582	0.61324	0.65763	0.69759	0.73251
23	0.45716	0.50478	0.55399	0.60235	0.64773	0.68868	0.72451
24	0.44448	0.49265	0.54259	0.59183	0.63815	0.68004	0.71676
25	0.43233	0.48100	0.53162	0.58167	0.62888	0.67166	0.70923
26	0.42071	0.46981	0.52104	0.57186	0.61991	0.66354	0.70193
27	0.40957	0.45905	0.51085	0.56237	0.61121	0.65566	0.69483
28	0.39890	0.44870	0.50101	0.55319	0.60279	0.64801	0.68793
29	0.38868	0.43875	0.49152	0.54431	0.59462	0.64059	0.68123
30	0.37887	0.42918	0.48236	0.53572	0.58670	0.63337	0.67471
35	0.33542	0.38634	0.44100	0.49663	0.55043	0.60019	0.64461
40	0.29976	0.35057	0.40593	0.46304	0.51894	0.57116	0.61813
45	0.27019	0.32040	0.37590	0.43392	0.49136	0.54552	0.59461
50	0.24542	0.29473	0.34997	0.40845	0.46698	0.52269	0.57356
60	0.20662	0.25358	0.30755	0.36605	0.42583	0.48373	0.53732
70	0.17791	0.22227	0.27442	0.33217	0.39234	0.45156	0.50710
80	0.15599	0.19775	0.24786	0.30445	0.36447	0.42442	0.48134
90	0.13879	0.17807	0.22609	0.28131	0.34083	0.40112	0.45902
100	0.12496	0.16196	0.20793	0.26167	0.32047	0.38082	0.43940

HARMONIC (B=1.0) DECLINE PRESENT VALUE FACTORS
CONTINUOUS INTEREST, J = 0.10

YEARS ON DECLINE	RATIO OF INITIAL TO FINAL RATE						
	1.0	2.0	4.0	8.0	16.0	32.0	64.0
1	0.95163	0.95708	0.96231	0.96709	0.97131	0.97494	0.97799
2	0.90635	0.91677	0.92681	0.93604	0.94423	0.95128	0.95723
3	0.86394	0.87887	0.89331	0.90666	0.91854	0.92880	0.93748
4	0.82420	0.84321	0.86168	0.87883	0.89415	0.90743	0.91867
5	0.78694	0.80965	0.83181	0.85247	0.87099	0.88709	0.90076
6	0.75198	0.77804	0.80357	0.82748	0.84898	0.86773	0.88369
7	0.71916	0.74824	0.77686	0.80376	0.82805	0.84928	0.86740
8	0.68834	0.72014	0.75158	0.78125	0.80813	0.83169	0.85185
9	0.65937	0.69362	0.72763	0.75986	0.78916	0.81491	0.83699
10	0.63212	0.66858	0.70493	0.73952	0.77107	0.79889	0.82279
11	0.60648	0.64491	0.68340	0.72017	0.75382	0.78358	0.80920
12	0.58234	0.62254	0.66297	0.70174	0.73736	0.76893	0.79619
13	0.55959	0.60137	0.64356	0.68419	0.72163	0.75492	0.78372
14	0.53815	0.58132	0.62512	0.66745	0.70660	0.74150	0.77176
15	0.51791	0.56233	0.60758	0.65149	0.69222	0.72864	0.76028
16	0.49881	0.54433	0.59089	0.63624	0.67846	0.71631	0.74926
17	0.48077	0.52725	0.57499	0.62167	0.66528	0.70447	0.73866
18	0.46372	0.51104	0.55985	0.60775	0.65264	0.69310	0.72848
19	0.44760	0.49564	0.54540	0.59443	0.64052	0.68217	0.71867
20	0.43233	0.48100	0.53162	0.58167	0.62888	0.67166	0.70923
21	0.41788	0.46708	0.51846	0.56945	0.61771	0.66155	0.70013
22	0.40418	0.45383	0.50588	0.55774	0.60697	0.65181	0.69135
23	0.39119	0.44121	0.49386	0.54651	0.59664	0.64242	0.68288
24	0.37887	0.42918	0.48236	0.53572	0.58670	0.63337	0.67470
25	0.36717	0.41771	0.47135	0.52536	0.57713	0.62464	0.66680
26	0.35605	0.40677	0.46080	0.51541	0.56790	0.61621	0.65916
27	0.34548	0.39632	0.45070	0.50584	0.55901	0.60806	0.65177
28	0.33542	0.38634	0.44100	0.49663	0.55043	0.60019	0.64461
29	0.32585	0.37680	0.43170	0.48776	0.54215	0.59258	0.63768
30	0.31674	0.36767	0.42276	0.47922	0.53415	0.58521	0.63096
35	0.27709	0.32748	0.38299	0.44083	0.49793	0.55164	0.60024
40	0.24542	0.29473	0.34997	0.40845	0.46698	0.52269	0.57356
45	0.21975	0.26765	0.32218	0.38079	0.44023	0.49742	0.55010
50	0.19865	0.24498	0.29853	0.35689	0.41684	0.47513	0.52927
60	0.16625	0.20930	0.26044	0.31766	0.37780	0.43745	0.49373
70	0.14273	0.18262	0.23116	0.28673	0.34640	0.40664	0.46433
80	0.12496	0.16196	0.20793	0.26167	0.32047	0.38082	0.43940
90	0.11110	0.14550	0.18905	0.24089	0.29861	0.35876	0.41789
100	0.10000	0.13209	0.17338	0.22335	0.27988	0.33962	0.39904

HARMONIC (B=1.0) DECLINE PRESENT VALUE FACTORS
CONTINUOUS INTEREST, J = 0.12

YEARS ON DECLINE	RATIO OF INITIAL TO FINAL RATE						
	1.0	2.0	4.0	8.0	16.0	32.0	64.0
1	0.94233	0.94882	0.95502	0.96071	0.96574	0.97006	0.97370
2	0.88905	0.90133	0.91317	0.92408	0.93377	0.94212	0.94917
3	0.83979	0.85722	0.87411	0.88977	0.90374	0.91582	0.92605
4	0.79420	0.81620	0.83765	0.85763	0.87552	0.89106	0.90425
5	0.75198	0.77804	0.80357	0.82747	0.84898	0.86771	0.88366
6	0.71284	0.74249	0.77169	0.79916	0.82398	0.84568	0.86421
7	0.67654	0.70935	0.74184	0.77256	0.80042	0.82488	0.84581
8	0.64282	0.67842	0.71386	0.74753	0.77820	0.80520	0.82838
9	0.61149	0.64954	0.68762	0.72396	0.75721	0.78658	0.81186
10	0.58234	0.62254	0.66297	0.70174	0.73736	0.76893	0.79617
11	0.55520	0.59727	0.63980	0.68078	0.71857	0.75218	0.78127
12	0.52991	0.57360	0.61800	0.66098	0.70077	0.73629	0.76710
13	0.50632	0.55141	0.59747	0.64225	0.68389	0.72117	0.75360
14	0.48430	0.53059	0.57811	0.62453	0.66787	0.70679	0.74074
15	0.46372	0.51104	0.55985	0.60775	0.65264	0.69309	0.72847
16	0.44448	0.49265	0.54259	0.59183	0.63815	0.68003	0.71675
17	0.42646	0.47535	0.52628	0.57672	0.62436	0.66757	0.70554
18	0.40957	0.45905	0.51085	0.56237	0.61121	0.65566	0.69482
19	0.39373	0.44368	0.49622	0.54872	0.59867	0.64427	0.68454
20	0.37887	0.42918	0.48236	0.53572	0.58670	0.63337	0.67470
21	0.36490	0.41548	0.46921	0.52334	0.57525	0.62293	0.66524
22	0.35176	0.40253	0.45671	0.51154	0.56431	0.61292	0.65617
23	0.33939	0.39028	0.44483	0.50027	0.55382	0.60331	0.64744
24	0.32773	0.37867	0.43353	0.48951	0.54378	0.59408	0.63904
25	0.31674	0.36767	0.42276	0.47922	0.53415	0.58520	0.63096
26	0.30636	0.35723	0.41250	0.46937	0.52490	0.57667	0.62316
27	0.29655	0.34732	0.40271	0.45994	0.51602	0.56845	0.61565
28	0.28728	0.33790	0.39337	0.45091	0.50748	0.56053	0.60839
29	0.27850	0.32894	0.38444	0.44224	0.49927	0.55289	0.60138
30	0.27019	0.32040	0.37590	0.43392	0.49136	0.54552	0.59461
35	0.23452	0.28329	0.33830	0.39688	0.45583	0.51219	0.56382
40	0.20662	0.25358	0.30755	0.36605	0.42583	0.48373	0.53732
45	0.18435	0.22937	0.28200	0.33999	0.40012	0.45907	0.51418
50	0.16625	0.20930	0.26044	0.31766	0.37780	0.43745	0.49373
60	0.13879	0.17807	0.22609	0.28131	0.34082	0.40112	0.45902
70	0.11902	0.15495	0.19993	0.25291	0.31130	0.37160	0.43044
80	0.10416	0.13715	0.17931	0.23003	0.28704	0.34697	0.40630
90	0.09259	0.12303	0.16263	0.21116	0.26669	0.32600	0.38551
100	0.08333	0.11156	0.14885	0.19530	0.24931	0.30786	0.36735

HARMONIC (B=1.0) DECLINE PRESENT VALUE FACTORS
CONTINUOUS INTEREST, J = 0.14

YEARS ON DECLINE	RATIO OF INITIAL TO FINAL RATE						
	1.0	2.0	4.0	8.0	16.0	32.0	64.0
1	0.93316	0.94065	0.94783	0.95441	0.96022	0.96522	0.96944
2	0.87220	0.88626	0.89984	0.91238	0.92352	0.93314	0.94126
3	0.81656	0.83634	0.85557	0.87343	0.88939	0.90322	0.91495
4	0.76570	0.79046	0.81467	0.83731	0.85763	0.87532	0.89036
5	0.71916	0.74824	0.77686	0.80375	0.82803	0.84925	0.86735
6	0.67654	0.70935	0.74184	0.77255	0.80042	0.82486	0.84578
7	0.63744	0.67347	0.70937	0.74350	0.77461	0.80201	0.82554
8	0.60154	0.64034	0.67923	0.71641	0.75046	0.78058	0.80652
9	0.56853	0.60969	0.65121	0.69111	0.72783	0.76044	0.78861
10	0.53815	0.58132	0.62512	0.66745	0.70660	0.74149	0.77173
11	0.51014	0.55501	0.60081	0.64530	0.68664	0.72363	0.75579
12	0.48430	0.53059	0.57811	0.62453	0.66786	0.70679	0.74073
13	0.46043	0.50789	0.55690	0.60504	0.65017	0.69087	0.72646
14	0.43834	0.48677	0.53705	0.58671	0.63348	0.67581	0.71294
15	0.41788	0.46708	0.51846	0.56945	0.61771	0.66154	0.70011
16	0.39890	0.44870	0.50101	0.55319	0.60279	0.64801	0.68791
17	0.38128	0.43154	0.48462	0.53784	0.58865	0.63515	0.67630
18	0.36490	0.41548	0.46921	0.52334	0.57525	0.62293	0.66524
19	0.34964	0.40045	0.45469	0.50962	0.56253	0.61128	0.65468
20	0.33542	0.38634	0.44100	0.49663	0.55043	0.60019	0.64460
21	0.32215	0.37310	0.42808	0.48431	0.53891	0.58960	0.63495
22	0.30975	0.36065	0.41587	0.47261	0.52794	0.57948	0.62572
23	0.29815	0.34894	0.40431	0.46149	0.51747	0.56979	0.61687
24	0.28728	0.33790	0.39337	0.45091	0.50748	0.56053	0.60839
25	0.27709	0.32748	0.38299	0.44083	0.49793	0.55164	0.60023
26	0.26751	0.31765	0.37314	0.43122	0.48878	0.54311	0.59239
27	0.25851	0.30835	0.36377	0.42205	0.48003	0.53493	0.58485
28	0.25004	0.29955	0.35487	0.41328	0.47163	0.52705	0.57758
29	0.24206	0.29121	0.34638	0.40490	0.46357	0.51948	0.57058
30	0.23452	0.28329	0.33830	0.39688	0.45583	0.51219	0.56382
35	0.20256	0.24921	0.30297	0.36141	0.42128	0.47938	0.53325
40	0.17791	0.22227	0.27442	0.33217	0.39234	0.45156	0.50709
45	0.15844	0.20052	0.25088	0.30764	0.36770	0.42758	0.48435
50	0.14273	0.18262	0.23116	0.28673	0.34640	0.40664	0.46432
60	0.11902	0.15495	0.19993	0.25291	0.31129	0.37160	0.43044
70	0.10204	0.13457	0.17630	0.22664	0.28341	0.34325	0.40262
80	0.08928	0.11895	0.15776	0.20558	0.26060	0.31968	0.37920
90	0.07936	0.10659	0.14281	0.18827	0.24153	0.29966	0.35907
100	0.07143	0.09656	0.13049	0.17376	0.22530	0.28240	0.34152

HARMONIC (B=1.0) DECLINE PRESENT VALUE FACTORS
CONTINUOUS INTEREST, J = 0.16

RATIO OF INITIAL TO FINAL RATE

YEARS ON DECLINE	1.0	2.0	4.0	8.0	16.0	32.0	64.0
1	0.92410	0.93259	0.94071	0.94817	0.95476	0.96043	0.96521
2	0.85578	0.87156	0.88682	0.90093	0.91349	0.92433	0.93350
3	0.79420	0.81620	0.83764	0.85761	0.87548	0.89099	0.90416
4	0.73861	0.76591	0.79270	0.81783	0.84045	0.86018	0.87699
5	0.68834	0.72014	0.75157	0.78124	0.80810	0.83164	0.85177
6	0.64282	0.67842	0.71386	0.74752	0.77818	0.80517	0.82833
7	0.60154	0.64034	0.67923	0.71640	0.75045	0.78056	0.80649
8	0.56403	0.60551	0.64736	0.68763	0.72471	0.75765	0.78612
9	0.52991	0.57360	0.61800	0.66097	0.70076	0.73627	0.76706
10	0.49881	0.54433	0.59089	0.63624	0.67845	0.71628	0.74922
11	0.47043	0.51742	0.56582	0.61324	0.65762	0.69757	0.73247
12	0.44448	0.49265	0.54259	0.59183	0.63815	0.68002	0.71672
13	0.42071	0.46981	0.52104	0.57185	0.61990	0.66353	0.70189
14	0.39890	0.44870	0.50101	0.55319	0.60278	0.64800	0.68790
15	0.37887	0.42918	0.48236	0.53572	0.58669	0.63336	0.67468
16	0.36043	0.41109	0.46497	0.51934	0.57155	0.61954	0.66216
17	0.34343	0.39429	0.44872	0.50397	0.55726	0.60646	0.65029
18	0.32773	0.37867	0.43353	0.48951	0.54378	0.59407	0.63903
19	0.31321	0.36413	0.41929	0.47589	0.53102	0.58232	0.62831
20	0.29976	0.35057	0.40593	0.46304	0.51894	0.57115	0.61811
21	0.28728	0.33790	0.39337	0.45091	0.50748	0.56052	0.60838
22	0.27568	0.32605	0.38155	0.43943	0.49659	0.55040	0.59909
23	0.26489	0.31494	0.37041	0.42856	0.48624	0.54074	0.59020
24	0.25482	0.30452	0.35990	0.41824	0.47638	0.53151	0.58170
25	0.24542	0.29473	0.34997	0.40845	0.46698	0.52269	0.57354
26	0.23663	0.28551	0.34057	0.39914	0.45801	0.51424	0.56572
27	0.22840	0.27683	0.33166	0.39027	0.44943	0.50614	0.55820
28	0.22068	0.26864	0.32321	0.38182	0.44123	0.49837	0.55098
29	0.21344	0.26090	0.31518	0.37375	0.43337	0.49090	0.54402
30	0.20662	0.25358	0.30755	0.36605	0.42583	0.48372	0.53731
35	0.17791	0.22227	0.27442	0.33217	0.39234	0.45156	0.50709
40	0.15599	0.19775	0.24786	0.30445	0.36447	0.42442	0.48134
45	0.13879	0.17807	0.22609	0.28131	0.34082	0.40112	0.45902
50	0.12496	0.16196	0.20793	0.26167	0.32047	0.38082	0.43940
60	0.10416	0.13715	0.17931	0.23003	0.28704	0.34697	0.40629
70	0.08928	0.11895	0.15776	0.20558	0.26060	0.31967	0.37919
80	0.07812	0.10503	0.14091	0.18604	0.23906	0.29704	0.35642
90	0.06944	0.09404	0.12736	0.17004	0.22109	0.27788	0.33690
100	0.06250	0.08513	0.11622	0.15666	0.20584	0.26139	0.31990

HARMONIC (B=1.0) DECLINE PRESENT VALUE FACTORS
CONTINUOUS INTEREST, J = 0.18

YEARS ON DECLINE	RATIO OF INITIAL TO FINAL RATE						
	1.0	2.0	4.0	8.0	16.0	32.0	64.0
1	0.91517	0.92462	0.93368	0.94199	0.94935	0.95568	0.96102
2	0.83979	0.85721	0.87409	0.88973	0.90365	0.91569	0.92588
3	0.77269	0.79678	0.82031	0.84228	0.86199	0.87912	0.89367
4	0.71234	0.74249	0.77168	0.79915	0.82394	0.84561	0.86410
5	0.65937	0.69362	0.72762	0.75984	0.78912	0.81484	0.83688
6	0.61149	0.64954	0.68761	0.72395	0.75718	0.78653	0.81178
7	0.56853	0.60969	0.65121	0.69110	0.72782	0.76041	0.78856
8	0.52991	0.57360	0.61800	0.66097	0.70076	0.73625	0.76704
9	0.49512	0.54084	0.58765	0.63327	0.67576	0.71387	0.74704
10	0.46372	0.51104	0.55985	0.60774	0.65263	0.69307	0.72842
11	0.43532	0.48387	0.53432	0.58418	0.63116	0.67371	0.71104
12	0.40957	0.45905	0.51084	0.56236	0.61121	0.65564	0.69478
13	0.38518	0.43633	0.48920	0.54214	0.59261	0.63874	0.67953
14	0.36490	0.41548	0.46920	0.52334	0.57525	0.62291	0.66521
15	0.34548	0.39632	0.45070	0.50584	0.55900	0.60805	0.65173
16	0.32773	0.37867	0.43353	0.48951	0.54378	0.59407	0.63901
17	0.31147	0.36239	0.41757	0.47424	0.52947	0.58088	0.62700
18	0.29655	0.34732	0.40271	0.45994	0.51602	0.56844	0.61562
19	0.28233	0.33336	0.38885	0.44653	0.50333	0.55666	0.60483
20	0.27019	0.32040	0.37590	0.43392	0.49135	0.54551	0.59459
21	0.25851	0.30835	0.36377	0.42205	0.48002	0.53492	0.58483
22	0.24771	0.29712	0.35240	0.41085	0.46929	0.52486	0.57554
23	0.23770	0.28664	0.34171	0.40027	0.45911	0.51527	0.56667
24	0.22840	0.27683	0.33166	0.39027	0.44943	0.50614	0.55820
25	0.21975	0.26765	0.32218	0.38079	0.44023	0.49742	0.55009
26	0.21169	0.25904	0.31324	0.37180	0.43145	0.48908	0.54231
27	0.20417	0.25094	0.30479	0.36325	0.42309	0.48110	0.53486
28	0.19713	0.24332	0.29678	0.35512	0.41509	0.47345	0.52769
29	0.19053	0.23614	0.28920	0.34738	0.40744	0.46612	0.52080
30	0.18435	0.22937	0.28200	0.33999	0.40012	0.45907	0.51417
35	0.15844	0.20052	0.25088	0.30764	0.36769	0.42758	0.48435
40	0.13879	0.17807	0.22609	0.28131	0.34082	0.40112	0.45901
45	0.12342	0.16015	0.20587	0.25942	0.31811	0.37846	0.43711
50	0.11110	0.14550	0.18905	0.24089	0.29861	0.35876	0.41788
60	0.09259	0.12303	0.16263	0.21116	0.26669	0.32600	0.38551
70	0.07936	0.10659	0.14281	0.18827	0.24153	0.29966	0.35907
80	0.06944	0.09404	0.12736	0.17004	0.22109	0.27788	0.33690
90	0.06173	0.08413	0.11496	0.15514	0.20409	0.25949	0.31793
100	0.05556	0.07612	0.10479	0.14273	0.18969	0.24368	0.30144

HARMONIC (B=1.0) DECLINE PRESENT VALUE FACTORS
CONTINUOUS INTEREST, J = 0.20

RATIO OF INITIAL TO FINAL RATE

YEARS ON DECLINE	1.0	2.0	4.0	8.0	16.0	32.0	64.0
1	0.90635	0.91675	0.92672	0.93588	0.94399	0.95097	0.95686
2	0.82420	0.84321	0.86165	0.87876	0.89402	0.90722	0.91841
3	0.75198	0.77803	0.80356	0.82744	0.84890	0.86758	0.88348
4	0.68834	0.72014	0.75157	0.78122	0.80807	0.83158	0.85168
5	0.63212	0.66858	0.70492	0.73950	0.77103	0.79880	0.82265
6	0.58234	0.62254	0.66297	0.70173	0.73733	0.76886	0.79606
7	0.53815	0.58132	0.62512	0.66744	0.70658	0.74144	0.77165
8	0.49881	0.54433	0.59089	0.63623	0.67844	0.71626	0.74916
9	0.46372	0.51104	0.55984	0.60774	0.65262	0.69306	0.72839
10	0.43233	0.48100	0.53162	0.58167	0.62887	0.67163	0.70915
11	0.40418	0.45383	0.50588	0.55774	0.60696	0.65178	0.69128
12	0.37887	0.42918	0.48236	0.53572	0.58669	0.63335	0.67464
13	0.35605	0.40677	0.46080	0.51541	0.56789	0.61618	0.65910
14	0.33542	0.38634	0.44100	0.49663	0.55042	0.60017	0.64456
15	0.31674	0.36767	0.42276	0.47922	0.53414	0.58519	0.63091
16	0.29976	0.35057	0.40593	0.46304	0.51893	0.57114	0.61808
17	0.28430	0.33486	0.39035	0.44798	0.50470	0.55794	0.60599
18	0.27019	0.32040	0.37590	0.43392	0.49135	0.54550	0.59457
19	0.25727	0.30706	0.36247	0.42077	0.47880	0.53377	0.58377
20	0.24542	0.29473	0.34997	0.40845	0.46698	0.52268	0.57353
21	0.23452	0.28329	0.33830	0.39688	0.45583	0.51218	0.56380
22	0.22448	0.27268	0.32738	0.38599	0.44528	0.50221	0.55454
23	0.21521	0.26280	0.31715	0.37573	0.43530	0.49274	0.54572
24	0.20662	0.25358	0.30755	0.36605	0.42583	0.48372	0.53730
25	0.19865	0.24498	0.29853	0.35689	0.41684	0.47512	0.52925
26	0.19125	0.23692	0.29002	0.34822	0.40828	0.46691	0.52155
27	0.18435	0.22937	0.28200	0.33999	0.40012	0.45907	0.51416
28	0.17791	0.22227	0.27442	0.33217	0.39234	0.45155	0.50708
29	0.17189	0.21559	0.26724	0.32474	0.38491	0.44435	0.50027
30	0.16625	0.20930	0.26044	0.31766	0.37780	0.43744	0.49372
35	0.14273	0.18262	0.23116	0.28673	0.34639	0.40663	0.46431
40	0.12496	0.16196	0.20793	0.26167	0.32046	0.38082	0.43939
45	0.11110	0.14550	0.18905	0.24089	0.29861	0.35876	0.41788
50	0.10000	0.13209	0.17338	0.22335	0.27988	0.33962	0.39903
60	0.08333	0.11156	0.14885	0.19530	0.24931	0.30786	0.36734
70	0.07143	0.09656	0.13049	0.17376	0.22530	0.28240	0.34152
80	0.06250	0.08513	0.11622	0.15666	0.20584	0.26139	0.31990
90	0.05556	0.07612	0.10479	0.14273	0.18969	0.24368	0.30143
100	0.05000	0.06884	0.09543	0.13113	0.17605	0.22850	0.28541

HARMONIC (B=1.0) DECLINE PRESENT VALUE FACTORS
CONTINUOUS INTEREST, J = 0.25

YEARS ON DECLINE	RATIO OF INITIAL TO FINAL RATE						
	1.0	2.0	4.0	8.0	16.0	32.0	64.0
1	0.88480	0.89750	0.90968	0.92089	0.93082	0.93937	0.94661
2	0.78694	0.80964	0.83176	0.85236	0.87078	0.88675	0.90031
3	0.70351	0.73398	0.76402	0.79230	0.81783	0.84014	0.85919
4	0.63212	0.66857	0.70492	0.73948	0.77098	0.79871	0.82251
5	0.57030	0.61180	0.65313	0.69283	0.72934	0.76171	0.78965
6	0.51791	0.56233	0.60758	0.65147	0.69217	0.72852	0.76008
7	0.47213	0.51904	0.56733	0.61462	0.65885	0.69863	0.73335
8	0.43233	0.48100	0.53162	0.58166	0.62885	0.67158	0.70907
9	0.39760	0.44744	0.49980	0.55206	0.60173	0.64700	0.68694
10	0.36717	0.41771	0.47135	0.52536	0.57710	0.62458	0.66668
11	0.34039	0.39128	0.44580	0.50118	0.55466	0.60404	0.64805
12	0.31674	0.36767	0.42276	0.47921	0.53413	0.58516	0.63086
13	0.29576	0.34652	0.40192	0.45917	0.51528	0.56774	0.61495
14	0.27709	0.32748	0.38299	0.44083	0.49791	0.55161	0.60016
15	0.26040	0.31030	0.36574	0.42397	0.48186	0.53663	0.58638
16	0.24542	0.29473	0.34997	0.40845	0.46698	0.52267	0.57349
17	0.23194	0.28057	0.33550	0.39409	0.45313	0.50962	0.56140
18	0.21975	0.26765	0.32218	0.38079	0.44022	0.49740	0.55005
19	0.20870	0.25583	0.30990	0.36842	0.42815	0.48592	0.53934
20	0.19865	0.24498	0.29853	0.35689	0.41683	0.47511	0.52922
21	0.18948	0.23499	0.28797	0.34612	0.40620	0.46491	0.51965
22	0.18108	0.22576	0.27816	0.33603	0.39618	0.45526	0.51056
23	0.17336	0.21722	0.26900	0.32656	0.38673	0.44612	0.50192
24	0.16625	0.20930	0.26044	0.31766	0.37780	0.43743	0.49370
25	0.15969	0.20193	0.25243	0.30926	0.36933	0.42918	0.48585
26	0.15361	0.19505	0.24490	0.30134	0.36130	0.42131	0.47835
27	0.14797	0.18863	0.23782	0.29384	0.35366	0.41380	0.47117
28	0.14273	0.18262	0.23116	0.28673	0.34639	0.40663	0.46430
29	0.13783	0.17697	0.22486	0.27999	0.33946	0.39976	0.45770
30	0.13326	0.17167	0.21891	0.27358	0.33284	0.39319	0.45135
35	0.11427	0.14929	0.19343	0.24575	0.30375	0.36397	0.42297
40	0.10000	0.13209	0.17338	0.22335	0.27988	0.33962	0.39902
45	0.08889	0.11846	0.15717	0.20490	0.25986	0.31890	0.37841
50	0.08000	0.10739	0.14378	0.18940	0.24279	0.30099	0.36041
60	0.06667	0.09049	0.12293	0.16475	0.21509	0.27142	0.33025
70	0.05714	0.07819	0.10743	0.14596	0.19347	0.24785	0.30579
80	0.05000	0.06884	0.09543	0.13113	0.17605	0.22850	0.28541
90	0.04444	0.06149	0.08586	0.11911	0.16167	0.21227	0.26806
100	0.04000	0.05556	0.07805	0.10915	0.14958	0.19842	0.25306

HARMONIC (B=1.0) DECLINE PRESENT VALUE FACTORS
CONTINUOUS INTEREST, J = 0.30

RATIO OF INITIAL TO FINAL RATE

YEARS ON DECLINE	1.0	2.0	4.0	8.0	16.0	32.0	64.0
1	0.86394	0.87882	0.89312	0.90628	0.91795	0.92803	0.93656
2	0.75198	0.77802	0.80351	0.82731	0.84867	0.86724	0.88303
3	0.65937	0.69361	0.72760	0.75976	0.78896	0.81457	0.83649
4	0.58234	0.62253	0.66295	0.70169	0.73722	0.76867	0.79577
5	0.51791	0.56233	0.60757	0.65145	0.69212	0.72843	0.75994
6	0.46372	0.51104	0.55984	0.60772	0.65256	0.69293	0.72818
7	0.41788	0.46708	0.51845	0.56943	0.61765	0.66141	0.69987
8	0.37887	0.42918	0.48236	0.53571	0.58665	0.63326	0.67448
9	0.34548	0.39632	0.45069	0.50583	0.55897	0.60797	0.65157
10	0.31674	0.36767	0.42276	0.47921	0.53411	0.58512	0.63079
11	0.29185	0.34255	0.39799	0.45537	0.51168	0.56438	0.61183
12	0.27019	0.32040	0.37590	0.43391	0.49133	0.54545	0.59447
13	0.25122	0.30078	0.35611	0.41451	0.47279	0.52811	0.57848
14	0.23452	0.28329	0.33829	0.39687	0.45581	0.51214	0.56371
15	0.21975	0.26765	0.32218	0.38079	0.44021	0.49738	0.55000
16	0.20662	0.25358	0.30755	0.36605	0.42582	0.48369	0.53723
17	0.19488	0.24088	0.29421	0.35249	0.41249	0.47094	0.52529
18	0.18435	0.22937	0.28200	0.33999	0.40011	0.45904	0.51410
19	0.17485	0.21888	0.27078	0.32841	0.38858	0.44789	C.50358
20	0.16625	0.20930	0.26044	0.31766	0.37779	0.43742	0.49366
21	0.15844	0.20052	0.25088	0.30764	0.36769	0.42756	0.48429
22	0.15131	0.19243	0.24202	0.29829	0.35819	0.41825	0.47541
23	0.14478	0.18498	0.23378	0.28953	0.34926	0.40945	0.46698
24	0.13879	0.17807	0.22609	0.28131	0.34082	0.40110	0.45897
25	0.13326	0.17167	0.21891	0.27358	0.33284	0.39318	0.45133
26	0.12815	0.16571	0.21218	0.26630	0.32528	0.38563	0.44404
27	0.12342	0.16015	0.20587	0.25942	0.31811	0.37845	0.43707
28	0.11902	0.15495	0.19993	0.25291	0.31129	0.37159	0.43039
29	0.11492	0.15008	0.19433	0.24674	0.30480	0.36503	0.42399
30	0.11110	0.14550	0.18905	0.24089	0.29861	0.35875	0.41785
35	0.09524	0.12628	0.16650	0.21556	0.27147	0.33094	0.39041
40	0.08333	0.11156	0.14885	0.19530	0.24931	0.30786	0.36732
45	0.07407	0.09992	0.13463	0.17867	0.23081	0.28828	0.34751
50	0.06667	0.09049	0.12293	0.16475	0.21509	0.27142	0.33025
60	0.05556	0.07612	0.10479	0.14273	0.18969	0.24368	0.30143
70	0.04762	0.06570	0.09135	0.12604	0.16998	0.22168	0.27814
80	0.04167	0.05779	0.08099	0.11292	0.15418	0.20371	0.25881
90	0.03704	0.05159	0.07276	0.10233	0.14120	0.18869	0.24242
100	0.03333	0.04658	0.06605	0.09358	0.13032	0.17592	0.22828

HARMONIC (B=1.0) DECLINE PRESENT VALUE FACTORS
CONTINUOUS INTEREST, J = 0.35

YEARS ON DECLINE	RATIO OF INITIAL TO FINAL RATE						
	1.0	2.0	4.0	8.0	16.0	32.0	64.0
1	0.84375	0.86070	0.87702	0.89204	0.90539	0.91693	0.92670
2	0.71916	0.74822	0.77677	0.80354	0.82763	0.84862	0.86651
3	0.61911	0.65657	0.69398	0.72960	0.76208	0.79068	0.81523
4	0.53815	0.58132	0.62510	0.66738	0.70642	0.74115	0.77120
5	0.47213	0.51904	0.56732	0.61458	0.65876	0.69845	0.73306
6	0.41788	0.46707	0.51845	0.56942	0.61761	0.66133	0.69973
7	0.37294	0.42338	0.47679	0.53046	0.58179	0.62878	0.67037
8	0.33542	0.38634	0.44100	0.49661	0.55037	0.60004	0.64431
9	0.30386	0.35471	0.41001	0.46696	0.52260	0.57445	0.62100
10	0.27709	0.32748	0.38299	0.44082	0.49788	0.55153	0.60001
11	0.25421	0.30389	0.35926	0.41761	0.47575	0.53086	0.58098
12	0.23452	0.28329	0.33829	0.39687	0.45580	0.51210	0.56364
13	0.21746	0.26520	0.31965	0.37824	0.43772	0.49499	0.54774
14	0.20256	0.24921	0.30297	0.36140	0.42126	0.47931	0.53310
15	0.18948	0.23499	0.28797	0.34612	0.40618	0.46487	0.51955
16	0.17791	0.22227	0.27442	0.33217	0.39233	0.45151	0.50697
17	0.16763	0.21084	0.26211	0.31939	0.37953	0.43910	0.49523
18	0.15844	0.20052	0.25088	0.30764	0.36768	0.42754	0.48425
19	0.15018	0.19115	0.24060	0.29678	0.35666	0.41673	0.47394
20	0.14273	0.18262	0.23116	0.28673	0.34638	0.40660	0.46423
21	0.13597	0.17481	0.22244	0.27739	0.33677	0.39707	0.45507
22	0.12981	0.16765	0.21438	0.26868	0.32775	0.38809	0.44640
23	0.12418	0.16105	0.20689	0.26054	0.31928	0.37961	0.43818
24	0.11902	0.15495	0.19993	0.25291	0.31129	0.37158	0.43036
25	0.11427	0.14929	0.19343	0.24575	0.30374	0.36395	0.42293
26	0.10988	0.14404	0.18735	0.23900	0.29660	0.35671	0.41583
27	0.10581	0.13915	0.18165	0.23265	0.28983	0.34981	0.40905
28	0.10204	0.13457	0.17629	0.22664	0.28341	0.34323	0.40257
29	0.09852	0.13029	0.17125	0.22095	0.27729	0.33695	0.39635
30	0.09524	0.12628	0.16650	0.21556	0.27147	0.33094	0.39039
35	0.08163	0.10943	0.14627	0.19230	0.24600	0.30437	0.36380
40	0.07143	0.09656	0.13049	0.17376	0.22529	0.28239	0.34149
45	0.06349	0.08641	0.11782	0.15861	0.20806	0.26381	0.32239
50	0.05714	0.07819	0.10743	0.14596	0.19346	0.24784	0.30578
60	0.04762	0.06570	0.09135	0.12604	0.16998	0.22168	0.27814
70	0.04082	0.05666	0.07949	0.11100	0.15185	0.20102	0.25589
80	0.03571	0.04980	0.07037	0.09923	0.13737	0.18421	0.23748
90	0.03175	0.04443	0.06314	0.08976	0.12551	0.17021	0.22191
100	0.02857	0.04010	0.05726	0.08196	0.11561	0.15834	0.20852

HARMONIC (B=1.0) DECLINE PRESENT VALUE FACTORS
CONTINUOUS INTEREST, J = 0.40

YEARS ON DECLINE — RATIO OF INITIAL TO FINAL RATE

YEARS ON DECLINE	1.0	2.0	4.0	8.0	16.0	32.0	64.0
1	0.82420	0.84313	0.86136	0.87817	0.89313	0.90606	0.91703
2	0.68834	0.72012	0.75147	0.78096	0.80758	0.83084	0.85069
3	0.58234	0.62253	0.66292	0.70159	0.73702	0.76833	0.79529
4	0.49881	0.54432	0.59086	0.63614	0.67822	0.71585	0.74854
5	0.43233	0.48100	0.53160	0.58161	0.62871	0.67131	0.70863
6	0.37887	0.42918	0.48235	0.53568	0.58657	0.63309	0.67419
7	0.33542	0.38634	0.44099	0.49660	0.55033	0.59996	0.64417
8	0.29976	0.35057	0.40592	0.46302	0.51886	0.57096	0.61774
9	0.27019	0.32040	0.37590	0.43390	0.49129	0.54535	0.59427
10	0.24542	0.29473	0.34996	0.40843	0.46693	0.52255	0.57325
11	0.22448	0.27268	0.32738	0.38598	0.44524	0.50209	0.55430
12	0.20662	0.25358	0.30755	0.36604	0.42579	0.48362	0.53708
13	0.19125	0.23692	0.29002	0.34821	0.40825	0.46682	0.52135
14	0.17791	0.22227	0.27442	0.33217	0.39231	0.45147	0.50690
15	0.16625	0.20930	0.26044	0.31765	0.37778	0.43737	0.49355
16	0.15599	0.19775	0.24785	0.30445	0.36444	0.42435	0.48118
17	0.14690	0.18740	0.23646	0.29238	0.35216	0.41229	0.46965
18	0.13879	0.17807	0.22609	0.28131	0.34081	0.40107	0.45888
19	0.13151	0.16963	0.21662	0.27110	0.33027	0.39059	0.44878
20	0.12496	0.16196	0.20793	0.26166	0.32045	0.38078	0.43928
21	0.11902	0.15495	0.19993	0.25291	0.31128	0.37156	0.43032
22	0.11362	0.14852	0.19254	0.24476	0.30269	0.36288	0.42185
23	0.10868	0.14261	0.18568	0.23715	0.29463	0.35469	0.41383
24	0.10416	0.13715	0.17931	0.23003	0.28703	0.34694	0.40620
25	0.10000	0.13209	0.17338	0.22335	0.27987	0.33959	0.39895
26	0.09615	0.12740	0.16783	0.21707	0.27310	0.33262	0.39204
27	0.09259	0.12303	0.16263	0.21116	0.26668	0.32598	0.38544
28	0.08928	0.11895	0.15776	0.20558	0.26060	0.31965	0.37912
29	0.08621	0.11514	0.15317	0.20030	0.25481	0.31361	0.37308
30	0.08333	0.11156	0.14885	0.19529	0.24931	0.30784	0.36728
35	0.07143	0.09656	0.13049	0.17376	0.22529	0.28238	0.34147
40	0.06250	0.08513	0.11622	0.15666	0.20583	0.26138	0.31986
45	0.05556	0.07612	0.10479	0.14273	0.18969	0.24367	0.30140
50	0.05000	0.06884	0.09543	0.13113	0.17605	0.22849	0.28539
60	0.04167	0.05779	0.08099	0.11292	0.15418	0.20371	0.25880
70	0.03571	0.04980	0.07037	0.09923	0.13737	0.18421	0.23747
80	0.03125	0.04376	0.06223	0.08855	0.12399	0.16840	0.21987
90	0.02778	0.03902	0.05578	0.07997	0.11307	0.15527	0.20503
100	0.02500	0.03521	0.05055	0.07293	0.10398	0.14417	0.19230

HARMONIC (B=1.0) DECLINE PRESENT VALUE FACTORS
CONTINUOUS INTEREST, J = 0.45

YEARS ON DECLINE	RATIO OF INITIAL TO FINAL RATE						
	1.0	2.0	4.0	8.0	16.0	32.0	64.0
1	0.80527	0.82607	0.84613	0.86465	0.88115	0.89543	0.90755
2	0.65937	0.69359	0.72750	0.75950	0.78848	0.81385	0.83555
3	0.54871	0.59120	0.63416	0.67553	0.71361	0.74739	0.77656
4	0.46372	0.51103	0.55981	0.60763	0.65235	0.69253	0.72758
5	0.39760	0.44744	0.49978	0.55199	0.60154	0.64663	0.68633
6	0.34548	0.39632	0.45068	0.50578	0.55885	0.60771	0.65113
7	0.30386	0.35471	0.41000	0.46694	0.52252	0.57429	0.62072
8	0.27019	0.32040	0.37589	0.43389	0.49126	0.54527	0.59413
9	0.24261	0.29179	0.34697	0.40546	0.46405	0.51981	0.57066
10	0.21975	0.26765	0.32218	0.38077	0.44016	0.49725	0.54973
11	0.20059	0.24707	0.30073	0.35912	0.41899	0.47709	0.53092
12	0.18435	0.22937	0.28200	0.33998	0.40008	0.45894	0.51388
13	0.17045	0.21398	0.26551	0.32293	0.38306	0.44248	0.49835
14	0.15844	0.20052	0.25088	0.30763	0.36766	0.42748	0.48411
15	0.14797	0.18863	0.23782	0.29383	0.35363	0.41372	0.47098
16	0.13879	0.17807	0.22609	0.28130	0.34080	0.40104	0.45881
17	0.13066	0.16863	0.21549	0.26988	0.32899	0.38930	0.44750
18	0.12342	0.16015	0.20587	0.25941	0.31809	0.37839	0.43694
19	0.11694	0.15247	0.19709	0.24978	0.30799	0.36822	0.42704
20	0.11110	0.14550	0.18905	0.24089	0.29859	0.35871	0.41774
21	0.10581	0.13915	0.18165	0.23264	0.28982	0.34978	0.40897
22	0.10101	0.13332	0.17482	0.22498	0.28162	0.34138	0.40069
23	0.09662	0.12797	0.16850	0.21784	0.27392	0.33345	0.39284
24	0.09259	0.12303	0.16263	0.21116	0.26668	0.32596	0.38540
25	0.08889	0.11846	0.15717	0.20490	0.25985	0.31887	0.37832
26	0.08547	0.11422	0.15206	0.19902	0.25341	0.31213	0.37157
27	0.08230	0.11027	0.14729	0.19349	0.24731	0.30573	0.36514
28	0.07936	0.10659	0.14281	0.18827	0.24152	0.29963	0.35898
29	0.07663	0.10315	0.13860	0.18333	0.23603	0.29382	0.35309
30	0.07407	0.09992	0.13463	0.17866	0.23081	0.28826	0.34745
35	0.06349	0.08641	0.11782	0.15861	0.20806	0.26380	0.32235
40	0.05556	0.07612	0.10479	0.14273	0.18969	0.24367	0.30139
45	0.04938	0.06803	0.09437	0.12982	0.17449	0.22674	0.28351
50	0.04444	0.06149	0.08586	0.11911	0.16167	0.21226	0.26803
60	0.03704	0.05159	0.07276	0.10233	0.14120	0.18869	0.24240
70	0.03175	0.04443	0.06314	0.08976	0.12551	0.17021	0.22190
80	0.02778	0.03902	0.05578	0.07997	0.11307	0.15527	0.20502
90	0.02469	0.03478	0.04996	0.07214	0.10294	0.14290	0.19083
100	0.02222	0.03138	0.04524	0.06571	0.09453	0.13247	0.17868

HARMONIC (B=1.0) DECLINE PRESENT VALUE FACTORS
CONTINUOUS INTEREST, J = 0.50

YEARS ON DECLINE	RATIO OF INITIAL TO FINAL RATE						
	1.0	2.0	4.0	8.0	16.0	32.0	64.0
1	0.78694	0.80952	0.83132	0.85147	0.86944	0.88501	0.89825
2	0.63212	0.66854	0.70477	0.73909	0.77025	0.79760	0.82103
3	0.51791	0.56232	0.60751	0.65125	0.69171	0.72773	0.75893
4	0.43233	0.48099	0.53158	0.58153	0.62853	0.67098	0.70814
5	0.36717	0.41771	0.47133	0.52527	0.57687	0.62411	0.66589
6	0.31674	0.36767	0.42275	0.47915	0.53395	0.58478	0.63019
7	0.27709	0.32748	0.38298	0.44078	0.49778	0.55129	0.59958
8	0.24542	0.29473	0.34996	0.40841	0.46686	0.52240	0.57298
9	0.21975	0.26765	0.32218	0.38076	0.44013	0.49717	0.54959
10	0.19865	0.24498	0.29852	0.35687	0.41676	0.47491	0.52882
11	0.18108	0.22576	0.27815	0.33601	0.39612	0.45509	0.51020
12	0.16625	0.20930	0.26044	0.31764	0.37774	0.43728	0.49337
13	0.15361	0.19505	0.24490	0.30132	0.36125	0.42118	0.47805
14	0.14273	0.18262	0.23115	0.28672	0.34635	0.40651	0.46402
15	0.13326	0.17167	0.21891	0.27357	0.33281	0.39308	0.45111
16	0.12496	0.16196	0.20793	0.26166	0.32043	0.38072	0.43915
17	0.11762	0.15329	0.19803	0.25081	0.30907	0.36929	0.42804
18	0.11110	0.14550	0.18905	0.24089	0.29858	0.35868	0.41768
19	0.10526	0.13847	0.18086	0.23176	0.28888	0.34880	0.40797
20	0.10000	0.13209	0.17338	0.22335	0.27986	0.33955	0.39885
21	0.09524	0.12628	0.16650	0.21556	0.27145	0.33089	0.39027
22	0.09091	0.12096	0.16016	0.20833	0.26359	0.32274	0.38216
23	0.08696	0.11607	0.15429	0.20159	0.25622	0.31507	0.37449
24	0.08333	0.11156	0.14885	0.19529	0.24930	0.30781	0.36721
25	0.08000	0.10739	0.14378	C.18940	0.24278	0.30095	0.36029
26	0.07692	0.10352	0.13905	0.18387	0.23662	0.29444	0.35370
27	0.07407	0.09992	0.13463	0.17866	0.23080	0.28825	0.34742
28	0.07143	0.09656	0.13049	0.17376	0.22529	0.28236	0.34141
29	0.06897	0.09343	0.12660	0.16913	0.22005	0.27675	0.33567
30	0.06667	0.09049	0.12293	0.16475	0.21508	0.27139	0.33016
35	0.05714	0.07819	0.10743	0.14596	0.19346	0.24782	0.30573
40	0.05000	0.06884	0.09543	0.13113	0.17604	0.22848	0.28535
45	0.04444	0.06149	0.08586	0.11911	0.16167	0.21226	0.26802
50	0.04000	0.05556	0.07805	0.10915	0.14958	0.19841	0.25302
60	0.03333	0.04658	0.06605	0.09358	0.13032	0.17591	0.22826
70	0.02857	0.04010	0.05726	0.08196	0.11561	0.15834	0.20851
80	0.02500	0.03521	0.05055	0.07293	0.10397	0.14417	0.19229
90	0.02222	0.03138	0.04524	0.06571	0.09453	0.13246	0.17868
100	0.02000	0.02830	0.04095	0.05981	0.08669	0.12261	0.16706

INDEX

M

N

O

P

S

T

U